U.T. TYLE
COLLEGE OF ENG

Heat and Fluid Flow in Microscale and Nanoscale Structures

WIT*PRESS*

WIT Press publishes leading books in Science and Technology.
Visit our website for new and current list of titles.
www.witpress.com

WIT*eLibrary*

Making the latest research accessible, the WIT electronic-library features papers presented at Wessex Institute of Technology's prestigious international conferences.
To access the library and view abstracts free of charge please visit www.witpress.com

International Series on Developments in Heat Transfer

Objectives

The Developments in Heat Transfer book Series publishes state-of-the-art books and provides valuable contributions to the literature in the field of heat transfer. The overall aim of the Series is to bring to the attention of the international community recent advances in heat transfer by authors in academic research and the engineering industry.

Research and development in heat transfer is of significant importance to many branches of technology, not least in energy technology. Developments include new, efficient heat exchangers, novel heat transfer equipment as well as the introduction of systems of heat exchangers in industrial processes. Application areas include heat recovery in the chemical and process industries, and buildings and dwelling houses where heat transfer plays a major role. Heat exchange combined with heat storage is also a methodology for improving the energy efficiency in industry, while cooling in gas turbine systems and combustion engines is another important area of heat transfer research.

To progress developments within the field both basic and applied research is needed. Advances in numerical solution methods of partial differential equations, high-speed, efficient and cheap computers, advanced experimental methods using LDV (laser-doppler-velocimetry), PIV (particle-image-velocimetry) and image processing of thermal pictures of liquid crystals, have all led to dramatic advances during recent years in the solution and investigation of complex problems within the field.

The aims of the Series are achieved by contributions to the volumes from invited authors only. This is backed by an internationally recognised Editorial Board for the Series who represent much of the active research worldwide. Volumes planned for the series include the following topics: Compact Heat Exchangers, Engineering Heat Transfer Phenomena, Fins and Fin Systems, Condensation, Materials Processing, Gas Turbine Cooling, Electronics Cooling, Combustion-Related Heat Transfer, Heat Transfer in Gas-Solid Flows, Thermal Radiation, the Boundary Element Method in Heat Transfer, Phase Change Problems, Heat Transfer in Micro-Devices, Plate-and-Frame Heat Exchangers, Turbulent Convective Heat Transfer in Ducts, Enhancement of Heat Transfer and other selected topics.

Series Editor
B. Sundén
Lund Institute of Technology
Box 118
22100 Lund
Sweden

Associate Editors

C.I. Adderley
Rolls Royce and Associates Limited
UK

E. Blums
Latvian Academy of Sciences
Latvia

C.A. Brebbia
Wessex Institute of Technology
UK

G. Comini
University of Udine
Italy

R.M. Cotta
COPPE/UFRJ,
Brazil

L. De Biase
University of Milan
Italy

G. De Mey
University of Ghent
Belgium

G. de Vahl Davies
University of New South Wales
Australia

S. del Guidice
University of Udine
Italy

M. Faghri
The University of Rhode Island
USA

P.J. Heggs
UMIST
UK

C. Herman
John Hopkins University
USA

D.B. Ingham
The University of Leeds
UK

Y. Jaluria
Rutgers University
USA

S. Kotake
University of Tokyo
Japan

M. Lamvik
Norwegian University of Science and Technology
Norway

P.S. Larsen
Technical University of Denmark
Denmark

B. Sarler
University of Ljubljana
Slovenia

D.B. Murray
Trinity College Dublin
Ireland

A.C.M. Sousa
University of New Brunswick
Canada

A.J. Nowak
Technical University of Silesia
Poland

D.B. Spalding
CHAM
UK

K. Onishi
Ibaraki University
Japan

J. Szmyd
University of Mining and Metallurgy
Poland

P.H. Oosthuizen
Queen's University Kingston
Canada

E. Van den Bulck
Katholieke Universiteit Leuven
Belgium

W. Roetzel
Universtaet der Bundeswehr
Germany

S. Yanniotis
Agricultural University of Athens
Greece

Heat and Fluid Flow in Microscale and Nanoscale Structures

Editors: M. Faghri & B. Sundén

Heat and Fluid Flow in Microscale and Nanoscale Structures

Series: Developments in Heat Transfer

M. Faghri

Department of Mechanical Engineering, University of Rhode Island, USA

B. Sundén

Division of Heat Transfer, Lund Institute of Technology, Sweden

Published by

WIT Press
Ashurst Lodge, Ashurst, Southampton, SO40 7AA, UK
Tel: 44 (0) 238 029 3223; Fax: 44 (0) 238 029 2853
E-Mail: witpress@witpress.com
http://www.witpress.com

For USA, Canada and Mexico

Computational Mechanics Inc
25 Bridge Street, Billerica, MA 01821, USA
Tel: 978 667 5841; Fax: 978 667 7582
E-Mail: info@witpress.com
http://www.witpress.com

British Library Cataloguing-in-Publication Data

A Catalogue record for this book is available
from the British Library

ISBN: 1-85312-893-7
ISSN: 1369-7331

Library of Congress Catalog Card Number: 2001088659

The text of the papers in this volume were set individually by the authors or under their supervision. Only minor corrections to the text may have been carried out by the publisher.

No responsibility is assumed by the Publisher, the Editors and Authors for any injury and/or damage to persons or property as a matter of products liability, negligence or otherwise, or from any use or operation of any methods, products, instructions or ideas contained in the material herein.

© WIT Press 2004.

Printed in Great Britain by Athenaeum Press, Gateshead.

All rights reserved. No part of this publication may be reproduced, stored in a retrieval system, or transmitted in any form or by any means, electronic, mechanical, photocopying, recording, or otherwise, without the prior written permission of the Publisher.

Contents

Preface .... **xiii**

Chapter 1 Miniature and microscale energy systems **1**
R.B. Peterson

1 Introduction 1
2 Overview 2
 2.1 Microscale energy systems 4
 2.2 Mesoscale energy systems 5
3 Scaling 12
 3.1 Scaling methodology 13
 3.2 Common phenomena important in energy systems 14
 3.3 TEC example 21
 3.4 Heat engine example 23
 3.5 Other thermal systems 24
4 Thermally based power systems 26
 4.1 A theoretical model for size limits 27
 4.2 Techniques for thermal management 33
5 Future directions 36
 5.1 Conventional mesoscopic devices 36
 5.2 High *ZT* thermal electric conversion 37

Chapter 2 Nanostructures for thermoelectric energy **45**
G. Chen, B. Yang & W. Liu

1 Introduction 47
2 Thermoelectric effects and devices with bulk materials 49
 2.1 Thermoelectric cooling devices 50
 2.2 Thermoelectric power generation devices 53
 2.3 Thermoelectric transport properties 53
3 Nanostructures for solid-state energy conversion 57
 3.1 Some recent experimental results on low-dimensional thermoelectrics 59
 3.2 General transport picture 63
 3.3 Coherent electron and phonon transport in nanostructures 66

3.4 Incoherent electron and phonon transport in nanostructures 73
3.5 Transport in the partially coherent regime 80
4 Summary 82

Chapter 3 Heat transport in superlattices and nanowires **93**
S.T. Huxtable, A.R. Abramson & A. Majumdar

1 Introduction 93
2 Superlattices 94
3 Nanowires and nanotubes 94
3.1 Nanowires 95
3.2 Nanotubes 96
4 Heat transport in bulk materials by phonons 97
4.1 Phonon scattering 100
5 Heat transport in low-dimensional structures 104
5.1 Acoustic impedance mismatch at a single interface 105
5.2 Phonon spectra mismatch 106
5.3 Phonon tunneling 107
5.4 Phonon wave interference and mini-bandgap formation 107
5.5 Interface scattering 110
6 Survey of previous work 111
6.1 Superlattices 112
6.2 Nanowires and nanotubes 119
7 Summary 122

Chapter 4 Thermomechanical formation and thermal detection of polymer nanostructures **131**
W.P. King & K.E. Goodson

1 Introduction 131
1.1 Motivation for AFM data storage 132
1.2 Review of thermomechanical data storage 134
1.3 Chapter overview 137
2 Relaxation kinetics in nanostructured polymer films 137
2.1 Fundamentals of mass transport in confined soft materials 138
2.2 Measuring flow characteristics in thin, nanostructured polymer films 140
3 Modeling and simulation of nanometer-scale thermomechanical data bit formation 146
3.1 Thermal analysis of the cantilever tip and polymer layer 148
3.2 Bit writing analysis 151
4 Thermal data reading and topography mapping 157
5 Summary and conclusions 161

Chapter 5 Two-phase flow microstructures in thin geometries: multi-field modelling 173

R. Kumar

1 Introduction ... 173
2 Global characteristics ... 175
2.1 Flow patterns in thin channels ... 175
2.2 Pressure drop and heat transfer ... 178
2.3 Summary ... 181
3 Local flow characteristics ... 182
3.1 Ensemble averaging approach ... 182
3.2 Multi-field modeling approach ... 183
3.3 Governing equations ... 184
3.4 Forces acting on a bubble in a narrow space ... 185
3.5 Annular flow forces (*cl-dv*) ... 192
3.6 Droplet models ... 201
3.7 Flow regime transition modeling ... 205
3.8 Heat transfer models ... 209
3.9 Turbulence models ... 212
3.10 Assessment of the multi-field model ... 213
4 Summary ... 216

Chapter 6 Radiative energy transport at the spatial and temporal micro/nanoscales 225

C.-H. Fan & J.P. Longtin

1 Introduction ... 225
2 Fundamentals ... 227
2.1 Properties of electromagnetic radiation ... 227
2.2 Sources of radiation in thermal engineering ... 230
2.3 Radiation-matter interactions ... 234
2.4 Characteristic length, time, and structure regimes for radiation-material interactions ... 247
3 Applications ... 251
3.1 Ultrafast laser materials processing ... 251
3.2 Laser scanning microscopy for biological systems ... 257
4 Future directions and concluding remarks ... 265

Chapter 7 Direct simulation Monte Carlo of gaseous flow and heat transfer in a microchannel 273

R. Sayegh, M. Faghri, Y. Asako & B. Sundén

1 Introduction ... 273
2 Description of the DSMC method ... 276
2.1 Grids ... 276
2.2 Molecular approximation ... 276

2.3 Time step 277
2.4 Molecular models 277
2.5 Molecular movement 277
2.6 Collisions between molecules 279
2.7 Boundary interaction 281
3 DSMC simulation of microchannel 282
3.1 Computational methodology 282
3.2 Description of the problem 284
3.3 Methodology of calculating important parameters 286
4 Results and discussion 288
4.1 Numerical criteria 288
4.2 Effects of cell size 289
4.3 Pressure distribution 289
4.4 Velocity profile 292
4.5 Slip velocity 292
4.6 Shear stress 293
4.7 Friction coefficient 293
4.8 Slip temperature 298
4.9 Nusselt number 299
5 Conclusions 299

Chapter 8 DSMC modeling of near-interface transport in liquid-vapor phase-change processes with multiple microscale effects **303**
V.P. Carey

1 Introduction 303
2 Phase equilibrium in microscale multiphase systems 304
2.1 Ultra-small bubbles and droplets 304
2.2 Ultra-thin liquid films 306
3 Molecular transport at interfaces 311
4 High Knudsen number and nonequilibrium effects 318
5 Variation of interfacial tension with interface curvature 320
6 Liquid phase and interfacial region effects 320
7 DSMC modeling of combined effects during vaporization and condensation 322
7.1 Post nucleation growth of microdroplets 323
7.2 Other processes involving multiple microscale effects 337
8 Concluding remarks 343

Chapter 9 Molecular dynamics simulation of nanoscale heat and fluid flow **349**
T. Ohara

1 Introduction 349
2 Basic equations and finite difference scheme 350
2.1 Basic equation for translational motion of molecules 350

2.2 Finite difference scheme 351
2.3 Rotational motion of polyatomic molecules 352
2.4 Deformation of molecules and intramolecular vibration 354
3 Intermolecular potential model 354
3.1 Lennard-Jones potential model for spherical molecules 355
3.2 Intermolecular potential model for a complex molecule: water 357
4 Macroscopic properties 359
4.1 Quantity of state 359
4.2 Transport properties 361
5 Boundary conditions and simulation system 363
5.1 Initial condition 363
5.2 Periodic boundary condition 364
5.3 Nonequilibrium system with a velocity and/or temperature gradient 366
5.4 Liquid-vapor coexistence system and determination of saturation curve 367
5.5 Solid wall and solid-liquid interface 368
6 MD application to heat and fluid flow 369
7 Future development 369

Preface

New developments in research and technology are needed which focus on the knowledge base at smaller time and length scales. This includes numerical methods such as Molecular Dynamics, and Direct Simulation Monte Carlo (DSMC) simulations for modeling at micro/nanoscale levels, experimental methods for high spatial-resolution and high time-resolution thermometry for measurements of thermal transport in solids, and developments of sensors for measurement of mass, pressure and temperature at micro/nanoscale levels in fluids.

Over the last 20 years, micro/nanoscale flow and heat transfer have been a most active area of research. The research is interdisciplinary involving scientists from various disciplines from engineering to physics, chemistry and materials science. Several books have been written on various aspect of this topic. Also, some journals have been created and various symposia have been held, and recently a special issue of the ASME *Journal of Heat Transfer* was completely devoted to this topic.

The objective of this book is to address the state-of-the-art knowledge in heat transfer and fluid flow in micro- and nanoscale structures. The chapters cover a wide range of topics and are invited contributions from some of the most prominent scientists who are authorities in this field.

The first chapter explores scaling laws and thermal issues associated with miniature and microscale energy systems. The scaling laws show processes that benefit from a reduction in scale and those that suffer from performance degradation as a result of miniaturization. The second chapter presents the principles of thermoelectric energy conversion in bulk materials and devices followed by a discussion on why nanostructures can be used to improve the energy conversion efficiency. It also discusses electron and phonon thermoelectric transport in nanostructures. Chapter 3 reviews heat transport mechanisms in nanostructures such as superlattices, nanowires and nanotubes. Also, a review of heat transport in bulk material by phonons is presented followed by an extension of heat conduction at micro- and nanoscales. A review of work on modeling and measurement of heat and mass transfer during thermomechanical formation and thermal detection of polymer nanostructures for data storage applications is presented in chapter 4. This is followed by a review article on the performance characteristics of two phase flow in microchannels in chapter 5. The nature and role of radiation heat transport in micro- and nanoscales are explored in chapter 6. This chapter provides an overview of several current research areas involving radiation energy exchange at micro and nanoscales levels.

The last three chapters deal with numerical issues related to modeling and numerical implementations of flow and heat transfer at micro- and nanoscales levels. Specifically, chapter 7 reviews numerical methodology and the application by direct simulation Monte Carlo for solving gaseous flow and heat transport in microchannels. Chapter 8 examines the DSMC method for modeling near-interface transport during vaporization and condensation in a microsystem. Finally, chapter 9 describes the basic theory and procedure for molecular dynamics simulation for modeling nanoscale flow and heat transfer problems.

All of the chapters follow a unified outline and presentation to aid accessibility and the book provides invaluable information for both graduate researchers and R & D engineers in industry and consultancy.

We are grateful to the authors and reviewers for their contributions. We also appreciate the cooperation and patience provided by the staff of WIT Press and for their encouragement and assistance in producing this volume. The editors would also like to thank the Wenner-Gren Center Foundation in Sweden for financial support.

M. Faghri and B. Sundén (Editors)
2003

CHAPTER 1

Miniature and microscale energy systems

R.B. Peterson
Department of Mechanical Engineering,
Oregon State University, USA.

Abstract

This work explores scaling laws and thermal issues associated with miniature and microscale energy systems. An overview is first provided of the possible technologies for these systems. Then scaling laws are introduced showing processes that benefit from a reduction in scale and those that suffer an "adverse" effect. Various thermally based technologies are also reviewed with the goal of identifying those that may be useful in the miniature and microscale size regimes. Although all thermally based systems suffer from performance degradation as their characteristic size is reduced, intermediate length scales (referred to here as mesoscopic) are fertile ground for developing a host of energy systems to power future miniature and microscale devices.

1 Introduction

The 21st century will see the development of a wide range of miniaturized power sources. Many different technologies will be incorporated into the design of these systems including (but not limited to) electrochemistry, turbo machinery, high-performance insulation, micro heat exchangers, micro reactors, and miniaturized fluidic components. Applications will be diverse and will span the size range from microscale to mesoscale systems. Definitions of exactly what length scales characterize these two size regimes will vary, but in the context used here, microscale spans the range between 1 and 100 microns while the mesoscale falls between about 100 microns and 1 cm. Larger devices will be called macroscale and smaller ones will be referred to as nanoscale.

Power will be the limiting factor determining the performance of future meso/microscale systems [1, 2]. Whether the area of interest is for extremely small applications (e.g., intracellular sensors or "larger" vascular-sized robotic devices), or for mesoscopic purposes such as man-portable power and cooling systems, power issues associated with each application must be addressed at the outset in order to develop a viable system. One consideration here is the primary source of energy. Choices are limited to one of five categories: chemical, nuclear, thermal, mechanical, or environmental – the latter term is used to describe sources harvested from the surroundings. Conversion of the primary energy source into a readily usable form (typically electrical or propulsive) is often the challenge of power systems design, and it becomes even more central to developing small-scale systems. Hence, the problem of powering meso/microscale devices is really one of energy conversion from one form to another. In this work, we will examine only a few of the unique challenges facing engineers as they develop these small-scale power systems. We will focus mainly on thermal energy systems, and then emphasize the power generation aspect of the topic. Note, however, that the term "energy systems" can be broadly interpreted to include other topics not covered here such as process heating, energy conversion of all types, heat pumping, cooling of high-flux devices, and chemical reforming, to name just a few.

2 Overview

To perform their function, all technologically useful devices require energy in one form or another. The most useful end-form will be either electrical or propulsive, although direct utilization of molecular bond energy for actuation is also of interest. This latter case occurs in biological systems at the molecular level (e.g., ATP mediated conformational changes). Energy sources and prime movers are well developed in the macroscopic size regime; however, there is an increasing need for better energy sources and power systems in the meso/microscale arena. Figure 1 gives an overview of length scales, energy sources, and power applications of miniature and microscale systems. Several of the listed sources and applications have not yet been realized. The characteristic length scale ranges from nanometers (10^{-9} m) to kilometers (10^{3} m). The top part of the figure gives common technological and natural systems. The bottom part of the figure first gives processes capable of supplying power to individual devices, followed by a selected number of applications that may be important in the coming decades. Note that although the listings are not exhaustive, an attempt has been made to provide important and timely examples in each case.

The nanoscale regime is the smallest length scale listed in Figure 1. This regime has received considerable attention over the past few years with increased funding being available to study both phenomena and applications. The vision for this area is to develop the mechanisms and devices that will permit new methodologies of material manipulation and provide a level of biological interaction hitherto unrealized. Molecular assemblers and cellular repair machines are just two of the many possible outcomes from this work. This vision is decades (if not longer) away, but biological systems can act as a proof-of-principle example of molecular machinery, especially the complex conformation changing molecules responsible for DNA repair and reproduction [3].

Technological Systems and Devices

Minimum e-beam Linewidths
Photolithography Limit (~2002)
Quantum Dots
Transistors
MEMS
Micro Channels
Micro Valves
MECS
Integrated Circuits
Miniature Aerial Vehicles
Gas Turbines and IC Engines
Autos
Aircraft
Ships
Solar Farms
Hydro Dams

Natural Systems

Simple Molecules
Cellular Organelles
Viruses
Bio Molecules
Most Cells
Capillaries
Human Hair
Beach Sand
Most Mites and Insects
Circulatory Valves
Land and Sea Mammals
Birds
Man
Trees
Armillaria Ostoyae (Fungi)
Coral

10^{-9} nm — 10^{-6} μm — 10^{-3} mm — 10^{0} m — 10^{3} km

Selected Energy Systems

Nanoscale Systems:
- -Molecular bond reactions.
- -Cluster-based reactions.
- -Photon processes.
- -Enzymatic reactions for molecular machines.

Microscale Systems:
- -Thin film fuel cells (room temp)
- -Thin film electrochemical cells.
- -Photon-to-electric devices.
- -Bio cell derived power (e.g., electric eel power cell).
- -Microscale radioisotope cells.

Mesoscale Systems:
- -Moderate temperature fuel cells.
- -Electrochemical cells.
- -Nuclear (beta cell, radioluminescence).
- -Selected combustion driven thermal systems (e.g., TPV, TEC, AMTEC, etc.).
- -Miniaturized traditional heat engines.

Macroscale Systems:
- -Large fuel cell and battery systems
- -Traditional prime movers, gas turbines, diesel engines, Stirling, etc.
- -AMTEC and TPV.
- -Wind, solar, hydroelectric, and nuclear sources.

Selected "Power" Applications

Nanoscale Systems:
- -Intracellular diagnostics.
- -Intracellular sensing and actuation.
- -Power for "nano" robots.
- -Energy for self-assembly of nano and micro structures.

Microscale Systems:
- -MEMS senors and actuators.
- -Microscale distributed sensor and monitoring networks.
- -Power for "micro" robots.
- -Implantable electronics.
- -Extra cellular, in vivo diagnostics and monitoring.

Mesoscale Systems:
- -Personal communication devices.
- -Handheld environmental monitoring units.
- -Portable and point application cooling.
- -Propulsion for miniature aerial vehicles.
- -Wearable electronics.
- -Power for "meso" robots, planetary rovers.
- -Remotely located, distributed sensing and monitoring networks.

Traditional Macroscopic Applications

Figure 1: Overview of the meso, micro, and nanoscale energy areas.

Energy systems relevant to this smallest scale of activity will be limited to the use of energy-rich molecules or photons [4]. The biological paradigm here is to immerse molecular machines in an energy-rich "soup" of feedstock molecules. Compounds such as ATP, NADH, and HNADA are used for this purpose. Also, a common theme in biological systems is to have both the feedstock material and energy source (e.g., a large macromolecule with an attached energy-rich active group on one end) combined together to form a participating species. Biological macromolecules are also capable of harvesting photons to drive molecular machinery as exemplified by chlorophyll in plants [5, 6]. Light has recently been employed to drive an artificial system of threading and rethreading [7]. In all likelihood, nanotechnology will operate under the same constraints, and use the same general concepts as demonstrated by biological systems.

2.1 Microscale energy systems

The next step up in length scale is characterized by the 1 μm to 0.1 mm size regime. As we shall see later in this chapter, the lower range of this length scale is still too small for maintaining a practical temperature difference for energy production, however, process heating in MEMS scale devices is feasible. Thus, power-producing concepts relying on ambient temperature processes will be the key to constructing practical microscale devices. Thin film structures will be important in this size range. Miniature thin film batteries based on several electrochemical couples have been designed and tested with "footprints" on the order of 100 microns on a side [8]. Also, PEM and solid oxide fuel cells have been demonstrated with a characteristic size of the order of 1 mm [9, 10, 11]. Reduction in size to even smaller dimensions appears technically feasible, but another route toward true microscale devices may be possible. Toward this end, a biological model may again lend some guidance. Cells tend to mix their reactants (both "fuel" and "oxidizer") in one homogenous soup and then depend on various organelles, macromolecules, and membranes to sort out the constituents for reaction. For a microscale fuel cell, this approach of combining the reactants has the advantage of eliminating manifolding and other ancillary components from the system. This may now be possible since mixed reactant fuel cells have been validated in the laboratory [12, 13].

Other microscale energy systems are possible. For example, a number of schemes have been proposed for utilizing the energy of radioactive decay [14, 15]. The so-called beta cell [16] could conceivably be reduced in size to qualify as a microscale device, or an interesting concept using high-energy decay products to produce radio luminescence can also be miniaturized [17]. The resulting light emissions can then be converted to electricity via photovoltaic cells. Other schemes are also possible; however, one problem plaguing radio decay concepts is their low conversion efficiency leading to microwatt power levels.

2.2 Mesoscale energy systems

The mesoscale regime will have the greatest opportunity for constructing viable miniature power sources. From about the 100 μm level up to a few centimeters, this size regime allows all major power producing concepts to be fundamentally realized. This includes thermally based systems. From Figure 1, the systems listed include elevated temperature fuel cells, standard electrochemical batteries, nuclear "batteries" (beta cells and radioluminescence cells), selected thermal systems including heat engines, and harvesting energy from environmental sources. The sections below provide additional details on a few of the mesoscopic energy systems listed in Figure 1.

2.2.1 Thin film moderate temperature fuel cells

Fuel cells are direct energy conversion devices that combine two reactants to produce electrical power [18]. The reactants are typically a fuel such as H_2, or methanol, and oxygen from the air. Fuel cells require an electrolyte capable of passing an ionic charge carrier across an electronic conduction barrier where the ions are driven by a concentration gradient. They also need a catalytic-based anode and cathode structure for reactant preparation. For meso/microscale systems, fuel cell structures will best be fabricated in thin film form in order to reduce the thickness of the device. However, depending on the desired power output of the system, the "footprint" may well be relatively large to supply the required power. Systems arranged in this manner are referred to as mixed scale systems on the basis that one critical dimension is small (the thickness) while the extent of the device (its footprint) can be the requisite size to satisfy a particular application.

Thin film fuel cells operate across a broad range of temperatures. Proton Exchange Membrane (PEM) cells based on Nafion or similar material can operate at room temperature but have better performance at elevated temperatures. The upper practical temperature limit for Nafion is approximately 100 °C, although pressurized systems can go higher. The reason for this limitation is the requirement for keeping the membrane saturated with water to promote ion passage. Fuels for PEM cells can include hydrogen and methanol, as well as other fuels if reforming takes place. Direct methanol fuel cells have received much current attention as a possible power source for portable electronics. Recent developments in PEM cells have resulted in new membrane materials operating near 200 °C where power densities can be higher and where catalysts on the fuel side (the anode) have less susceptibility to carbon monoxide poisoning [19]. This is critical for cells consuming a reformer gas since CO in low concentrations is usually present even after processing.

Higher temperature systems are also a possible choice for mesoscale power systems. Solid Oxide Fuel Cells (SOFC) have traditionally been made with electrolytes of yittria-stabilized zirconia having a thickness greater than 100 μm. This has dictated operating temperatures approaching 1000 °C. Research over the past decade on thin film SOFC has shown the possibility of operating at temperatures as low as 500 °C [see, e.g., 20, 21, 22]. This makes them attractive for small-scale systems. Attributes for the thin film SOFC at these lower temperatures include tolerance to many types of fuels including CO, no water management issues, and the possibility of operating with either internal reforming or direct fuel oxidation. If the power density of the thin film SOFC can be maintained at the lower temperatures, then practical small-scale systems may result.

2.2.2 Nuclear batteries

Considerable work has been accomplished on miniaturized radioisotope power sources over the last few decades [23, 24]. NASA and DOE have provided support to develop power generators running off plutonium with direct energy conversion provided by thermoelectric elements. Deep space missions, because they go beyond the point where solar energy is feasible, all rely on radioisotope power generators. Most of this work has focused on larger systems generating continuous electrical power in the 100–1000 watt range. Smaller systems, especially for implantable applications, have also been developed, but not to the degree of the larger systems [25].

Radioisotope power sources, or "nuclear batteries," have one overriding advantages: the *energy* density is orders of magnitude higher than any conventional source. However, this advantage is tempered by the reality that this energy is metered out over a long period of time at a rate that cannot be changed by any known method. Thus, the *power* density of nuclear batteries often suffers considerably when compared to conventional sources. Nevertheless, where long-term power is needed at relatively low levels, radioisotope systems represent a viable choice in certain situations. These include power generation in remote locations, space based missions where reliability and weight issues are paramount, and implantable "micropower" applications where recharging is difficult. When the danger of using a nuclear source of power is either low or mitigated by other considerations (such as an emergency or wartime), relatively long-term shelf life and storage capability are also advantages.

Plutonium, an alpha emitter, is the material of choice for long-term missions where moderate power outputs are needed for extended periods. The use of this isotope for interplanetary missions has been studied by NASA for decades. It has a half-life of 87 years, the energy density (both per unit mass and volume) is high compared with other possible choices, and the alpha particles are easily stopped thus turning the kinetic energy of alpha emission into thermal energy. NASA has supported work on a general heat source module generating 1 watt of thermal energy and fully encased for protection [26]. Other isotopes are also being studied as potential power sources. The thermal electric elements for energy conversion can be miniaturized and are discussed in a later section of this chapter.

Beta emitters can be used to create secondary particles such as photons and electrons. The generated particles can then go on to produce an electrical current by way of a suitably arranged PN junction. The former of these processes is classified as a radio luminescence device, while the latter is a beta cell: see [16] and [17]. One interesting development in the radio luminescence area is the microencapsulation of the radioactive material [27]. In one proposed design, tritium is contained in small glass spheres where the inner surface of the sphere is coated with a phosphor. The light thus generated falls on a photovoltaic cell to produce microwatts of power.

Nuclear battery development at the microscale has been relatively modest compared to systems intended to power interplanetary probes. The small systems to date have been restricted to microwatt power levels and poor power densities due to low conversion efficiencies. Beta cells and radio luminescence devices have efficiencies in the 0.1% range and are currently only acceptable for systems requiring extremely low power levels. Their simplicity and near ambient temperature operation, however, afford them great reliability and the capability of being produced in small sizes. Furthermore, new developments in this area could result in higher-efficiency devices with higher power generation. Radioisotope power supplies using Thermal Electric Conversion (TEC) can boost conversion efficiency into the 1 to 5% range with an accompanying power increase. Along with this comes the need for rather aggressive thermal management techniques to allow the low heat generation levels to develop an acceptably high temperature for conversion. As described in a later section, if TEC material can be developed having *ZT* factors exceeding about 2 to 3, higher-performance systems can be developed. Even with this future prospect, a thermally based system will require the use of low thermal conductivity materials and high temperature multifoil vacuum insulation in order to achieve acceptable temperatures at the hot end of the device. Unless future developments result in new, microtechnology-based insulation designs, the heat management issues of these thermally based nuclear systems will result in bulky designs.

2.2.3 Combustion driven thermal systems

Of all the *conventional* sources of energy, liquid hydrocarbons have the highest energy densities. Compared with electrochemical cells, these fuels have energy densities of between 35 and 300 times greater than present battery technology. This assumes a liquid fuel energy density of 42 kJ/gm (with air coming from the surroundings) compared to a zinc–air battery at 1.2 kJ/gm or a lead–acid battery at 0.125 kJ/gm [28]. With such potential, liquid fuel-based power cells (for applications such as personal electronic devices and remotely located unattended sensors) would be an important enabling technology for a variety of applications. Realization of these power cells will be accomplished by development of microengines, moderate temperature (450–800 °C depending on the type of fuel) microcombustors combined with an efficient thermal-to-electric converter [29], or, perhaps, using direct energy conversion devices such as PEM or solid oxide fuel cells (SOFC). The former type of fuel cell would require fuel processing for hydrogen generation unless direct methanol technology is used. Note that SOFCs have features in common with combustion, but conversion efficiencies are not limited by the Carnot ideal.

Combustion systems have a number of challenges associated with miniaturization. Thermal management of the generated heat is one important aspect and is a common theme throughout most of this chapter. Another is the quenching issue. Gas phase (i.e., flame-based or flameless) combustion will become difficult to achieve except at the highest possible temperatures, thus requiring the so-called excess enthalpy burners developed by Weinberg: see [79] and [29]. Quench distance depends on temperature, pressure, and the type of fuel used. For hydrocarbon–air combustion at atmospheric pressure and room temperature walls, this distance is approximately 3 mm. As the temperature is increased at atmospheric pressure, the distance can fall below 1 mm at 600–800 °C. With catalytic combustion, surface reactions dominate and much lower combustor sizes can result, especially at the lower end of the temperature range which starts at about 450 °C for hydrocarbon–air combustion. Thus, for miniaturized combustion systems, catalytic combustion and the associated problems of catalyst attachment and light-off become important issues to consider.

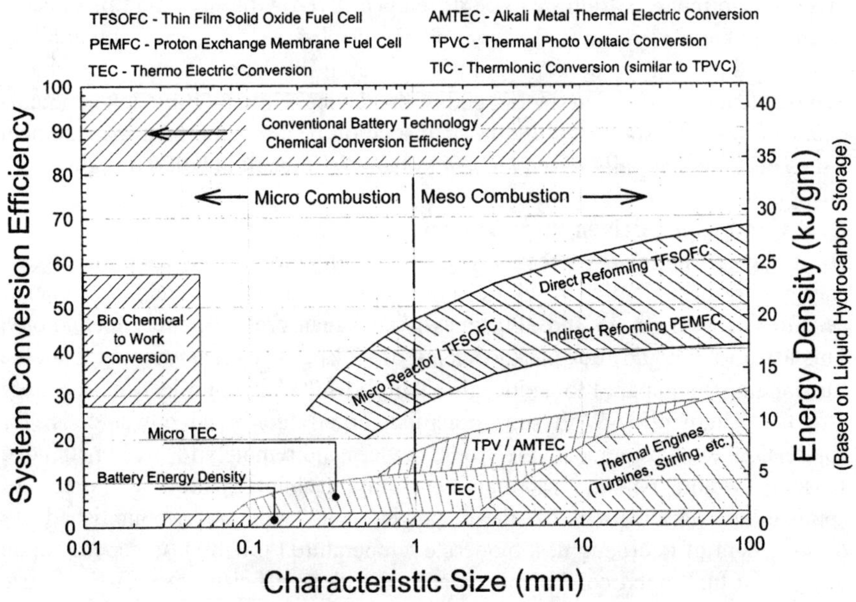

Figure 2: Regime map for meso/microscale energy systems.

Several methods exist for converting chemical energy to electricity. Some of the methods are true heat engines requiring reactants to be burned first to produce thermal energy, followed by heat-to-electric conversion. Other techniques rely on direct energy conversion, which theoretically, side steps the inherent thermodynamic restrictions on heat engines. Figure 2 shows the author's best estimates of the performance for various conversion technologies plotted as a function of characteristic size. The curves are based on a number of considerations including current operating efficiencies, theoretical performance, number of adverse scaling issues involved, and the potential for miniaturization. The various conversion processes shown are not all true thermal engines (fuel cells operate on a direct energy conversion process). Also, batteries are included for comparison, but their performance is split on the graph since they have high chemical conversion efficiency but low energy density. The right vertical axis of the plot has a maximum value based on storing a liquid hydrocarbon fuel with an energy density of 42 kJ/gm. Thus the estimated energy density of each conversion process (except for batteries and biological systems) is determined by an overall estimated conversion efficiency multiplied by 42 kJ/gm. The upper bound for each method should be considered the maximum obtainable performance for mature technology, which could be several years in the future. However, the figure does provide a view as to where the most important areas are for development of thermally based meso/microscale power cells. Note that thin film Solid Oxide Fuel Cells (SOFCs) can convert chemical reactants directly to electricity, but the cells work at elevated temperatures and can have a significant internal irreversibility (i.e., heat producing). In this respect, it practically behaves like a combustion driven heat engine, but theoretically its efficiency is not Carnot-limited.

One of the easiest approaches to producing electrical power from liquid hydrocarbons would be to integrate a thermoelectric device into a package containing a microcombustor. Most of the commercial thermoelectric devices are discrete components with a length scale of about 1 mm [30]. Thin film devices are currently being developed [31]. The temperature needs for a thermoelectric converter are modest and can easily be met. However, an efficiency of approximately 5% (limited by the current "*ZT* factors") seriously impacts the useful energy density. For example, using 42 kJ/gm for hydrocarbon–air combustion, 5% of this is 2.1 kJ/gm. This is a little higher than zinc–air batteries, but probably not high enough to provide a clear motivation for development. Also, when an overall system is considered which includes air delivery to the combustor, fuel metering and control, and thermal insulation, larger physical sizes result. Note, however, if high *ZT* materials are discovered and developed, thermoelectrics could become very attractive. Thermal photovoltaics (TPV), thermionics, and AMTEC could provide higher efficiency (with much developmental effort) in the 10–20% range [32, 33, 34, 35], but the operating temperature of the microcombustor would probably have to exceed 1200 °C for these concepts to be practical. At these high combustor temperatures, serious reliability issues exist. Also, a demonstrated miniature and/or microscale TPV or thermionic device has not yet been developed. Micro thermal engines may also play a role in combustion powered mesoscopic energy systems, but heat leakage and isothermal problems remain to be solved. This latter topic is covered in the next section.

2.2.4 Miniaturized heat engines

Several groups are currently involved in miniaturizing heat engines for power production and propulsion [see, e.g., 36, 37, 38]. Most of this work is focused on reducing the scale of a traditional prime mover. However, at least one program [39] is developing a unique engine designed from the start with miniaturization in mind. A review of this subject area is provided by Fernandez-Pello [29].

Many characteristics of traditional engines make them attractive for use in power generation and propulsion. They tend to be self-aspirating and rely on combustion, which at the macroscale is a very robust form of heat generation. Fuel is plentiful and inexpensive, with storage easily realized. The energy density of the fuel (or fuel plus container), when compared to electrochemical sources, is high. Along with these advantages come a number of drawbacks especially where miniaturization is concerned. A few of these are explained below.

A thermal engine can only convert a part of the fuel's chemical energy to useful work due to the intermediate step of converting the chemical energy first to heat. This is in contrast to direct energy conversion, of which fuel cells are an example. However, conversion efficiency is respectable in macroscopic engines and can approach the 30–40% range at design speed and power output. Note that the *overall* conversion efficiency of fuel cells rarely exceeds 50% due to the cell's internal electrochemical irreversibilities, power conversion electronics, and fuel utilization. Furthermore, if reforming of a hydrocarbon fuel is necessary, additional thermal and chemical losses occur from the high temperature reactor needed. Thus, thermal engines provide a practical and reliable way of extracting chemical energy bound up in hydrocarbon fuels.

Most common thermal engines (Otto, Diesel, Brayton, Stirling, and Rankine cycles) rely on compressing a cooler, or condensed, working fluid and expanding a hot working fluid. This is often accomplished through mechanisms that have rubbing or clearance seals, hot regions separated from cold structures, and bearings to allow differential motion between parts. All three of these defining characteristics can suffer adverse scaling effects. Although scaling laws will be covered in more detail in the next section, consider the case of maintaining the required temperature difference for the Brayton cycle (gas turbine). By their very nature, these engines exploit a difference in temperature to generate work from heat. This temperature difference is the driving force for engine operation, and as the temperature difference increases, better thermal efficiency results. However, as the size of the engine is reduced, the internal structure of the engine acts as a thermal pathway shunting a portion of the useable heat through two unproductive pathways. First, some of the heat is transferred to the surroundings without producing work. Second, heat is conducted through the connection between the turbine and the compressor, thus heating the inlet air stream. At smaller sizes, it becomes progressively harder to insulate the hot section of the engine to prevent heat flow through the two leakage paths. Even maintaining a workable temperature difference becomes a challenge. According to Kaplat [40], the ratio of "parasitic" heat loss to overall power output scales as the inverse of the linear dimension squared, assuming the same specific power and the same operating temperatures are maintained. Simple calculations for palm-sized gas turbines having a 300 K inlet temperature and a turbine operating at 1800 K show the parasitic heat loss to be comparable to the power output of the device. The parasitic effects become more pronounced as the engine is further miniaturized until no practical conventional design is possible. Note that Otto and Diesel cycles, because the hot and cold sections are coincident, are even more sensitive to the impact of size reduction. Although these latter comments concerning conventional prime movers may be on the pessimistic side and are based on accepted macroscopic heat transfer arguments, there is much room for new ideas and approaches to the problem. If novel and innovative ways of applying the power cycles are developed, and new methods of thermal management and implementing combustion come about, then the above mentioned limitations can be "shifted" to a smaller size regime allowing true microscale engine development.

Micro thermal engines as characterized by miniature Wankel engines, Stirling engines, and micro gas turbines are all practical macroscopic engines, but they suffer from significant sealing problems, reduced subcomponent efficiency, friction, and thermal management issues when scaled to the mesoscopic size regime. To date, the MIT micro gas turbine project [41, 42, 43] is the most advanced. The contemplated device has a rotor diameter of approximately 5 to 10 mm, a combustor temperature in excess of 1200 °C, and rotation rates up to 2 million RPMs. A working device has yet to be made and conversion efficiencies are not expected to exceed those of thermoelectric devices. A notable aspect of the MIT work is the thorough exploration of problems facing miniaturization of prime movers. Bearings and fluid friction through small passages are two important issues being examined by the project.

Wankel engines and compressors have received considerable recent interest. At the University of California, Berkeley, a project has been studying the miniaturization of a Wankel rotary engine powered by internal combustion [44]. The particular design features making this approach attractive are simple overall design and no valves. Furthermore, a near two-dimensional layout of the engine would permit MEMS-type fabrication if workable designs emerge from the laboratory. However, as with most miniature internal combustion engines, the Wankel has serious questions regarding seals, internal heat transfer, and combustion in small spaces. A Wankel compressor is also being examined by researchers at the University of Central Florida in their Energy Systems Miniaturization program [45]. This device may find itself a component in a larger power generating system relying on Brayton-like cycle characteristics. The compressor is also meant for two-dimensional manufacturing. An interesting feature of the design when used solely as a compressor is that six compression "strokes" can be achieved for every revolution of the power input shaft. Seals, lubrication, and friction are problems for the compressor group to solve.

Thermal engines designed from the start for miniaturization have been studied by at least one group. Washington State University has a program to develop a heat engine-on-a-chip device that uses flexible membranes, phase change of a working fluid, and the piezoelectric effect to create a Carnot-like cycle [46]. This approach is different from other groups – a traditional engine cycle is not being used as a model for a reduced size version. Rather, the unique features of high heat transfer rates and single chamber two-phase flow are being exploited for engine operation. Washington State University researchers characterize their engine as a "mixed-length scale" device since a large planar design results with a small thickness.

For practical micro thermal engines, it will be important to exploit all options available for enhancing performance. Two techniques useful in this context are heat recovery and isothermal expansion. The first of these techniques uses regeneration (or recuperation) wherever feasible. Brayton cycle engines are potential candidates for heat recovery where a portion of the thermal exhaust energy is used to preheat the compressed air stream. Meso/microscale counterflow heat exchangers are needed for this purpose. The second technique, isothermal expansion by heat addition (or "reheating"), would be used in conjunction with recuperation. Isothermal expansion leads to higher performance and efficiency if heat recovery can be effectively employed. These concepts can be applied to other cycles, although it may be more difficult to do so. These two techniques are well known, but rarely used in macroscopic engines due to the added complexity. Note that from a cycle pressure ratio point of view, using heat recovery and isothermal expansion makes sense because miniature heat engines will probably use lower pressure ratios than macroscopic engines.

3 Scaling

In a landmark talk given in 1959 entitled *There's Plenty of Room at the Bottom*, Richard P. Feynman [47] introduced the field of microscale and nanoscale engineering by describing a number of different scenarios and approaches to making things very small. He likened the potential of this new technology to early physics research at low temperatures, or at high pressures, where discoveries led to important advances in both science and technology. Feynman's premise was that no fundamental physical laws limit the size of a machine down to the microscopic level. Different effects may become important, but it is a matter of understanding how to harness those different effects that determine how fast microscale technology is developed. Of the several examples given in the talk, scaling of an automobile stands out from an engineering point of view. Problems such as tolerances, force scaling, electrical effects, lubrication, and heat transfer were briefly touched upon as areas where microscale behavior would be radically different from macroscale behavior. Scaling studies of this nature can provide great insight into the fundamental challenges that take place when attempting to miniaturize a macroscopic device. Feynman did identify heat transfer as one fundamental mechanism that would prevent an internal combustion engine from working at small length scales. But, at what length scale does this occur? Rapid thermal losses from a nascent ignition kernel inside a small cylinder would quench exothermic reactions. Increasing the pressure or temperature of the reactant mixture would help, but the scaling consequences are inevitable when it comes to the type of combustion found in the common automobile IC engine. Feynman's example provides insight and understanding into some of the challenges confronting miniaturization of energy systems. It is also of interest to examine the means to determine ultimate size limits of relevant power producing concepts. This section examines selected scaling laws for miniature energy systems.

3.1 Scaling methodology

Humans are accustom to viewing the world at a particular length scale. Based on this view, intuition has developed regarding how our surroundings work. Outside this "natural" size level, our intuition can lead us astray. People are often amazed at how ants can lift several times their weight, or how a grasshopper can jump an incredible distance compared to its size. However, knowing that mass scales with the characteristic length cubed and muscle power scales with the same length squared leads to the conclusion that these super feats of performance are a direct result of being small.

Much of our insight into scaling first came by observing natural systems and phenomena. Thompson [48] has explored the scaling issue of natural systems in a book entitled *On Growth and Form*. The cube law of mass with "size" can explain obvious behavioral characteristics of large and small animals. The larger animals typical move slowly and ponderously as they travel and forage. However, small animals live a fast, agile existence. Also pointed out by Thompson are less obvious effects such as the diameter of trees growing with the 3/2 power of their height. Other observations include warm-blooded animals having a size limit due to surface area decreasing with the square of length, while mass, and hence internal heat generation, decreasing with the cube. Shrews are noted for their voracious appetites since they burn proportionally more energy in staying warm than larger animals. Hayashi [49] has examined a number of natural and technological systems for their scaling characteristics. An interesting derivation made was to show why bigger fish and ships travel faster than smaller ones, but only with the square root of their characteristic size. This is a result of the available energy for motion scaling with the length cubed while the resistance to motion, i.e., skin friction, scaling with the surface area and realizing that the power dissipated in movement goes with the velocity squared. One other interesting result from scaling analysis is an estimate of the ultimate size of walking land animals. Benton [50] shows that if legs are required to both support the weight of the animal and walk at the same time, the weight of the animal cannot exceed approximately 140 tons. Although a number of assumptions about structure and behavior went into the analysis, it is interesting to note how scaling has been applied to this type of estimate.

Scaling of technological systems has been examined by Trimmer [51] who has developed an elegant vertical bracket notation for representing how systems behave. As a classic example, Trimmer shows how a simple parallel plate capacitor behaves upon isometric (or, isomorphic) scaling, i.e., all dimensions are reduced by the same proportion giving a self-similar geometry upon reduction in size. Madou [52], who provides an excellent coverage of scaling and its impact on the operation of microscale components, provides a description of Trimmer's method. The basic scaling operation is to generate a governing equation for a particular effect which could be a force, power output, heat transfer rate, efficiency, etc., and to identify the functionality of the variables on characteristic size. Material properties such as density, viscosity, and thermal conductivity are treated as invariant with scale. This is an appropriate assumption for mesoscopic systems as well as the higher ranges of the microscale regime. It obviously breaks down as the continuum limit is reached, or when device size begins to approach the graininess of the materials used. With this simple approach to scaling, a number of functionalities have been identified and discussed in previous works. Table 1 shows some of the common physical, or technological, effects and their dependencies on length as given by Madou [52] and others. It should be noted that a number of ways exist for creating a scaling equation. It can depend on the type of scaling, either isomorphic or restricting the reduction in size to only one dimension. It can also depend on the constraints imposed on the scaling process. For example, in scaling magnetic forces, if the current density is kept constant, the force between two wires or coils is dependent on L^4. If the interaction is taken between a coil and a permanent magnet, the force scales as L^3. Since heat can more easily be transferred from a coil to the surroundings at small dimensions, a constant heat flow per unit area leads to L^3 for two coils or $L^{2.5}$ for a coil and a permanent magnet [53]. Hence, functionality can be dependent on how a system is analyzed.

3.2 Common phenomena important in energy systems

Miniaturizing thermally based energy systems can involve many of the phenomena listed in Table 1. In addition to the more obvious effects such as heat transfer, gas expansion, and fluid flow, a number of mechanical effects may also be necessary to consider such as friction. Several of the above mentioned effects are explored in the following.

Table 1: Scaling of physical effects or processes.

Process	Scaling	Comments
Time	L^0	Invariant with scaling
Distance or length	L	Scales proportionally with length
Area	L^2	Dependent on the linear dimension squared
Mass	L^3	Mass is proportional to volume
Velocity	L	Units of m/s, hence proportional to length
Power	L^3	Power often scales with the volume
Energy of capacitor	L^3	Energy of capacitor tied to volume
Stored energy, fuel	L^3	Liquid fuel storage is tied to volume
Surface tension	L	Surface tension important at small scales
Inertia forces	L^3	Dependent on the mass of the object
Electrostatic force	L^2	Force generated by parallel plates
Magnetic force	L^3	Magnetic force can scale with L^2, L^3, or L^4
Muscle force	L^2	Muscle force scales with area
SMA force	L^2	SMA actuators depend on area
Kinetic friction	L^2 or L^3	Dependent on applied normal force
Frequency	L^{-1}	Frequency of operation increases on scaling

Differential movement between mechanical components will be accompanied by friction. A classic example at the macroscale would be a piston rubbing along the inside surface of a cylinder. A seal, such as a piston ring, in contact with the cylinder generates a frictional force as the piston executes its cycle. This frictional effect could be dry or lubricated, depending on the application. Another comparable system would be the seal on a Wankel rotor spinning inside its contoured chamber. The constitutive relationship governing kinetic friction can be describe as

$$F_k = \mu_k N \tag{1}$$

where μ_k is the coefficient of kinetic friction and N is the magnitude of the applied normal force. This frictional force, assumed here to be between dry unlubricated surfaces, is approximately independent of the contact area and is proportional to the normal force. Additionally, kinetic friction is relatively independent of the rubbing speed at the contact surface as long as the temperature is constant [54]. Thus, in order to determine frictional scaling effects, one needs to examine how N would vary as a system is reduced in size. If the normal force were produced by weight or inertia, then the magnitude of the frictional force would be dependent upon L^3, since the mass of an object varies with the cube of its characteristic size. If, however, the normal force were dependent on a pressure (such as occurs at the face of a piston), then the dependency would scale with the area, or L^2. It is interesting to place the frictional force in the context of an operating, power producing device. It is often stated that microscale devices suffer from enhanced frictional effects. This may be true in instances where the normal force scales with surface tension or surface attractive forces that may cause sticking. Static friction may also be important to consider in this case. At the mesoscale level, if the normal force on a friction-bearing component is produced ultimately by a pressure acting over a piston area, then the frictional effects and the motive force for power production scale comparably. Also, the work produced during a stroke depends on $P\Delta V$, or directly on the change in gas volume. This scales with L^3. The work dissipated in overcoming kinetic friction would depend on the stroke length and piston area, thus also leading to a dependency of L^3. The conclusion here is that work done by an expanding working fluid decreases as fast as the work dissipated in overcoming friction, hence kinetic friction does not appear to dominate piston-type mechanisms upon scaling.

Clearance seals may appear to provide a way to avoid contact friction. Clearance seals have been considered previously in Stirling engine design [55] and are important in other energy producing devices such as compressor and turbine rotors of gas turbines, and labyrinth seals on rotating machinery. Analysis of clearance seals is more involved than examining the relationship governing friction – a model is needed for describing the behavior of the seal in the context of an expansion process. Consider a piston inside a cylinder having a piston-to-cylinder wall clearance of h, as shown in Figure 3. At the start of the expansion, the motive gas is confined to the volume of the cylinder to the right of the piston. As the gas expands, some of it escapes through the clearance space between the piston and cylinder. A simple model for this situation assumes the space is very narrow such that laminar flow occurs. Also, for estimation purposes, only average quantities are used in the formulation of a descriptive relation. Thus, the mass of gas undergoing expansion is given by

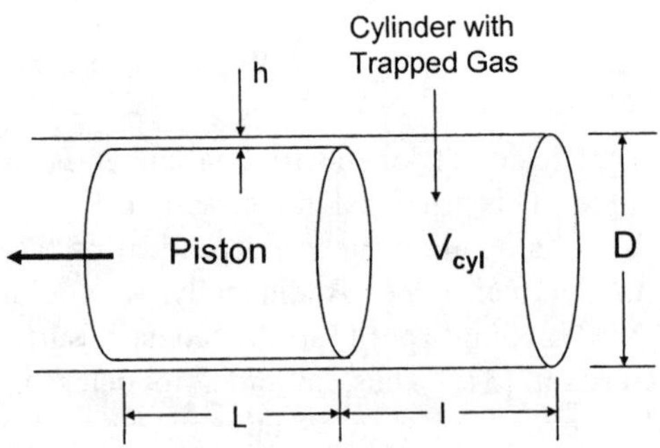

Figure 3: Model for a clearance seal in a piston–cylinder arrangement.

$$m_g = \rho_g V = \rho_g \frac{\pi}{4} D^2 l \tag{2}$$

where ρ is the density of the gas, D is the diameter of the cylinder, and l is the length of the region containing the gas. Now, if laminar flow is assumed in the annular gap between the piston and cylinder walls, an expression for the mass flow rate through the annular gap takes the form

$$\dot{m} = \frac{\pi}{8} \frac{\rho D \Delta p h^3}{\mu L} \tag{3}$$

where ρ in this case is the density of the gas flowing through the annular gap, Δp is the pressure difference across the length of the piston, h is the gap width, μ is a representative viscosity, and L is the length of the piston. One way to characterize the performance of the clearance seal is to form a quantity that describes the fraction of mass leaving the expansion volume during a characteristic time period (e.g., the time it takes to expand the gas). Introducing τ as this characteristic time, the performance metric becomes

$$\frac{\dot{m}\tau}{m_g} = \frac{\rho \Delta p}{2\mu \rho_g} \frac{h^3 \tau}{LDl} \tag{4}$$

For proper functioning of the seal, this ratio should be much smaller than unity. The right hand side of the expression is separated into two parts: the first being a quantity that is assumed invariant upon scaling while the other is dependent on the characteristic size of the piston-cylinder arrangement. Upon isomorphic scaling, the functionality of the expression shows the clearance seal maintains its performance if τ remains constant. This is an important statement because it shows clearance seals apparently do not suffer any adverse scaling effects. However, practical considerations tell a different story. For a 1 cm diameter piston operating in the 100 to 1000 Hz frequency range, a practical clearance seal is on the order of 10 μm. Upon isomorphic scaling, a reduction in size by a factor of 10 brings the clearance gap down to 1 micron, and further reductions push the gap into the nanoscale regime. This gap must be maintained during the motion of the piston over relatively long travel distances. The challenge of maintaining isomorphic dimensions in scaling the clearance seal may be insurmountable for small-scale gas expansion devices. This also includes rotary gas dynamic seals on compressor and turbine rotors. One helpful consideration here is that τ often decreases as the characteristic size of the device is reduced, as discussed next. If this is the case, then isomorphic scaling is actually favorable for a clearance seal. However, if due to fabrication constraints the gap width remains constant, the h^3 term in the numerator of eqn. (4) is potentially a severe disadvantage during scaling.

The cycle frequency of a gas expander or compressor will often scale favorably. Again, the piston–cylinder arrangement is examined as a model. Assume a piston divides a cylinder into two spaces where one space undergoes a gas expansion or compression process, and the other contains a spring for restoring the position of the piston after the process. This arrangement is often used in free piston Stirling engines where operation at the resonant frequency occurs. The spring in the case of Stirling machines can be a gas volume trapped on one side of the piston. For this case, a first order approximation of the spring constant is

$$k_s = C\frac{PA_p}{V} \tag{5}$$

where C is a constant, P is an average gas pressure in the bounce space, A_p is the cross-sectional area of the piston, and V is an average volume for the bounce space. The frequency of operation for the piston can be approximated by

$$f_n = \frac{1}{2\pi}\left(\frac{k_s}{m_p}\right)^{1/2} \tag{6}$$

where m_p is the mass of the piston. Substituting into this expression for the spring constant and piston mass gives

$$f_n = \left(\frac{CP}{4\pi^2\rho}\right)^{1/2}\left(\frac{D_P^2}{VL}\right)^{1/2} \tag{7}$$

where D_P and L are the diameter and length of the piston, respectively. The second term on the right hand side of the expression contains the dependency on the dimensions of the system. Upon isomorphic scaling, the frequency is inversely related to the characteristic size of the system. This dependency will also hold for mechanical springs as well as the rotating mechanisms such as turbine and compressor disks (if the peripheral speed remains constant). Hence, small-scale systems will have a higher frequency of operation than larger scale devices.

Heat transfer plays *the* primary role in thermally based energy systems. However, a number of different ways exist for examining the scaling characteristics of heat transfer systems. Different phenomena are relevant for the three different modes of transfer, namely conduction, convection, and radiation. An in-depth scaling analysis of all relevant problems will not be given here, but a sampling of the types of scaling important to miniature energy systems will be presented. The basic expression for one-dimensional heat conduction is

$$\dot{Q} = -kA\frac{dT}{dx} = -kA\frac{T_2 - T_1}{L} \quad (8)$$

where k (assumed constant here) is the thermal conductivity of the material. The second equivalence on the right hand side of eqn. (8) assumes constant properties where a slab of material of thickness L has a temperature difference imposed across it. There are many ways of looking at scaling in this problem. For isomorphic cases, conduction heat transfer scales with the characteristic length, L, of the problem (this assumes $A \sim L^2$). Another important view is to look at "mixed length scale" systems. This type of heat transfer configuration could take the form of a large plane surface area and a much smaller thickness dimension characterizing the heat transfer. If only the thickness dimension is scaled, then the heat transfer varies with the inverse of the characteristic length assuming the temperature difference can be maintained. This type of scaling may be important if microscale devices are arrayed together in a planar configuration to provide macroscale effects.

In some specific cases, the scaling is even stronger. Consider the situation where internal heat generation exists in a slab of material with a half thickness of L. The internal heat generation rate can be expressed as

$$\dot{Q} = \frac{(T_{max} - T_s)2k}{L^2} \quad (9)$$

where T_{max} is the centerline temperature and T_S is the surface temperature of the slab. Considering just the heat generation rate, it is obvious that it scales with the inverse square of the characteristic length if the temperature difference remains constant.

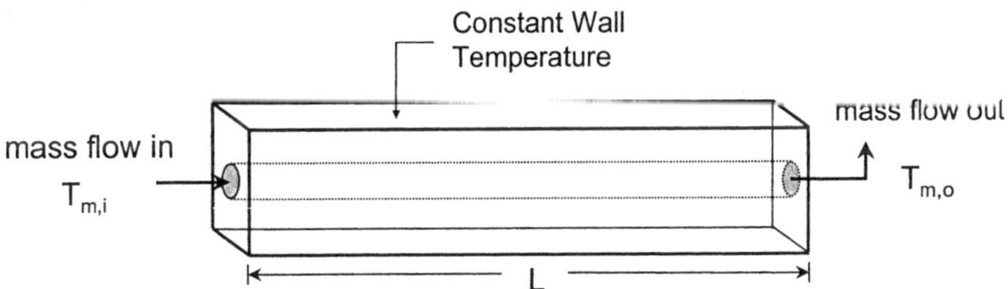

Figure 4: Flow of fluid through a constant wall temperature structure.

One other important case is considered here. Figure 4 shows the classic heat transfer problem where an internal flow is maintained along a constant wall temperature duct. The nondimensional fluid temperature difference between the inlet and outlet describes the performance of the system from the point of view of transferring heat to the fluid. This quantity is defined as (see, e.g., [56])

$$\frac{T_s - T_{m,o}}{T_s - T_{m,i}} = \exp\left[-\frac{P_C L}{\dot{m}c_p}\bar{h}\right] \tag{10}$$

where $\dot{m}c_p$ is the mass flow rate multiplied by the constant pressure specific heat. Also, P_C is the inside perimeter of the flow conduit, h is the convective heat transfer coefficient, and L is the length of the duct where the outlet temperature, $T_{m,o,}$ is measured. Keeping the average flow velocity, U, constant upon scaling and introducing the Nusselt number, this expression can be recast in the form

$$\frac{T_s - T_{m,o}}{T_s - T_{m,i}} = \exp\left[-\frac{4Nu_D k_f}{\rho U c_p}\frac{L}{D^2}\right] \tag{11}$$

where k_f and ρ are the fluid thermal conductivity and density, respectively. The first term in the brackets is assumed invariant upon scaling while the second term contains the quantities dependent on the characteristic length of the system (note that L in this case is the length of the conduit and not a scaling parameter). If the nondimensional outlet temperature is to remain constant upon scaling, L/D^2 must also remain constant. Thus, reducing the conduit diameter by a factor of two reduces the length by a factor of four. This is an inverse square relationship and it results in a strong performance increase (i.e., decrease in length or residence time of the fluid) as the size of the system is reduced. If the same mass flow as some original larger device is to be matched, several smaller conduits must be used in place of the single larger one, but the length of the scaled device will always be shorter.

The cost of this increase in performance is a higher pressure drop across the smaller flow passage. This pressure drop is given by

$$\Delta P = f\frac{\rho U^2}{2D}L \tag{12}$$

whereupon, assuming laminar flow,

$$\Delta P = 32U\mu\frac{L}{D^2} \tag{13}$$

For the situation where flow through the passage is driven by a pressure drop (under laminar conditions), the mass flow rate can be written as

$$\dot{m} = \frac{\pi\rho\Delta P}{128\mu}\frac{D^4}{L} \tag{14}$$

Thus for smaller flow conduits, the pressure drop upon isomorphic scaling increases with $1/D$ whereas the mass flow rate decreases with D^3. This functionality will eventually dictate a loss of performance due to excessive pressure drops and small flow rates. However, for mesoscopic systems, pressure drops can still be acceptable.

3.3 TEC example

All heat engines are governed by specific physical laws. These laws can often be simplified to express a particular physical process critical to the operation of the engine. One particular type of heat engine operating with an unconventional working fluid is a thermoelectric generator. In this case, electrons execute a cycle governed by the laws of thermodynamics.

Consider a Peltier generator used for power production from a low-grade source of heat. The analysis presented here is similar to that given by Min and Rowe [57] where a Peltier thermoelectric device is positioned between a heat source at T_H and a heat sink at T_L, as shown in Figure 5. Examining first the ideal case where the surfaces of the generator are maintained at $\Delta T_0 = T_H - T_L$, the power output would be

$$P_i = \left(\frac{\alpha^2}{\rho}\right)\left(\frac{\Delta T_0^2}{2}\right)\left(\frac{A_0}{L_0}\right) \tag{15}$$

where α is the Seebeck coefficient, ΔT_0 is the temperature difference across the device, ρ is the electrical resistivity, and L_0/A_0 is the ratio of the thermoelectric element length to cross-sectional area. This expression predicts an increasing power output with decreasing element thickness (assuming A_0 remains constant, i.e., nonisomorphic scaling). Thus, as L_0 goes to zero, the ideal case predicts an unrealistically large power output.

Looking at the efficiency, heat transferred from the high-temperature reservoir can be approximated as $\dot{Q}_H = k_p A_0(\Delta T / L_0)$. If eqn. (15) is divided by this quantity, the resulting expression yields,

$$\eta = \frac{P_i}{\dot{Q}_H} = \frac{\left(\frac{\alpha^2}{\rho}\right)\left(\frac{\Delta T_0^2}{2}\right)\left(\frac{A_0}{L_0}\right)}{k_p A_0\left(\frac{\Delta T_0}{L_0}\right)} = \left(\frac{\alpha^2}{\rho}\right)\left(\frac{\Delta T_0}{2k_p}\right) \tag{16}$$

where k_P is the thermal conductivity of the generator material. Equation (16) shows that the ideal thermal efficiency of a thermoelectric generator is invariant with scaling. Thus, as the device becomes smaller, the increasingly higher rates of heat flowing through the generator will be converted into electricity with the same thermal efficiency.

This ideal analysis, however, does not accurately portray the actual heat transfer situation that accompanies the operation of the generator. Any real thermoelectric device, as shown in Figure 5, has a layer of insulating ceramic on both sides. In addition, thin copper straps are incorporated into the design to provide a current flow through the junctions. Finally, the temperatures T_H and T_L are typically provided by gaseous or liquid heat transfer fluids where a convective coefficient, h, is needed to describe the heat transfer between the reservoirs and the top and bottom ceramic plates. With this situation, a thermal resistance is present between the hot and cold junctions and their respective reservoirs. This fundamentally changes the heat transfer, and hence the scaling laws.

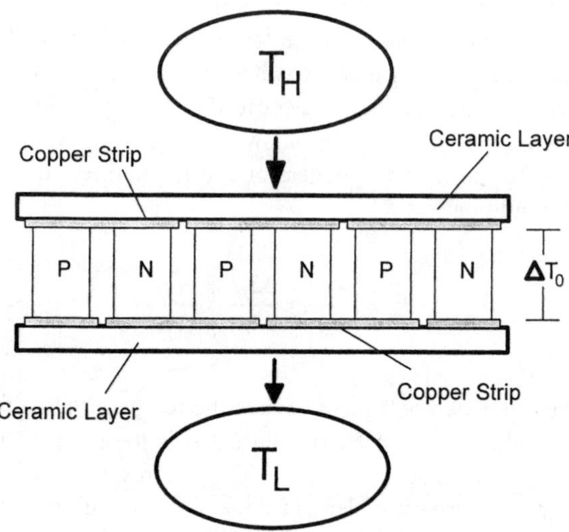

Figure 5: Schematic diagram of a Peltier thermoelectric generator.

If the total thermal resistance between the hot and cold reservoirs can be represented by three series resistances composed of two overall heat transfer coefficients and a single conduction term, then it can easily be shown that,

$$\Delta T_0 = \frac{L_0}{k_p A_0}\left(\frac{T_H - T_L}{\dfrac{1}{h_H A_0} + \dfrac{1}{h_L A_0} + \dfrac{L_0}{k_p A_0}}\right) \tag{17}$$

where h_H and h_L are overall heat transfer coefficients for the hot and cold sides of the device, respectively. If the efficiency is assumed small enough such that Q_H is approximately equal to Q_L, then substituting eqn. (17) into eqn. (16) gives

$$\eta = \left(\frac{\alpha^2}{\rho}\right)\left(\frac{\Delta T_{H-L}}{2k_p}\right)\left(\frac{1}{1+\left(k_p / L_0 \quad h_{tot}\right)}\right) \propto \left(\frac{1}{1+\left(1/L_0\right)}\right) \tag{18}$$

where the final proportionality assumes that α, ρ, $\Delta T_{H\text{-}L}$, k_P, and h_{tot} are invariant with scaling. Finally, as the characteristic size of the device is reduced, $1/L_0$ becomes much larger than unity and eqn. (18) reduces to the efficiency being proportional to L_0. This exercise has demonstrated that the thermal efficiency of TEC generators suffers as the characteristic size is reduced.

3.4 Heat engine example

Can the same analysis be applied to a conventional heat engine? Peterson [58] presented work on scaling heat engines by developing a generalized model, which included both intrinsic and extrinsic heat conduction terms. Early in the paper, a simplified model was presented to make clear the effect of size reduction on the thermal efficiency.

Figure 6 shows a simple model of a heat engine operating on a Stirling cycle with a thermal shunt resistance connecting the two reservoirs. This model includes an intrinsic source of irreversibility that is tied to a heat conductance shunt path. The model also contains a characteristic size parameter, L. High and low temperature reservoirs are denoted by T_H and T_L, respectively.

For a simple approach to scaling, let the shunt establish a thermal conductance between the high and low temperature reservoirs. The cross-sectional area and length of this shunt path are $A_C = \pi / 4\, D^2$ and L, respectively. Further, assume that during scaling, D is proportional to L. Note that heat flow through the shunt, Q_S, is unavailable for work production. The engine generates work from another heat flow path, Q_{HE}. An estimate of the output power uses the difference between isothermal expansion of the working fluid at T_H and isothermal compression at T_L. For a regenerative engine cycle such as a Stirling engine,

$$\dot{W}_{out} = R\dot{m}T_H \ln(r) - R\dot{m}T_L \ln(r) = R\dot{m}\ln(r)\left(T_H - T_L\right) \tag{19}$$

where R is the gas constant, $\dot{m}$ is the mass flow rate through the engine, and r is the volume ratio characterizing the compression and expansion processes. The terms R, r, T_H, and T_L are assumed invariant with scaling, but $\dot{m}$ is not. It is assumed that $\dot{m} = m_f\, f$ where m_f is the mass of working fluid within the engine and f is the frequency of operation. Now, m_f will scale with L^3, but f is proportional to $(1/L)$ due to the fact that smaller engines operate at higher frequencies. Also, it can easily be shown that $\dot{W}_{out} \propto \dot{Q}_{HE}$ for an ideal heat engine.

With these quantities and eqn. (19), the thermal efficiency of the heat engine shown in Figure 6 can be expressed as

$$\eta = \frac{\dot{W}_{out}}{\dot{Q}_S + \dot{Q}_{HE}} \propto \frac{L^3(1/L)}{L + L^2} \propto L \tag{20}$$

For the proportionality terms of eqn. (20), the shunt conductance was modeled as purely conductive, i.e., $\dot{Q}_S = k_p A_C (\Delta T / L)$. Upon scaling, k_P and ΔT are assumed invariant. Also, in the denominator of eqn. (20), as L goes to zero, L^2 can be neglected. This simple exercise shows that even without external sources of thermal resistance, the engine efficiency decreases as its characteristic size is reduced. This is primarily caused by the intrinsic shunt conductance, which is made up of all structural components connecting the high-temperature reservoir with the low-temperature one. As L approaches zero, the shunt conductance acts as a dead short between the two reservoirs causing the system efficiency to approach zero.

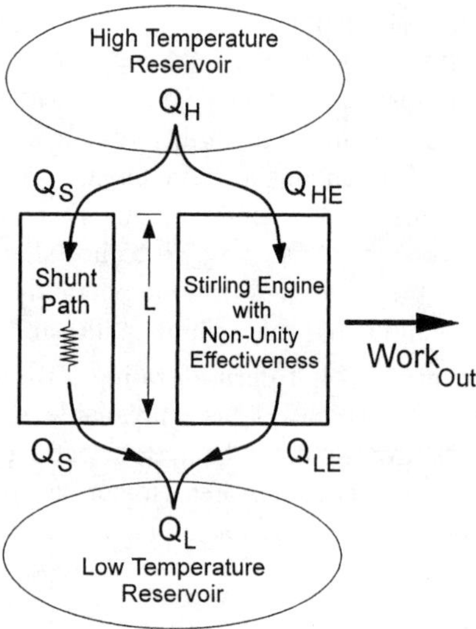

Figure 6: Schematic diagram of Stirling engine model with thermal shunt path.

3.5 Other thermal systems

Thermally based MEMS are devices that rely on the transfer, or generation, of thermal energy for their function. Relatively little work has been accomplished in this area to date, but with increasing interest in the topic of microtechnology-based energy and chemical systems (MECS), thermally based MEMS are beginning to receive attention. In order for these devices to function properly, a small region within the overall device must be thermally isolated from the surrounding ambient environment so that high or low temperatures can be maintained. The isolated region, however, may still require the capability of exchanging energy with the surrounding structure and hence must be physically connected to it – this is where energy transfer processes create significant challenges for the designer. That is, as the characteristic size of the device is reduced, heat transfer through the structure is enhanced making it increasingly more difficult to maintain a temperature difference. Consequently, device performance will degrade as the length scale decreases unless special provisions are made to minimize the effects of heat transfer.

Many of these thermally based systems could be evaluated for their scaling laws, and for their potential applications in the meso/microscale size regime. One final topic discussed in this section concerns heat exchangers; specifically counterflow devices for use in isolating a high-temperature region from ambient conditions. Such an arrangement could be used for micro-refrigeration based on the vapor compression cycle, recuperative micro-combustors, small-scale chemical reactors, or elevated temperature fuel cells.

The design of counterflow heat exchangers is mature from the standpoint of engineering large-scale devices for industry. The mechanism of heat transfer between the two fluid streams can be expressed by an overall transfer coefficient while heat conducted in the streamwise direction can be neglected. Meso/microscale devices differ from their macroscale counterparts in that streamwise heat conduction must be considered. This is because of the small length scales involved and the fact that streamwise, or axial, conduction rivals the transverse flow of heat in the microscale regime.

Early past work in this area includes Meissner [59], Hausen [60], Landau and Hlinka [61], and Kroeger [62]. Only the latter two provided quantitative treatments of the differential equations describing the problem. The analysis by Landau and Hlinka was important for the way the boundary conditions were treated – at both ends of the heat exchanger, insulated wall conditions were used. This choice guided later similar studies of the problem including Kroeger's work, which greatly extended the applicable range of previous results.

Modern day treatments of the problem [63, 64, 65] examined thermally isolated systems where insulated boundary conditions were used. This approach essentially neglects conductive heat loss to the surroundings as would occur at the base of a microscale heat exchanger where it connects to an ambient temperature structure. The author has examined this problem with a focus on determining the heat loss from small-scale heat exchangers. The nonzero temperature gradients at the attachment point of the device was explicitly taken into account in a series of papers designed to look at various heat loss mechanisms and quantify the total overall loss [66, 67, 68]. As one example of the consequences of scaling counterflow heat exchangers, Figure 7 shows the loss of heat from a counterflow device as it is scaled through the mesoscale regime. Energy losses due to conduction, radiation, and nonunity effectiveness are plotted as a function of characteristic length of the device. The details of the analysis can be found in the above referenced papers, but for the case examined in Figure 7, the heat exchanger has a geometry as shown in the inset to the figure, and the end temperature was set to 1200 K. Thermal conductivity, external surface emissivity, and other important parameters can be found in the cited work. The curves clearly show the dominant heat loss mechanisms as the device is scaled. Ultimately, conductive losses prevail – they are determined by the thermal conductivities of the fluid and wall material, and the characteristic length of the heat exchanger. This type of analysis is useful for determining the ultimate practical size limitations on counterflow heat exchangers.

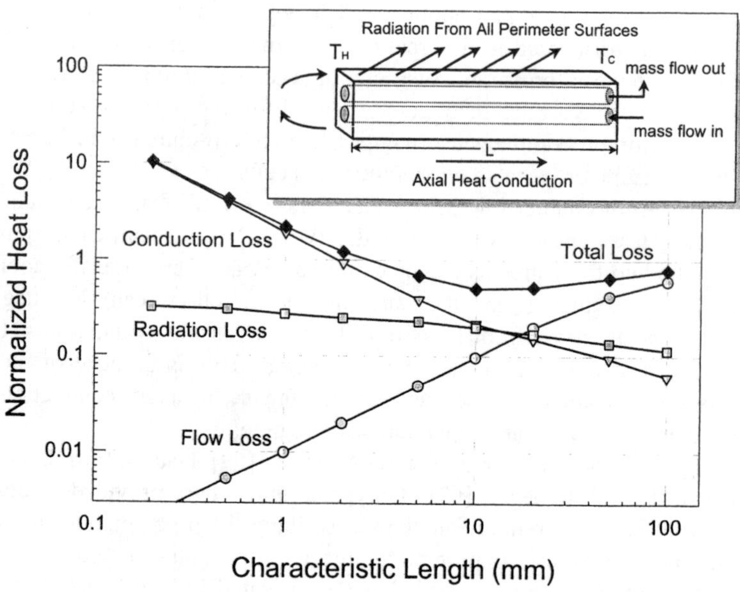

Figure 7: Behavior of the primary heat loss mechanisms in a counterflow heat exchanger upon scaling.

4 Thermally based power systems

From the concepts covered so far, heat transfer plays *the* dominant role in determining the physical size of small-scale, thermally based power systems. The key design characteristic here is the close proximity of the hot section to the ambient temperature section. Upon a reduction in size, heat transfer between the two sections is enhanced thus lowering performance. Of the thermally based devices listed in Figure 2, all will suffer this "degradation" upon scaling. We have already examined the TEC concept and found its performance to depend on heat transfer effects. Engines built from mechanical components (such as pistons and cylinders, or turbines and compressors) have physical structures that also conduct heat throughout the device. At first glance, concepts such as AMTEC, thermionic, and thermophotovoltaic energy systems may be less susceptible to the thermal degradation problem. This conclusion may result from realizing that the "working fluids" (sodium atoms, electrons, or photons) of these engines pass across a "gap" with little parasitic heat transfer occurring. But, how is the heating accomplished in these devices? If combustion is used, the reactants must be delivered through ducting to the hot section and exhaust flow must also be removed. If a nuclear source of heat is utilized, the hot section still cannot be "free-floating" – it must be structurally tied to the rest of the system. Even if conceptually the physical tie could be severed between the hot and ambient temperature sections, thermal radiation would take place. This decoupled configuration would then be susceptible to surface area scaling (proportional to L^2) while the heat generation scaled with L^3. Based on this, the elevated temperature of a free-floating structure would be progressively more difficult to maintain upon size reduction. Thus, all practical thermally based energy systems will have the problem of thermal degradation of performance upon scaling.

Can a model be developed that captures this degradation effect while providing an estimate for the limits on physical size? Any calculation will necessarily require some physical configuration and material properties to estimate the ultimate size. A true theoretical limit would be difficult to set. However, the following describes a methodology, and provides a numerical example, of what can be accomplished in this area.

4.1 A theoretical model for size limits

A model for predicting the performance of small-scale power systems is shown in Figure 8. Fundamental processes captured by the model are: (1) a combustor at a temperature elevated from the surroundings, (2) a support structure for isolating the hot region from ambient, (3) a counterflow heat exchanger, as part of the support structure, for supplying the combustor with reactants, and (4) an ideal Carnot heat engine for converting heat to work. The model takes into account a number of system characteristics, including:

- Heat loss due to radiation, conduction, and nonunity effectiveness (the latter from a counterflow heat exchanger)
- Variation in the relative size of the combustor with respect to the device length
- Variation in the length, diameter, and wall thickness of the delivery tubes
- Thermal conductivity and surface emissivity as system input parameters
- Mass flow rates and other thermal properties, including heat of combustion, can be set for the reactant mixture entering the combustor
- The model is restricted to constant properties and steady state operation.

Heat loss mechanisms are central to this model. Using simple expressions for the losses, an energy balance is performed on the combustor yielding its temperature. Included in this balance is heat supplied to the ideal heat engine to produce work. This heat loss term is critically important for determining the overall operating performance of the combined system. The computational model treats the heat flow through the engine as another loss from the hot region. Thus, if nearly all of the heat from combustion is routed through the engine, the combustor would operate at an unrealistic temperature only slightly above ambient. At the other extreme, if little heat is sent through the ideal engine, the temperature of the combustor is high and is determined by the major heat losses, which are balanced by the mass flow and heating value of the reactants. There is a maximum operating point between these two extremes giving the most favorable operating point for the engine.

An energy balance on the combustor requires an operating temperature. Although this temperature is found through solving a set of algebraic equations, for the sake of discussion, it will be assumed to have a value of T_H. The configuration chosen for the model is a reaction chamber operating at temperature T_H and connected to ambient conditions through two tubes – one tube for the delivery of reactants and one for removing products. The tubes are arranged in a counterflow configuration to recover some of the thermal energy leaving the system with the products. Because the surroundings are at ambient temperature, it is anticipated that heat loss from the combustor will be from radiation to the surroundings and conduction through the delivery tubes. The tubes also lose heat by radiation and conduction, the latter being through axial heat conduction down the length of the tubes. In addition, the counterflow heat exchanger introduces a loss due to nonunity effectiveness.

For the combustor, gray body radiation to the surroundings is modeled by

$$\dot{Q}_{rad} = \varepsilon\sigma F_{1-2} A_{surf}\left(T_{surf}^4 - T_{surr}^4\right) \tag{21}$$

where ε is the surface emissivity, σ is the Stefan Boltzmann constant, F_{1-2} is the radiation shape factor (assumed to be unity in this work), A_{surf} is the surface area of the hot region, and T_{surr} is assumed to be 300 K. Conduction losses from the combustor are also present because of the temperature gradient that exists at the hot end of the delivery tubes, as shown in Figure 8. As shown in the figure, there is a direct coupling of the conduction losses from the combustor and the radiation and conduction along the length of the counterflow heat exchanger. Another way of looking at the problem is to realize that any energy loss due to radiation from, and conduction within, the heat exchanger must come from the combustor by way of conduction. Using an energy balance, the conduction loss from the combustor is set to the sum of the radiation and conduction losses from the heat exchanger. Thus, the conduction loss from the combustor can be written as

$$\dot{Q}_{cond} = \varepsilon\sigma F_{1-2} P_{HEX} \int_0^L \left(T_s^4(x) - T_{surr}^4\right)dx + k_{eff} A_{tot}\left(\frac{T_H - T_{surr}}{L}\right) \tag{22}$$

where P_{HEX} is the combined delivery tube (or heat exchanger) perimeter, $T_s(x)$ is the temperature gradient along the length of the delivery tubes, k_{eff} is the effective thermal conductivity of the tube wall and gas flow area, A_{tot} is the total cross sectional area of the delivery tubes, and L is the length of the delivery tubes. In this expression, the temperature profile is assumed linear. Although this only approximates the true profile, it is adequate for estimating the radiation and conduction losses. The last heat loss term for the combustor is associated with the entering reactants. Due to nonunity effectiveness, the reactants are not at T_H but are at a temperature somewhat below this value. This heat loss can be determined by

$$\dot{Q}_{eff} = \dot{m}C_p\left(T_H - T_{surr}\right)\left(1 - E\right) \tag{23}$$

where $\dot{m}$ is the reactant mass flow rate, C_p is the constant pressure specific heat of the reactants, and $\underline{E}$ represents the effectiveness of the counterflow delivery tube arrangement. With these expressions, and considering the useful energy supplied to the ideal heat engine, the energy balance on the combustor is

$$\dot{Q}_{tot} = \dot{Q}_{rad} + \dot{Q}_{cond} + \dot{Q}_{eff} + \dot{Q}_{useful} \tag{24}$$

Since the first three terms on the right hand side all increase as T_H increases, the energy balance is attained when Q_{tot}, calculated by eqn. (24), equals the heat generated by combustion of the reactants (at a mass flow of $\dot{m}$). Finding this temperature requires solving a system of algebraic equations. Note that Q_{useful} is set to progressively larger values starting from zero and going to the total heat of combustion for the reactant flow. Throughout this range, the combustor temperature goes from its highest values to near ambient. Within this range, a maximum power point exists where both Q_{useful} and the heat engine efficiency (assumed to be equal to the Carnot efficiency) maximizes the engine power output. This is taken as the natural operating point of the combined system. The computational routine calculates this point for a given set of input parameters.

From Figure 8, the geometry of the combustor is shown to be spherical with delivery tubes depicted as passages in a cylinder. This baseline configuration can be changed through parameter ratios incorporated into the model. For example, the scaling parameter for the model is based on L (the length of the delivery tubes). The combustor diameter is tied to L through a diameter-to-length ratio. Other geometric parameters for the delivery tube diameter and wall thickness are also tied to L. With this method of connecting the geometrical constraints of the problem together, isomorphic scaling studies can be done by changing the magnitude of L.

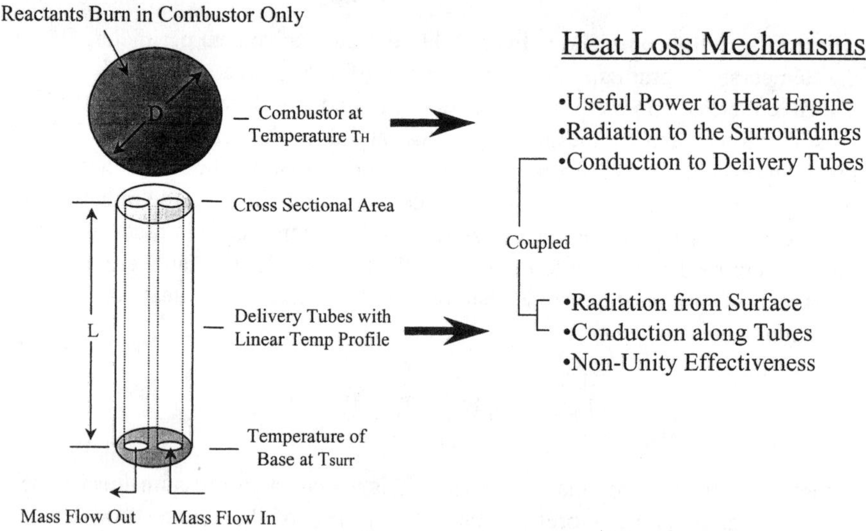

Figure 8: Diagram of the microcombustor thermal loss model.

The computational model has been run for two cases to explore the size limits imposed on the system due to heat transfer effects. Figure 9 shows the results of the computations for a microcombustor having a length-to-combustor diameter ratio of 10. It was assumed that the heat of reaction of the incoming flow was similar to stoichiometric hydrogen–air (3.0×10^6 J/gm) and that the flow velocity through the inlet tube was 5 m/s. The corresponding flow rate gives progressively shorter residence times in the combustor, and for a length scale of 1 mm, the residence time falls below 1 msec. It is normally necessary to have a residence time of between 1 and 10 msec for high combustion efficiency, thus assuming 100% heat release at the smaller length scales is certainly an idealization. Practical devices may have lower flow velocities and hence larger ultimate sizes. The model results are given in terms of overall thermal efficiency as a function of the characteristic length of the combustor, *L*. Also, the power output as a function of characteristic length is provided. It should be noted that the overall system efficiency is the relevant quantity to plot. It is calculated from the power generated by the thermal engine divided by the heat release rate from combustion. To assess the impact of two important system properties on performance, the effective thermal conductivity of the support structure is varied from 1 to 100 W/m-K and different surface emissivities are used.

Each curve in Figure 9 holds these parameters constant as the characteristic length is varied. The results of this study are interesting. The key findings show that surface emissivity controls the overall thermal efficiency while effective thermal conductivity dictates the ultimate size limitations. When low thermal conductivity (1–10 W/m-K) materials are used for constructing the device, thermal losses will limit the characteristic size to a range between 0.1 and 1 mm. However, practical considerations will probably dictate sizes closer to 1 mm and up since many of the model assumptions were idealizations. This includes using the Carnot efficiency for the heat engine.

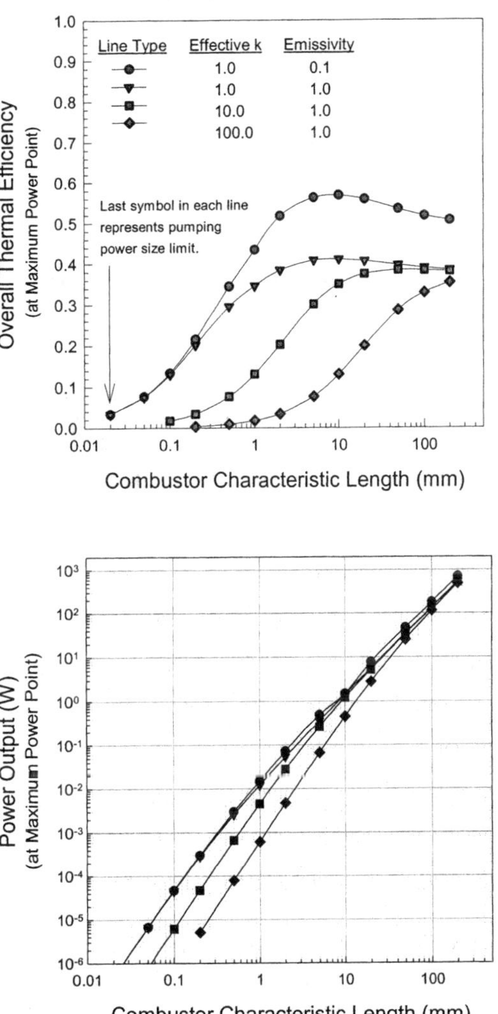

Figure 9: Scaling curves for a microcombustor with an aspect ratio of 10 to 1.

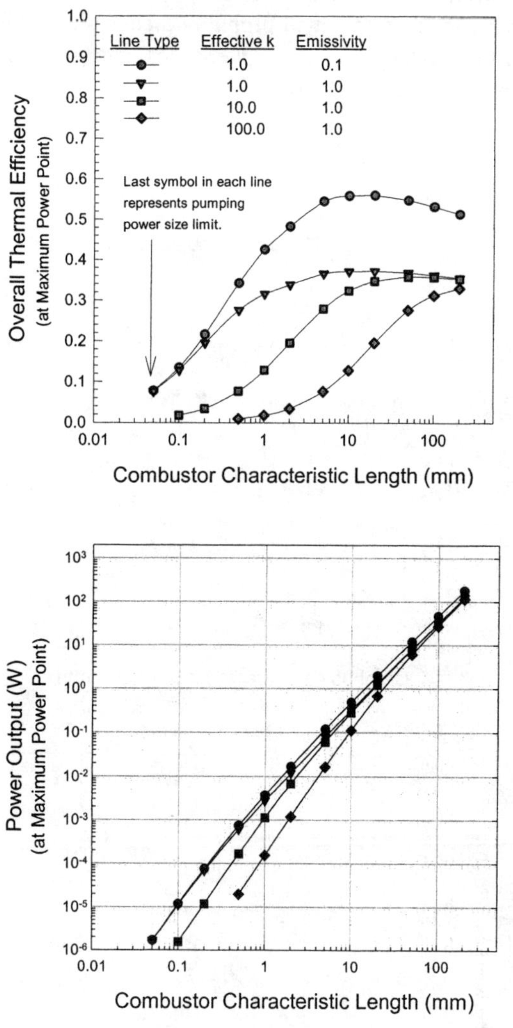

Figure 10: Scaling curves for a microcombustor with an aspect ratio of 20 to 1.

Figure 10 shows a similar scaling study where the aspect ratio was changed from 10 to 20. This is meant to examine the effects of different system geometries on performance. Although the quantitative nature of each curve changes slightly, the overall trends remain intact. There is no large numerical difference in system performance. The decrease in conduction heat loss by going to a longer, thinner structure appears to be offset by additional radiative losses from a higher surface area. With this model, the geometry of the combustor could be examined in more detail in order to determine if some optimal configuration exists. Note that in each curve, a lower size limit is placed on the characteristic length. This is determined by the point at which the heat engine can no longer produce enough power to pump the reactants through the combustor.

4.2 Techniques for thermal management

Heat loss is the primary concern in reducing the scale of thermally based energy systems. Various loss mechanisms are relevant to meso/microscale devices; however, they are typically grouped into: (1) conductive, (2) radiative, and (3) flow losses. Techniques are being developed to help managed these losses, but even the most aggressive means inevitably fail at the smaller length scales. A brief discussion follows of some of the more important considerations.

Conductive heat loss ultimately limits the size of thermally based systems. For a general heat engine model, conduction will impact performance in two ways. First, heat flow around the power generating section of the engine is a parasitic loss to the surroundings. This directly affects the overall efficiency since a portion of the engine's heat flow does not produce work. This loss mechanism continues to grow as the characteristic size decreases. The second conductive effect is more fundamental: isothermal conditions throughout the cycle lead to an inherent thermodynamic loss of efficiency. That is, heat conduction through the engine structure tends to heat the compression cycle and cool the expansion cycle leading to a loss of efficiency. This is clearly present in engines that separate compression and expansion processes such as gas turbines. The compressor of the device is directly linked to the turbine through a shaft (as well as through other engine structures such as ducting). The turbine must run hot while the compressor stays cool. As the separation distance between these two engine components shrinks, heat from the turbine increasingly flows to the compressor, thus heating up the working fluid at the precise point in the cycle where lower temperatures are needed. Although material choice and engine design can help delay the effects of conduction, as the device is scaled down, conduction ultimately prevents the engine from working.

To help mitigate conduction effects in mesoscopic heat engines, low thermal conductivity materials are needed. Silicon, a choice for most MEMS, is a particularly poor choice for small, thermally based devices. The room temperature thermal conductivity is approximately 150 W/m-K. At temperatures above 1000 °C, it falls to below 25 W/m-K [69]. Various metallic alloys and intermetallics, including common stainless steels, are potential fabrication materials for small-scale heat engines. Most stainless steels have a room temperature thermal conductivity between 11 and 15 W/m-K [70]. In contrast to silicon, the thermal conductivity of metals increases with increasing temperature. It is interesting to note that between about 1000 and 1200 °C, the thermal conductivities of silicon and stainless steel become similar. The best materials will probably turn out to be ceramics and glasses with amorphous structures.

Fused silica is at the low end of the thermal conductivity range with a room temperature value of 1.4 W/m-K. This climbs to approximately 4 W/m-K at 1200 °C. Polymeric materials are not often considered for elevated temperature use. However, polyimide (and other similar polymers) can be used up to approximately 400 °C, which could make them attractive in certain applications.

In most conceivable thermal engine concepts, the hot section is structurally connected to cooler sections of the engine and ultimately to ambient conditions. The structure can be engineered to reduce thermal conduction, but heat flow from around the hot section must also be minimized. Insulation is needed to prevent undesirable heat loss to the surroundings. Since meso/microscale devices imply little room available for insulation, high performance materials (or techniques) are called for. The three important cases considered here are aerogel, vacuum gap, and multi-foil insulation.

Aerogel insulation was first developed in 1931 by Kistler [71]. Although not produced in bulk quantities until relatively recently, this material is finding important commercial application in the area of high-performance vacuum insulation. The special feature of aerogel giving it high insulation capacity is a low density, open cellular structure at the nanoscale level. This permits restrictive conduction through the cellular structure itself and through trapped gas if present. Much work has been done recently at making the material commercially viable. One particular formulation available on the market is a "reconstituted" aerogel that is essentially composed of compacted aerogel particles [72]. Other, near aerogel-like substances can also be found commercially including submicron particle compresses [73]. Most of the silica-based gels can have operating temperatures in excess of 500 °C. Also, IR absorbers are typically incorporated into the compacted material to reduce radiation heat transfer. When evacuated, the effective thermal conductivity at room temperature can be below 0.010 W/m-K [74]. However, the thermal conductivity can exceed 0.2 W/m-K at temperatures above 500 °C. The unique feature of the nanoscale cellular structure permits low thermal conductivity to be realized at rough vacuum conditions, typically 20 mbars and below. The commercial potential for this unique material currently resides in Vacuum Insulation Panels (VIPs) [75]. Although designs vary, the major components of VIPs are the core insulation material, an outer barrier to maintain vacuum conditions, and getters to remove background gases from the interior. Technology is rapidly progressing in this area, but applications are typically for temperatures near ambient. This is evident by the types of barrier materials being considered which are multi-layer polymeric/metal film compositions. Getters have also been specifically developed for this application [76]. They are designed to remove the major background gases remaining after pump down and low temperature bake-out. These gases include water vapor, oxygen, nitrogen, hydrogen, carbon dioxide, and hydrocarbons. With state-of-the-art barrier materials and getters, rough vacuum conditions (< 20 mbars) can be maintained in VIPs for many years.

Vacuum gaps are also effective insulation. If the two inward facing surfaces of a gap are coated with a low emissivity metal (with a typical $\varepsilon = 0.05$) then the "effective" thermal isolation exceeds that of aerogel by a factor of approximately three. Gap thickness need not be a concern until dimensions start to approach the wavelength of the thermal radiation passing across the gap. However, other practical concerns must be considered. First, high vacuum conditions must be maintained so that the mean free path of the residue gas between the surfaces exceeds the gap dimension. Secondly, structural support of the gap may dictate thick material walls or internal support points spanning the gap. Both can lead to heat loss exceeding that of the radiation across the gap. Thus, even though a vacuum gap appears promising as an insulation technique, practical considerations at the meso/microscale level often result in compromised performance.

Multi-foil vacuum insulation has been developed for cryogenic use [77]. It is termed "super insulation" and has very good performance. A similar approach for high temperature systems has been developed by Advanced Modular Power Systems, Inc. [78]. Many layers (up to 50) of stainless steel foil (approximately 10 μm thick) are alternated with submicron sized zirconia particles. Again, vacuum conditions are required for the multi-layered system to work, but claimed performance is many times that of aerogel for the same thickness. One design constraint with multi-foil systems is the nonisotropic insulation capability. Typically, the in-plane effective thermal conductivity is much larger than the cross-plane value. Thus, end effects become problematic in the design of a fully insulated system.

One final topic in this section relates to the recovery of waste heat. Effective thermal management of a small-scale energy system will often require the recovery of heat from an exhaust stream emerging from the power-producing device. This is evident, for example, when miniature gas turbines and simple combustors are considered. Both cases have significant sensible heat capacity in their exhaust streams and recuperation of this energy will be important. Focusing on the combustor example, recuperative burners have been studied for decades as a way of enhancing thermal performance, or for burning low-energy-content fuel (see, e.g., Weinberg [79]). Recently, the author [80, 81] and other groups [82, 83] have been developing microcombustors for small-scale power generation. One of the features of these combustors is to use the thermal energy in the exhaust stream to heat the incoming reactants. Counterflow heat exchange appears to be the most straightforward way of accomplishing this. Some of Weinberg's configurations based on conduction may also be practical. It should be noted that, as the size of the counterflow heat exchanger is reduced, streamwise conduction grows and becomes a major contributor to heat loss from the system. Determining the characteristic size when this occurs has been the subject of past studies by the author [see refs. 66– 68]. With the use of heat recovery, low thermal conductivity materials, and high performance insulation, small-scale energy systems based on thermal processes become practical in the mesoscale size regime.

5 Future directions

Power generation for small-scale devices will be dominated by battery technology in the near term. This mature form of energy storage has been developed into a reliable, if not high energy density, form of power for meso/microscale devices. PEM fuel cells will also play an important role in energy production for small-scale devices. Direct methanol cells based on low temperature membrane materials are being developed for meso/microscale systems and significant contributions should result from the current effort in this area. The potential for much higher energy densities in the form of conventional stored hydrocarbon fuels and nuclear sources of heat will continue to be studied and developed into useful systems, first in niche areas, and then into more general energy delivery devices.

5.1 Conventional mesoscopic devices

Conventional mesoscopic *thermal energy systems*, as considered in this section, are combustion driven heat engines and direct energy conversion devices, the latter further being restricted to elevated temperature systems such as solid oxide fuels cells. This area of development will show important new technologies emerging from the laboratory such as microcombustors and mesoscopic thermal engines with respectable efficiencies, perhaps at the 10% level. Better performance in the 1 mm to 1 cm size range will hinge on thermal management of the heat used to drive the system as well as combining techniques together for different temperature ranges. For example, perhaps a TEC can generate electricity from a source at 1000 K and deliver the rejected heat at 500 K to a small-scale Rankine cycle device. This combined cycle approach may yield efficiencies approaching 15–20% in the nearer term. If high *ZT* materials are developed, efficiencies could reach the 25–30% range. In addition, thin film SOFC operating between 500 and 600 °C have the potential of revolutionizing small-scale power generation. This temperature range in small packages is achievable with the newest generation of insulating materials, and this range is also practical from a thermal management point of view. Another possible candidate for small-scale thermal-to-electric conversion is AMTEC, but not the traditional devices now being developed. Rather, thin film AMTEC holds the same promise as thin film SOFC did a decade ago. That is, lower operating temperatures and higher efficiencies (as well as the potential for mass production). To the author's knowledge, little if any work has been reported on thin film AMTEC devices. This should be a fertile ground for developing small-scale energy conversion devices.

An overriding issue with all of the conventional mesoscopic energy systems is the size and complexity of the "balance-of-plant" components, such as air movers, fuel delivery components, valves, and other associative concerns such as power for start-up. These issues can be explored and resolved over the next few years with dedicated effort, especially in the development of microscale engineering and MEMS-type fabrication techniques. From both practical work and theoretical modeling, it appears that the size limit for thermally activated energy systems is in the 1 mm region. Breakthrough concepts will be needed to push the technology to smaller sizes.

5.2 High *ZT* thermal electric conversion

If high *ZT* materials are developed for thermal-to-electric conversion, then most of the other heat engine concepts considered here will become obsolete. The required range for a breakthrough in this area is a *ZT* factor of between 3 and 5 throughout the temperature range for conversion, perhaps between 800 K and ambient. This one breakthrough would have important ramifications throughout the energy conversion field, but especially for power generation and refrigeration. However, the development of high *ZT* materials is today a matter of much speculation and little encouraging results. Work is being focused on materials such as skutterudite and nano-engineered materials [84] with atomic scale layering and/or reduced dimensional transport of electricity and heat. For these latter materials, even if substantial evidence existed for high performance materials, economical production would be a daunting task.

Assuming a high *ZT* material can be developed, it could essentially act as a solid-state heat engine for the conversion of heat-to-electricity at length scales from the 1 cm range to the sub mm range. Certainly microcombustors could provide the required heat source for a high performance TEC device. Nuclear sources could also be used. Thermal management would remain a very important consideration in such devices, but the heat-to-electric conversion step would be a known factor in overall device design. Engineers could then concern themselves with issues such as packaging, balance of plant, and thermal isolation of the hot section. With demonstrated microcombustors in the sub-mm^3 range, heat engines-on-a-chip also become feasible with the potential for mass production.

References

[1] Goemans, P.A.F.M., Microsystems and Energy: The Role of Energy. In: *Microsystem Technology: Exploring Opportunities*, ed. G.K. Lebbink, Samsom BedrijfsInformatie bv, pp. 50–64, 1994.

[2] Koeneman, P.B., Busch-Vishniac, I.J. & Wood, K.L., Feasibility of Micro Power Supplies for MEMS, *Journal of MicroElectroMechanical Systems*, **6(4)**, pp. 355–362, 1997.

[3] Alberts, B., Bray, D., Johnson, A., Lewis, J., Raff, M., Roberts, K. & Walter, P., *Essential Cell Biology, An Introduction to the Molecular*

Biology of the Cell, Garland Publishing, Inc.: New York & London, pp. 189–205, 1998.

[4] Ballardini, R., Balzani, V., Credi, A., Gandolfi, M.T. & Venturi, M., Artificial Molecular-Level Machines: Which Energy to Make Them Work? *Accounts of Chemical Research*, **34(6)**, pp. 445–455, 2001.

[5] Stowell, M.H.B., McPhillips, T.M., Rees, D.C., Soltis, S.M., Abresch, E. & Feher, G., Light-Induced Structural Changes in Photosynthetic Reaction Center: Implications for Mechanism of Electron-Proton Transfer, *Science*, **276(5313)**, pp 512–817, 1997.

[6] Alberts, B., Bray, D., Johnson, A., Lewis, J., Raff, M., Roberts, K. & Walter, P., *Essential Cell Biology, An Introduction to the Molecular Biology of the Cell*, Garland Publishing, Inc.: New York & London, pp. 430–442, 1998.

[7] Balzani, V., Credi, A, Marchioni, F. & Stoddart, J.F., Artificial Molecular-Level Machines: Dethreading-Rethreading of a Pseudorotaxane Powered Exclusively by Light Energy. *Chemical Communications*, **18**, pp. 1860–1861, 2001.

[8] Ryan, D. M., LaFollette, R. M. & Salmon, L., Microscopic Batteries for MicroElectroMechanical Systems (MEMS). *Proceeding of the 32nd Intersociety Energy Conversion Engineering Conference (IECEC)*, Vol. 1, pp. 77–82, Honolulu, HW, July 27 – August 1, 1997.

[9] Kelly, S.C., Deluga, G.A. & Smyrl, W.H., A Miniature Methanol/Air Polymer Electrolyte Fuel Cell. *Electrochemical and Solid-State Letters*, **3(9)**, pp. 407–409, 2000.

[10] Wainwright, J., Savinell, R., Dudik, L. & Liu, C.C., A microfabricated Hydrogen/Air Fuel Cell. *Proceedings of the 195th Electrochemical Society Meeting*, Seattle, WA, May 2–6, 1999.

[11] Morse, J. D. & Jankowski, A.F., A Novel Thin Film Solid Oxide Fuel Cell for Microscale Energy Conversion. *Proceedings of the International Mechanical Engineering Congress and Exposition (ASME)*, Nashville, TN, November 15–20, 1999.

[12] Dyer, C.K., A Novel Thin-Film Electrochemical Device for Energy Conversion. *Nature*, **343**, pp. 547–548, 1990.

[13] Hibino, T., Hashimoto, A., Inoue, T., Tokuno, J-I., Yoshida, S.-I. & Sano, M., A Low-Operating-Temperature Solid Oxide Fuel Cell in Hydrocarbon-Air Mixtures. *Science*, **288**, pp. 2031–2033, 2000.

[14] Angrist, S.W., *Direct Energy Conversion*, Allyn and Bacon, Inc.: Boston & London, pp. 420–428, 1982.

[15] Miley, G.H., *Direct Conversion of Nuclear Radiation Energy*, Monograph Series on Nuclear Science and Technology, American Nuclear Society, pp. 1–89, 1970.

[16] Olsen, L.C., Advanced Betavoltaic Power Sources. *Proceeding of the 9th Intersociety Energy Conversion Engineering Conference (IECEC)*, pp. 754–762, San Francisco, CA, August 26–30, 1974.

[17] US Patent 5124610, Tritiated Light Emitting Polymer Electrical Energy Source, Issued June 23, 1992.

[18] Larminie, J. & Dicks, A., *Fuel Cell Systems Explained*, John Wiley and Sons: New York, 2000.

[19] Steel, C.H. & Heinzel, A., Materials for Fuel-Cell Technologies. *Nature*, **414**, pp. 345–352, 2001.

[20] Tsai, T., Perry, E. & Barnett, S., Low-Temperature Solid-Oxide Fuel Cells, Utilizing Thin Bilayer Electrolytes. *J. Electrochem. Soc.*, **144(5)**, pp. L130–L132, 1997.

[21] Park, S., Vohs, J.M. & Gorte, R.J., Direct Oxidation of Hydrocarbons in a Solid-Oxide Fuel Cell. *Nature*, **404**, pp. 265–267, 2000.

[22] Steel, C.H. & Heinzel, A., Materials for Fuel-Cell Technologies. *Nature*, **414**, pp. 345–352, 2001.

[23] Fleurial, J.P., Snyder, G.J., Patel, P., Herman, J.A., Caillat, T., Nesmith, B. & Kolawa, E.A., Miniaturized Radioisotope Solid State Power Sources. *AIP Conference Proceeding: Space Technology and Applications International Forum*, Vol. 2, ed. M. S. El-Genk, pp. 1500–1507, Albuquerque, NM, January 2000.

[24] Allen, D.T., Bass, J.C., Elsner, N.B., Ghamaty, S. & Morris, C.C., Milliwatt Thermoelectric Generator for Space Applications, *AIP Conference Proceeding: Space Technology and Applications International Forum*, Vol. 2, ed. M. S. El-Genk, pp. 1476–1481, Albuquerque, NM, January 2000.

[25] Rowe, D.M., Miniature Semiconductor Thermoelectric Devices. In: *CRC Handbook of Thermoelectrics*, ed. D.M. Rowe, CRC Press: New York, pp. 441–458, 1995.

[26] El-Genk, M.S., Tournier, J.-M., Sholtis, J.A. & Lipinski, R.J., Coated Particle Fuel for Radioisotope Power systems and Heater Units: Status and Future Research Needs. *AIP Conference Proceeding: Space Technology and Applications International Forum*, Vol. 2, ed. M.S. El-Genk, pp. 1466–1475, Albuquerque, NM, January 2000.

[27] US Patent 5443657, Power Source Using a Photovoltaic Array and Self-Luminous Microspheres, Issued August 22, 1995.

[28] Crompton, T.R., *Battery Reference Book*, Butterworth-Heinemann, Ltd.: Oxford, pp. 2/1–2/23, 1995.

[29] Fernandez-Pello, A.C., Micro-Power Generation Using Combustion: Issues and Approaches. *Proceedings of the 29th International Symposium on Combustion*, In Press, Supporo, Japan, July 21–26, 2002.

[30] McNaughton, A.G., Commercially Available Generators. In: *CRC Handbook of Thermoelectrics*, ed. D.M. Rowe, CRC Press: New York, pp. 459–469, 1995.

[31] Fleurial, J.P., Borshchevsky, A., Caillat, T. & Ewell, R., New Materials and Devices for Thermoelectric Applications, *Proceeding of the 32nd*

Intersociety Energy Conversion Engineering Conference (IECEC), Vol. 2, pp. 1080–1085, Honolulu, HW, July 27 – August 1, 1997.
[32] Hatsopoulos, G.N. & Gyftopoulos, E.P., *Thermionic Energy Conversion*, Vol. II, The MIT Press: Cambridge, MA, pp. 591–646, 1979.
[33] Coutts, T.J. & Fitzgerald, M.C., Thermophotovoltaics. *Scientific American*, pp. 90–95, September, 1998.
[34] Lamp, T.R., Applications for Advanced Energy Conversion Applications. *Proceedings of the International Mechanical Engineering Congress and Exposition (ASME)*, AESD Vol. 38, Anaheim, CA, November 15–20, pp. 455–460, 1998.
[35] Ramalingam, M., Doyle, E., Shukla, K. & Donovan, B., Advanced Energy Conversion Technologies Based Small Power Systems for Remote Site Applications. *Proceedings of the 36th Intersociety Energy Conversion Engineering Conference*, Vol. 1, pp. 373–378, Savannah, Georgia, July 29 – August 2, 2001.
[36] Fu, K., Knobloch, A.J., Martinez, F.C., Walther, D.C., Fernandez-Pello, C., Pisano, A.P. & Liepmann, D., Design and Fabrication of a Silicon-based MEMS Rotary Engine. *Proceedings of the International Mechanical Engineering Congress and Exposition (ASME)*, New York, NY, November 11–16, 2001.
[37] Lipkin, R., Micro Steam Engine Makes Forceful Debut. *Science News*, **144(13)**, p. 197, September 25, 1993.
[38] Kirtas, M. & Menon, S., Simulation of a Combustion Driven Micro-Engine, *Proceedings of the Aerospace Sciences Meeting and Exhibit (AIAA)*, paper no. 2001-0944, Reno, NV, January 8–11, 2001.
[39] Xu, C., Hall, J., Richards, C., Bahr, D. & Richards, R., Design of a Micro Heat Engine. *Proceedings of the International Mechanical Engineering Conference and Exposition (ASME)*, MEMS Symposium Vol. 2, pp. 261–267, Orlando, FL, November, 2000.
[40] Kapat, J. & Chow, L., Scaling Laws in Miniature Heat Engines. *Advanced Energy Systems Division Newsletter*, ed. S. Garimella, pp. 5–6, Fall, 2001.
[41] Epstien, A.H. & Senturia, S.D., Macro Power from Micro Machinery. *Science*, **276**, p. 1211, 1997.
[42] Dornheim, M., Turbojet on a Chip to Run in 2000. *Aviation Week and Space Technology*, pp. 50–52, July 12, 1999.
[43] Mehra, M., Zhang, X., Ayon, A.A., Waitz, I.A., Schmidt, M.A. & Spadaccini, C.M., A Six-Wafer Combustion Systems for a Silicon Micro Gas Turbine. *Journal of Microelectromechanical Systems*, **9(4)**, pp. 517–527, 2000.
[44] Fu, K., Knobloch, A., Cooley, B., Walther, D., Liepmann, D., Fernandez-Pello, A.C. & Miyasaka, K., Micro-Scale Combustion Research for Applications to MEMS Rotary IC Engines. *Proceedings of the 35th National Heat Transfer Conference (ASME)*, paper no. NHTC2001-11212, Anaheim, CA, 2001.

[45] Finger, G.W., Kapat, J.S. & Chow, L.C., Design and Analysis of a Miniature Rotary Wankel Compressor. *Proceedings of the International Mechanical Engineering Congress and Exposition (ASME)*, New York, NY, November 11–16, 2001.

[46] Richards, C.D., Bahr, D.F., Xu, C-G. & Richards, R.F., MEMS Power: The P3 System, *Proceedings of the 36th Intersociety Energy Conversion Engineering Conference (IECEC)*, Savannah, GA, July 30–August 3, 2001.

[47] Feynman, R.P., "There's Plenty of Room at the Bottom," talk given at the 1959 annual meeting of the American Physical Society. Reprinted in the *Journal of Microelectromechanical Systems*, Vol. 1, pp. 60–66. March 1992.

[48] Thompson, D.W., *On Growth and Form*, University Press: Cambridge, MA, 1992.

[49] Hayashi, T., Micromechanism and Their Characteristics. *IEEE International Workshop on MicroElectroMechanical Systems, MEMS '94*, Oiso, Japan, pp. 39–44, 1994.

[50] Benton, M., Dinosaur Summer. In: *Book of Life*, ed. S.J. Gould, W.W. Norton & Company: New York & London, pp. 127–167, 2001.

[51] Trimmer, W.S.N., Microrobots and Micromechanical Systems. *Sensors and Actuators*, **19**, pp. 267–287, 1989.

[52] Madou, M., *Fundamentals of Microfabrication*, CRC Press: New York, pp. 405–447, 1997.

[53] Trimmer, T., Micromechanical Systems. *Proceedings of the Integrated Micro-Motion Systems: Micromaching, Control, and Applications* (3rd Toyota Conference), Aichi, Japan, October, pp. 1–15, 1990.

[54] Resnick, R. & Halliday, D., *Physics, Part I*, John Wiley & Sons: New York, pp. 111–114, 1966.

[55] White, M.A., Colenbrander, K., Olan, R.W. & Penswick, L.B., Generators That Won't Wear Out. *Mechanical Engineering*, pp. 92–96, February, 1996.

[56] Incropera, F.P. & DeWitt, D.P., *Fundamentals of Heat and Mass Transfer*, 4th edn, John Wiley & Sons: New York, pp. 435–439, 1996.

[57] Min, G. & Rowe, D.M., Peltier Devices as Generators. In: *CRC Handbook of Thermoelectrics*, ed. D.M. Rowe, CRC Press: New York, pp. 479–488, 1994.

[58] Peterson, R.B., Micro Thermal Engines: Is There Any Room at the Bottom? *Proceedings of the International Mechanical Engineering Congress and Exposition (ASME)*, Nashville, TN, November 15–20, 1999.

[59] Meissner, W., Ueber die Vorgaenge in den Gegenstromapparaten der Gasverfluessiger. *Zeitschrift fuer technische Physik*, **7**, pp. 235–238, 1926.

[60] Hausen, H., *Waermeuebertragung im Gegenstrom, Gleichstrom und Kreuzstrom*, Springer Verlag: Berlin, p. 186, 1950.

[61] Landau, H.G. & Hlinka, J.W., Steady-State Temperature Distribution in a Counterflow Heat Exchanger Including Longitudinal Conduction in the Wall, ASME Paper No. 60-WA-236, 1960.
[62] Kroeger, P.G., Performance Deterioration in High Effectiveness Heat Exchangers Due to Axial Heat Conduction Effects. In: *Advances in Cryogenic Engineering,* ed. K.D. Timmerhaus, Plenum Press: New York, pp. 363–372, 1967.
[63] Kays, W.M. & London, A.L., *Compact Heat Exchangers*, 2nd edn, McGraw-Hill Book Company: New York, 1964.
[64] Bahnke, G.D. & Howard, C.P., The Effect of Longitudinal Heat Conduction on Periodic-Flow Heat Exchanger Performance. *J. Eng. Power*, pp. 105–120, April, 1964.
[65] Shah, R.K., A Review of Longitudinal Wall Heat Conduction in Rucuperators. *J. Energy, Heat, and Mass Transfer*, **16**, pp. 881–888, 1994.
[66] Peterson, R.B., Numerical Modelling of Conduction Effects in Microscale Counterflow Heat Exchangers. *Microscale Thermophysical Engineering*, **3**, pp. 17–30, 1999.
[67] Peterson, R.B. & Vanderhoff, J.A., High Temperature Microscale Reactor Analysis Using a Counterflow Heat Exchanger Model. *Proceedings of the International Mechanical Engineering Congress and Exposition (ASME)*, ASME, Orlando, FL, November 5–10, 2000.
[68] Peterson, R.B. & Vanderhoff, J.A., Analysis of a Bayonet-Type Counterflow Heat Exchanger with Axial Conduction and Radiative Heat Loss. *Numerical Heat Transfer, Part A*, **40**, pp. 203–219, 2001.
[69] Glassbrenner, C.J. & Slack, G.A, Thermal Conductivity of Silicon and Germanium from 3°K to the Melting Point. *Phys. Rev.*, **134**, pp. A1058–A1069, 1964
[70] Incropera, F.P. & DeWitt, D.P., *Fundamentals of Heat and Mass Transfer*, 4th edn, John Wiley & Sons: New York, pp. 827–830, 1996.
[71] Kistler, S.S., Coherent Expanded Aerogels and Jellies. *Nature*, **127**, p. 741, 1931.
[72] Nanogel Advanced Insulation, product literature from Cabot Corporation, Billerica, MA, 2002
[73] Microtherm Insulation, product literature from Microtherm International Ltd., Upton, U.K., 2002.
[74] Fricke, J. & Tillotson, T., Aerogels: Production, Characterization, and Application. *Thin Solid Films*, **297**, pp. 212–223, 1997.
[75] Stovall, T.K., An Introduction to VIP Technology. *Proceedings of the Vacuum Insulation Panel Symposium*, Baltimore MD, May 3– 4, 1999.
[76] Mazza, F., Present, Future, and Exotic Applications of Gettering Materials. *Proceedings of the 23rd IUVSTA Workshop on Gettering Materials*, Bonassola, Italy, June 5–10, 1999.

[77] Molnar, W., Insulation. In: *Cryogenic Fundamentals*, ed. G.G. Haselden, Academic Press: London & New York, pp. 199–234, 1971.
[78] Personal communications with Advanced Modular Power Systems, Inc., Ann Arbor, Michigan, March 8, 2002.
[79] Weinberg, F.J., Combustion in Heat-recirculating Burners. In: *Advanced Combustion Methods*, ed. F.J. Weinberg, Academic Press: London, pp. 183–234, 1986.
[80] Peterson, R.B. & Vanderhoff, J.A., A Catalytic Combustor for Microscale Applications. *Combustion Science and Technology Comm.,* **1**, pp. 10–13, 2000.
[81] Hatfield, J.M. & Peterson, R.B., A Catalytically Sustained Microcombustor Burning Propane. *Proceedings of the International Mechanical Engineering Congress and Exposition (ASME)*, ASME, New York, NY, November 11–16, 2001.
[82] Sitzki, L., Borer, K., Wussow, S., Schuster, E., Maruta, K. & Ronney, P., Combustion and Power Generation in Microscale Excess Enthalpy Burners, 2nd Joint Meeting of the US Section of the Combustion Institute, Oakland, CA, March 26–29, 2001.
[83] Brooks, K.P., Call, C.J. & Drost, M.K., Integrated Microchannel Combustor/Evaporator Development. *Proceedings of the Process Miniaturization 2nd International Conference on Microreaction Technology* (AIChE), New Orleans, LA, March 9–12, pp. 196–200, 1998.
[84] Nolas, G.S., Sharp, J. & Goldsmid, H.J., *Thermoelectrics, Basic Principles and New Materials Development*, Springer: Berlin & New York, pp. 178–191, 235–254, 2001.

CHAPTER 2

Nanostructures for thermoelectric energy conversion

G. Chen[1,2], B. Yang[3] & W. Liu[2]
[1]*Mechanical Engineering Department,*
Massachusetts Institute of Technology, USA
[2]*Mechanical and Aerospace Engineering Department,*
University of California at Los Angeles, USA
[3]*Mechanical Engineering Department,*
University of Maryland at College Park, USA

Abstract

Thermoelectric energy conversion employs electrons and holes as the working fluids to carry energy from one place to another. Highly efficient thermoelectric cooling and power generation devices require materials with high electrical conductivity and Seebeck coefficient but a low thermal conductivity, which are difficult to find in bulk form. Nanostructures provide additional parameter space that can be engineered to steer these properties towards desired directions. This chapter discusses electron and phonon thermoelectric transport in nanostructures. It starts with an introduction of thermoelectric effects and thermoelectric devices, followed by discussion of various size effects in nanostructures on electron and phonon transport that can be exploited for improving the thermoelectric figure-of-merit. Recent experimental results on promising materials systems are summarized but emphasis is placed on the mechanisms governing the electron and phonon transport in nanostructures.

Nomenclature

- A cross-section area in eqn. (6)
  Richardson constant in eqn. (43)
- B B factor
- b microscopic quantity of interest
- C specific heat

D density of states per unit volume
$\boldsymbol{E}$ energy or energy function
F Fermi–Dirac integral
$\boldsymbol{F}$ force vector
f Bose–Einstein or Fermi–Dirac distribution function
g distribution function for particles
$\hbar$ Planck constant divided by 2π
I electrical current
J electrical or heat current density
$\boldsymbol{J}$ current density vector
K thermal conductance
k thermal conductivity in eqns. (6), (11), (32), and (36)
wave vector in eqns. (23) and (40)
k_B Boltzmann
$\boldsymbol{k}$ wavevector
L transport coefficient in eqns. (27)–(32)
length in eqn. (6) and eqn. (40).
m electron mass
N phonon distribution function
P fraction of specularly scattered phonons
q cooling power in eqn. (5), (8), (15), and (18)
elementary charge elsewhere
R electrical resistance
$\boldsymbol{r}$ coordinator
S Seebeck coefficient
t transmissivity of carriers
U electrical potential
$\boldsymbol{v}$ velocity vector
W electrical power
x coordinate
Z thermoelectric figure-of-merit
ZT dimensionless thermoelectric figure-of-merit
ε electrical field
Φ electrochemical potential
ϕ coefficient of performance
η efficiency of power generation
Λ phonon mean free path
μ electron mobility
Π Peltier coefficient
π Thomson coefficient
ρ electrical resistivity
σ electrical conductivity
τ relaxation time
ω frequency
ξ chemical potential

Subscripts

0	at equilibrium state
1	medium 1
2	medium 2
21	from medium 2 to medium 1
12	from medium 1 to medium 2
C	cold side
c	conduction band
e	electrical current
f	Fermi level
H	hot side
L	electrical load
n	n-type semiconductor
p	phonon in eqns. (35) and (36) polarization in eqns. (41) and (42) p-type semiconductor elsewhere
q	heat flux
r	coordinate vector
x	coordinate
y	coordinate
z	coordinate

1 Introduction

Using electrons as carriers for thermal to electric energy conversion is not a new idea [1]. Thermocouples converting thermal energy to electrical energy are based on the Seebeck effect that was discovered in 1821. The Peltier effect, which was discovered in 1834, has been used to make commodities such as electrically cooled picnic boxes and water chillers. Despite their many applications, energy conversion technologies using electrons in solid-state materials have not been able to compete against mainstream energy conversion technologies based on mechanical machinery, because of their relatively low efficiency. For thermoelectric devices, the maximum theoretical efficiency (for both cooling and power generation) is determined by a combination of materials properties called figure-of-merit, which is defined as $Z = S^2\sigma/k$, where S $(= -V/\Delta T)$ is the Seebeck coefficient defined as the voltage difference generated in a material per unit temperature difference at the corresponding two points, σ the electrical conductivity and k the thermal conductivity. The figure-of-merit has units of inverse temperature and thus the dimensionless figure-of-merit ZT, where T is in Kelvin, is used more often. The Seebeck coefficient is a measure of the thermal energy carried per electron. The reason that the electrical conductivity σ enters Z is due to the Joule heating in the thermoelectric element. Naturally, the Joule heat should be minimized through increasing the electrical conductivity. The thermal conductivity k appears in the denominator of Z because the thermoelectric elements also act as the thermal insulation between the hot and the cold sides. A high thermal conductivity causes too much heat leakage through heat conduction. The best *ZT* materials are found in heavily doped semiconductors. Insulators have poor electrical conductivities. Metals have relatively low Seebeck coefficients. In addition, the thermal conductivity of a metal, which is dominated by electrons, is proportional to the electrical conductivity, as dictated by the Wiedmann–Franz law. It is thus hard to realize high *ZT* in metals. In semiconductors, the thermal conductivity consists of contributions from electrons (k_e) and phonons (k_p), with the majority contribution coming from phonons. The phonon thermal conductivity can be reduced without causing too much reduction in the electrical conductivity. A proven approach to reduce the phonon thermal conductivity is through alloying [2]. The mass difference scattering in an alloy reduces the lattice thermal conductivity significantly without much degradation to the electrical conductivity. The traditional cooling materials are alloys of Bi_2Te_3 with Sb_2Te_3 (such as $Bi_{0.5}Sb_{1.5}Te_3$, p-type) and Bi_2Te_3 with Bi_2Se_3 (such as $Bi_2Te_{2.7}Se_{0.3}$, n-type), with a *ZT* at room temperature approximately equal to one. A typical power generation material is the alloy of silicon and germanium, with a $ZT \sim 0.6$ at 700 °C [3].

While the search for high *ZT* materials before the 1990s was mostly limited to bulk materials, there has been extensive research in the area of artificial semiconductor nanostructures in the last 30 years. Various means of producing ultra-thin and high-quality crystalline layers (such as molecular beam epitaxy and metal organic chemical vapor deposition) have been used to alter the "bulk" characteristics of the materials. Even though electrical and optical properties of these artificial crystalline structures have been extensively studied, much less attention was paid to their thermal and thermoelectric properties. Thermoelectric properties of low-dimensional structures started to attract attention in the 1990s, in parallel to renewed interest in certain bulk thermoelectric materials, such as skutterudites [4]. Compared to the research in bulk materials that emphasizes reducing the thermal conductivity, nanostructures offer a chance of improving both the electron and phonon transport through the use of quantum and classical size and interface effects. Several directions have been explored such as quantum size effects for electrons [5, 6], thermionic emission at interfaces [7, 8], and interface scattering of phonons [9, 10]. Impressive *ZT* values have been reported in some low-dimensional structures [11, 12, 13]. Comprehensive reviews on the progress of thermoelectric materials research is presented in recently published reviews [3, 4, 14–17] and in the proceedings of the various International Conferences on Thermoelectrics held in recent years.

In this article, we will first give a brief introduction to the basic theories of thermoelectric devices and materials and then focus on discussing why nanostructures can lead to a potentially higher figure-of-merit, with an emphasis on the physical mechanisms of thermoelectric transport for electrons and phonons in nanostructures.

2 Thermoelectric effects and devices with bulk materials

There exist three basic thermoelectric effects: the Seebeck effect, the Peltier effect, and the Thomson effect, as illustrated in Figure 1. The Seebeck effect describes the phenomenon of voltage generation when a material is subject to a temperature gradient under the open circuit condition, as shown in Figure 1(a). Under a temperature gradient, electrons diffuse from the hot side to the cold side. At a steady state, the internal field (Seebeck voltage), caused by the electron concentration difference, balances the driving force for diffusion. The Seebeck coefficient is defined as

$$S = -\frac{1}{q}\frac{dU/dx}{dT/dx} = \frac{\varepsilon}{dT/dx} \tag{1}$$

where U is the electric potential and ε the electric field.

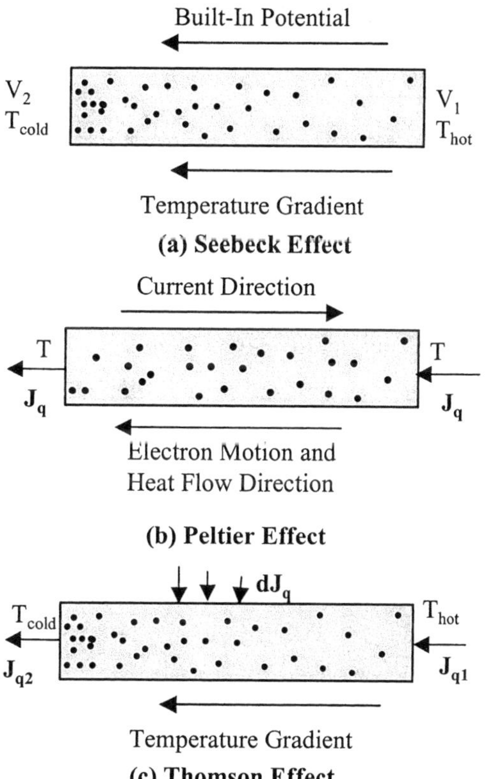

Figure 1: Three basic thermoelectric effects: (a) Seebeck effect, (b) Peltier effect, and (c) Thomson effect.

Accompanying the electron flow is also the heat flow. The Peltier effect measures the heat carried by electrons when a current flows through a material maintained at a constant temperature, as shown in Figure 1(b). The Peltier coefficient Π defines the relationship between the heat flux (J_q) and the current density (J_e),

$$J_q = \Pi \times J_e \tag{2}$$

In the case shown in Figure 1(b), the Peltier heat enters from one end of the conductor and is rejected at the other end. When two materials of different Peltier coefficients are joined together, the imbalance of the heat in and out at each junction creates cooling or heating effects that can be used for refrigeration and heat pumping.

The Thomson effect occurs when current flows through a conductor under a temperature gradient [Figure 1(c)]. It was found by Thomson (Lord Kelvin) that heat could be released or absorbed along the conductor, depending on the current and temperature gradient directions. The Thomson coefficient is defined as

$$\pi = \frac{1}{J_e} \frac{dJ_q / dx}{dT / dx} \tag{3}$$

Typically, the Thomson effect is small and often neglected in device analysis.

All three effects are due to electron flow and their interactions with the lattice and thus these coefficients are interrelated through the Kelvin relations,

$$\Pi = S \times T \qquad \pi = T \frac{dS}{dT} \tag{4}$$

2.1 Thermoelectric cooling devices

Thermoelectric devices are typically made of multiple p-type and n-type semiconductor elements and connected such that the current flow is in series while the heat flow is in parallel, as shown in Figure 2(a). The reason for the use of both p and n type elements (legs) is because the Seebeck and Peltier coefficients are usually of opposite sign, such that both types of elements contribute to the desired thermoelectric effect. The reason that the legs are electrically in series is due to the small electrical resistance of each element. Putting many legs in series makes the power supply easier. Despite such a configuration, the total resistances of bulk thermoelectric devices are usually small, and typical thermoelectric devices use currents of the order of amps.

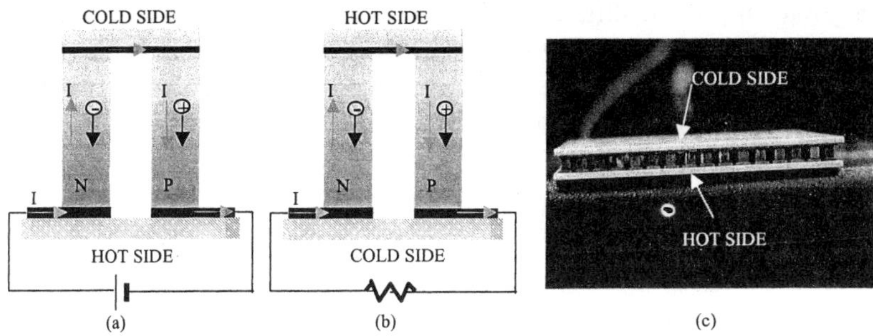

Figure 2: Thermoelectric devices: (a) cooling mode, (b) power generator mode, and (c) an actual device made of many pairs of p-n junctions electrically in series and thermally in parallel.

We consider a pair of thermoelectric legs as shown in Figure 2(a) used for cooling. Current is passed from the n-type leg to the p-type leg such that both electrons in the n-type leg and holes in the p-type leg move away from the cold junction, thus carrying heat out of the cold junction. The Peltier cooling is proportional to $(\Pi_2 - \Pi_1)I$, where I is the current. In addition to the heat taken out by the current flow, there is also reverse heat flow from the hot side to the cold side due to heat conduction by the solid. In addition, some of the Joule heat generated inside the thermoelectric element is also conducted back to the cold junction. The total cooling power of the couple is thus

$$q_C = (S_p - S_n) \times I \times T_C - K(T_H - T_C) - \frac{1}{2} I^2 R \tag{5}$$

where

$$K = \frac{k_p A_p}{L_p} + \frac{k_n A_n}{L_n} \quad \text{and} \quad R = \frac{L_p \rho_p}{A_p} + \frac{L_n \rho_n}{A_n} \tag{6}$$

are the thermal conductance and electrical resistance of both branches, respectively. The one-half factor in eqn. (5) means that half of the Joule heat conducts back to the cold side. The electrical power W consumed is

$$W = (S_p - S_n) \times I \times (T_H - T_C) + I^2 R \tag{7}$$

The coefficient of performance (COP) used to describe the efficiency of refrigeration is

$$\phi = \frac{q_C}{W} = \frac{(S_p - S_n) \times I \times T_C - K(T_H - T_C) - \frac{1}{2} I^2 R}{(S_p - S_n) \times I \times (T_H - T_C) + I^2 R} \tag{8}$$

At a given temperature difference, the COP is dependent on the current I. For the refrigeration application, two special cases are of interest. One is the maximum cooling power. The other is the maximum COP at a given temperature difference. For the first case, the current I_q can be determined by solving $dq_C/dI = 0$, which gives

$$I_q = \frac{(S_p - S_n)T_C}{R} \tag{9}$$

$$\phi_q = \frac{0.5ZT_C^2 - (T_H - T_C)}{ZT_H T_C} \tag{10}$$

where $Z = (S_p - S_n)^2/(KR)$. The product $K \times R$ is a minimum when $[(L_n A_p)/(L_p A_n)] = [(\rho_p k_n)/(\rho_n k_p)]^{1/2}$, which gives the maximum Z as

$$Z = \frac{(S_p - S_n)^2}{\left[(k_p \rho_p)^{1/2} + (k_n \rho_n)^{1/2}\right]^2}. \tag{11}$$

The device figure-of-merit thus depends on the properties of both legs. The figure-of-merit of individual materials $Z = S^2\sigma/k$ is nevertheless a useful measure of the material's potential for making a useful device.

For the second case, the current at maximum COP is determined by $d\phi/dI = 0$, which leads to

$$I_\phi = \frac{(S_p - S_n)(T_H - T_C)}{R\left(\sqrt{1 + ZT_M} - 1\right)} \tag{12}$$

$$\phi_{\max} = \frac{T_C\left(\sqrt{1 + ZT_M} - T_H/T_c\right)}{(T_H - T_C)\left(\sqrt{1 + ZT_M} + 1\right)} \tag{13}$$

where $T_M = (T_H + T_C)/2$ is the average temperature of heat source and heat sink. It is clear that as Z goes to infinity, ϕ_{max} approaches the COP of a Carnot cycle.

Another parameter that is useful for refrigeration performance is the maximum temperature difference that a given system can ideally reach. The maximum temperature difference is reached when there is no net cooling power taken from the source. Therefore, the COP is zero at the maximum temperature difference. The corresponding maximum temperature difference can be obtained from eqn. (10) as

$$(T_H - T_C)_{\max} = \frac{1}{2} Z T_C^{\ 2} \tag{14}$$

2.2 Thermoelectric power generation devices

The power generation mode can be similarly analyzed. In a power generation mode as shown in Figure 2(b), the heat supplied to the hot side should be

$$q_H = (S_p - S_n) I T_H - K(T_H - T_C) - \frac{1}{2} I^2 R \tag{15}$$

and the power output is

$$W = I^2 R_L \tag{16}$$

where R_L is the external load resistance of the output circuit. The output current is given by

$$I = \frac{(S_p - S_n)(T_H - T_C)}{R + R_L} \tag{17}$$

Therefore, the thermal efficiency is

$$\eta = \frac{W}{q_H} = \frac{I^2 R_L}{(S_p - S_n) I T_H + K(T_H - T_C) - \frac{1}{2} I^2 R} \tag{18}$$

It can be shown that the maximum power is obtained with a matched load $R_L = R$. The maximum efficiency, which does not have to occur at the maximum power, is determined by setting $d\eta/dR_L = 0$,

$$\eta_{\max} = \frac{(T_H - T_C)\left(\sqrt{1 + ZT_M} + 1\right)}{T_H\left(\sqrt{1 + ZT_M} + T_c / T_H\right)} \tag{19}$$

2.3 Thermoelectric transport properties

The figure-of-merit includes three transport properties, S, σ and k. Both S and σ are determined by electrons, although electron–phonon scattering can have a significant impact on their actual values. The thermal conductivity consists of contributions from both electrons and phonons. The electron transport properties are typically derived from the Boltzmann equation under the relaxation time approximation,

$$\mathbf{v} \bullet \nabla_{\mathbf{r}} f + \frac{\mathbf{F}}{\hbar} \bullet \nabla_k f = -\frac{f - f_o}{\tau} \quad (20)$$

where $f(\mathbf{r},\mathbf{k})$ is the distribution function in phase space that includes both coordinates $\mathbf{r}$ and wavevector $\mathbf{k}$, and $\hbar$ is the Planck constant divided by 2π. The force $\mathbf{F}$ acting on a charge in the absence of a magnetic field is given by

$$\mathbf{F} = q\varepsilon = q \times \left(-\frac{1}{q}\frac{dE_c}{dx} \right) = -\frac{dE_c}{dx} \quad (21)$$

where q is the unit charge and E_c is the conduction band edge of a semiconductor. The equilibrium distribution f_o for electrons obeys the Fermi–Dirac distribution,

$$f_o = \frac{1}{\exp\left(\frac{E - E_f}{k_B T} \right) + 1} \quad (22)$$

where the Fermi level E_f is a function of $\mathbf{r}$ and the carrier concentration. The electron energy consists of a potential energy and a kinetic energy part. Under the spherical and parabolic band approximation, it can be expressed as

$$E = E_c + \frac{\hbar^2 k^2}{2m^*} \quad (23)$$

where m* is the electron effective mass. In bulk materials, the first order approximation to the Boltzmann equation is to replace the real distribution function on the left hand side of eqn. (20) by the equilibrium distribution function, which leads to the following solution of the distribution function,

$$f = f_o + \tau\mathbf{v} \bullet \left(\nabla E_f + \frac{E - E_f}{T} \nabla T \right) \frac{\partial f_o}{\partial E} \quad (24)$$

With this distribution function, we can evaluate heat and current flux. The current density $\mathbf{J}$ carried by electrons can be expressed as [18]

$$\mathbf{J}(\mathbf{r}) = \frac{1}{4\pi^3} \iiint q\mathbf{v}(\mathbf{k}) f(\mathbf{r},\mathbf{k}) d^3\mathbf{k} \quad (25)$$

To find a proper expression for the heat flux carried by electrons, we must consider the fact that the local particle number density is not a constant. For a Lagrangian thermodynamic system, the first law of thermodynamics can be expressed as $dq = du - E_f dN$, where E_f is the chemical potential (or Fermi level). Translating this expression into an Euler form through the use of the Reynolds equation as often done in fluid mechanics, we get the following equation for evaluating the heat flux carried by the charge

$$\mathbf{J}_q(\mathbf{r}) = \frac{1}{4\pi^3} \iiint [E(\mathbf{k}) - E_f(\mathbf{r})] \mathbf{v}(\mathbf{k}) f(\mathbf{r}, \mathbf{k}) \, d^3\mathbf{k} \tag{26}$$

Substituting eqn. (24) into eqns. (25) and (26) leads to

$$\mathbf{J}(\mathbf{r}) = q^2 L_0 \left(-\frac{1}{q} \nabla\Phi \right) + \frac{q}{T} L_1 (-\nabla T) \tag{27}$$

$$\mathbf{J}_q(\mathbf{r}) = q L_1 \left(-\frac{1}{q} \nabla\Phi \right) + \frac{1}{T} L_2 (-\nabla T) \tag{28}$$

where Φ is the electrochemical potential ($-\nabla\Phi / q = \boldsymbol{\mathcal{E}} + \nabla E_f / q$). The transport coefficients L_n (for $n = 0, 1, 2$) are defined by the following integrals

$$L_n = \frac{1}{4\pi^3} \iiint \tau(\mathbf{k}) v(\mathbf{k}) v(\mathbf{k}) (E(\mathbf{k}) - E_F)^n \left(-\frac{\partial f_o}{\partial E} \right) d^3\mathbf{k} \tag{29}$$

From the expressions for $\mathbf{J}$ and $\mathbf{J}_q$, various materials parameters, such as the electrical conductivity, thermal conductivity due to electrons, and the Seebeck coefficient can be calculated. For simplicity we assume that both the current flow and the temperature gradient are in the x-direction. For bulk materials, the transport coefficients can be expressed as

$$\sigma = \left. \frac{J}{-\nabla\Phi / q} \right|_{\nabla T = 0} = q^2 L_0 = \frac{q^2}{3} \int \tau v^2 \left(-\frac{\partial f_o}{\partial E} \right) D(E) dE \tag{30}$$

$$S = \left. \frac{-\nabla\Phi / q}{\nabla T} \right|_{J=0} = \frac{1}{qT} L_0^{-1} L_1 = \frac{1}{qT} \frac{\int \tau v^2 (E - E_f)(\partial f_o / \partial E) D(E) dE}{\int \tau v^2 (\partial f_o / \partial E) D(E) dE} \tag{31}$$

$$k_e = \left. \frac{J_q}{\nabla T} \right|_{J=0} = \frac{-1}{T} L_1 L_0^{-1} L_1 + \frac{1}{T} L_2 \tag{32}$$

The phonon thermal conductivity is also modeled from the Boltzmann equation under the relaxation time approximation,

$$\mathbf{v} \bullet \nabla_{\mathbf{r}} N = -\frac{N - N_o}{\tau} \tag{33}$$

where N is the phonon distribution function and N_o is the Bose–Einstein distribution. More detailed modeling separates the Umklapp and normal scattering processes [19] but at room temperature, the later is often negligible. Again, for bulk materials, one can replace N in the left hand side of eqn. (33) by N_o and obtain the first order approximation for N,

$$N = N_o - \tau \mathbf{v} \bullet \nabla_{\mathbf{r}} N_o \tag{34}$$

The phonon heat flux can thus be calculated from

$$\mathbf{J}_p = -\nabla T \iiint \tau_p \mathbf{v}\mathbf{v}(\partial N_o / \partial T) d^3k = -k \nabla T \tag{35}$$

For bulk materials, the thermal conductivity can be expressed as

$$k_p = \frac{1}{3} \sum \int C(\omega) v_{\mathrm{p}}(\omega) \Lambda(\omega) d\omega \tag{36}$$

where C is the specific heat of phonons at frequency ω, v_p the phonon group velocity, and Λ the phonon mean free path.

Using the above transport property expressions, ZT can be further written as [6]

$$Z_{3D} T = \frac{\left[(5F_{3/2} / 3F_{1/2}) - \xi^*\right]^2 (3F_{1/2} / 2)}{1 / B_{3D} + 7F_{5/2} / 2 - (25F_{3/2}^2 / 6F_{1/2})} \tag{37}$$

where

$$B_{3D} = \frac{m^{*3/2}}{3\pi^2} \left(\frac{2k_B T}{\hbar^2} \right)^{3/2} \frac{k_B^{\,2} T \mu}{e k_p} \tag{38}$$

and $m^* = (m_x m_y m_z)^{1/3}$ is the effective density-of-states mass of electrons in the band, μ the electron mobility, ξ* the chemical potential normalized by $k_B T$, and F_i is the Fermi–Dirac integral defined as

$$F_i(\xi^*) = \int_0^\infty \frac{x^i dx}{\exp(x - \xi^*) + 1} \tag{39}$$

In eqns. (37) and (38), we have used the subscript 3D to denote that those expressions are derived considering the density-of-states of three-dimensional bulk crystals. In low-dimensional structures, these expressions must be reformulated [6]. In eqn. (37), the reduced chemical potential ξ^* is a free variable that can be controlled by doping. The optimum value for the chemical potential is chosen to maximize *ZT*. Therefore, thermoelectric materials development involves careful control and optimization of doping.

Some strategies can be developed by examining the above expressions for their materials properties and *ZT*. Equations (30) and (31) show that the Seebeck coefficient and the electrical conductivity depend on the density of states and the relaxation time. Because the derivative of the Fermi-Dirac distribution is nonzero only in a region of the order of k_BT near the Fermi level, one should increase the density of states near the Fermi level to increase the electrical conductivity, and increase the average energy of electrons relative to the Fermi level [see eqn. (31)] to increase the Seebeck coefficient. Another line to develop strategies for increasing *ZT* is to increase the *B* factor, which depends on the electron effective mass, the carrier mobility, and the phonon thermal conductivity. Thus thermoelectric materials research is often guided by finding materials that have a large *B* factor, which include a large electron (hole) effective mass and high mobility, and a low lattice thermal conductivity. Such materials are succinctly called phonon glass-electron-crystal materials by Slack [20]. To reduce the thermal conductivity, materials with a small phonon group velocity and a short relaxation time are desired. Roughly speaking, the phonon group velocity is proportional to $(K/m)^{1/2}$ where *K* is the spring constant between the atoms and m is the mass of the atom. Thus, materials with high atomic mass are often used for thermoelectric materials. The phonon relaxation time can be reduced by scattering, such as through alloying and adding phonon rattlers.

While the general *ZT* formulation for bulk materials has played and will continue to play an instrumental role in developing strategies in the search of highly efficient thermoelectric materials, it should be kept in mind that these expressions are derived by using a set of approximations. We will mention the following that are related to our discussion: (1) bulk density of states for electrons and holes, (2) local equilibrium approximation, and (3) isotropic relaxation time approximation. Many of the developments in nanostructures in recent years can be attributed to relaxing one or several of these approximations.

3 Nanostructures for solid-state energy conversion

We define nanostructures as inhomogeneous materials with at least one length scale in the nanometer range. Figure 3 shows a few examples of nanostructures, including a regular [3(a)] and a quantum dot [3(b)] superlattice that are grown by molecular beam epitaxy and two bismuth nanowires [3(c)] obtained by filling anodized alumina templates with desired materials. Thermoelectric transport in such nanostructures can differ from the description for bulk materials for the following reasons.

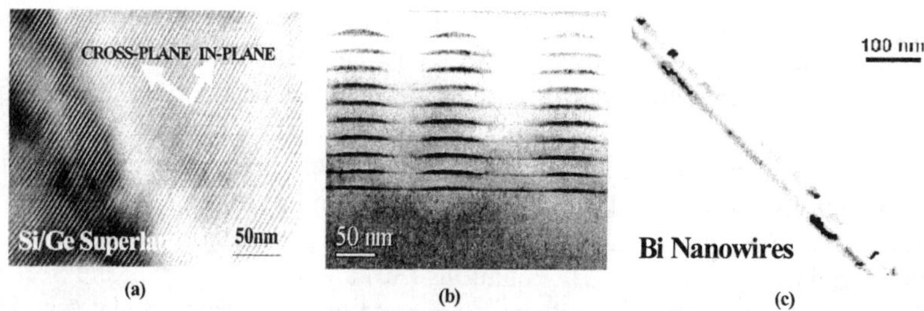

Figure 3: Examples of nanostructures for potential thermoelectric applications: (a) superlattices, (b) quantum dot superlattices, and (c) quantum wires (Courtesy of Professor K.L. Wang and Professor M.S. Dresselhaus).

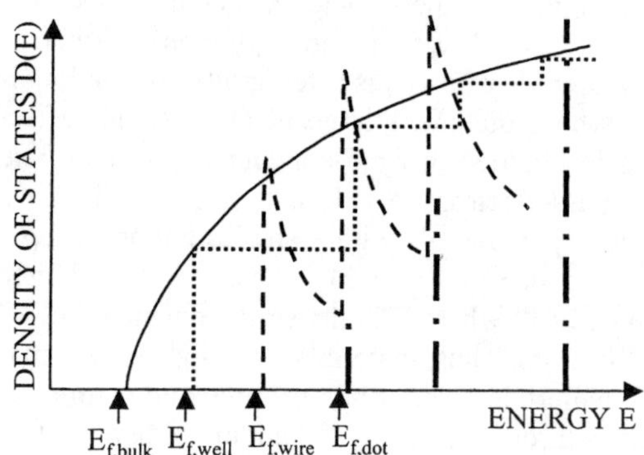

Figure 4: Schematic illustration of the electron density of states in bulk materials, quantum wells, quantum wires, and quantum dots.

(1) Interfaces and boundaries of nanostructures impose constraints on the electron and phonon waves, which lead to a change in their energy states and correspondingly, their density of states and group velocity. Figure 4 shows schematically the density of states of electrons in bulk form, in a quantum well, a quantum wire, and a quantum dot. Taking a quantum well as an example, the quantization of electrons leads to a staircase form of the density of states. By placing the Fermi level within k_BT of the first staircase, one can have a larger density of states for those carriers that contribute to the electrical conductivity, and a larger asymmetry in the electron energy that leads to an increase in the

Seebeck coefficient, as suggested by eqn. (31). A similar effect can be realized by exploring the energy filtering function of a potential barrier at an interface [7]. (2) The phonon thermal conductivity can be reduced through interface scattering and through the alteration of the phonon spectrum in low-dimensional structures [10]. Our plan for this section is to start with a summary of recent experimental results and then discuss in depth possible mechanisms of electron and phonon transport in nanostructures that can be exploited for improving the thermoelectric energy conversion efficiency and that are responsible for some of the recent experimental observations.

3.1 Some recent experimental results on low-dimensional thermoelectrics

There are several key ideas in using nanostructures to improve the energy conversion efficiency. For electron transport, one is the use of quantum size effects on electrons to increase the power factor [5, 6]. This mainly explores the electron transport along the confinement plane such as along the film plane of a quantum well or superlattice, or along the axis of a nanowire. Another idea for electron transport is to use interfaces to filter low energy electrons, based on electron transport perpendicular to the interface or the film plane [21]. This has been further developed into a thermionic emission approach [7, 8]. For phonon transport, a key idea is to use interface effects in superlattices to reduce the thermal conductivity [9, 10]. These ideas will be discussed in subsequent sections.

A proof-of-principle demonstration of the quantum size effect on electron transport for improving the power factor was first reported in 1996 on PbTe/Eu_xPb_{1-x}Te based quantum wells [22] and later on Si/SiGe quantum wells [23]. These experiments focus on the transport inside the quantum well only and thus the reported ZT is not for the whole structure that includes the barriers. Later, Harman and co-workers discovered that PbTe quantum dot superlattices [11] have a much improved power factor over the whole structure. Their most recent experimental cooling data from a single p-type PbTe/PbSeTe quantum-dot superlattice along the in-plane direction exceeds the best observed in bulk Bi_2Te_3 systems and an equivalent ZT of such structures at room temperature of 1.3 [24]. Part of the improvement in ZT comes from the increase in the power factor while another part comes from the reduction of the lattice thermal conductivity. A conservative estimate of the ZT of this material system, based on the thermal conductivity value of equivalent alloys, shows ZT at higher temperatures reaching 2.

The thermal conductivity reduction strategy was pursued by Venkatasubramanian [9]. The key assumption at the beginning was that interfaces reduce the thermal conductivity significantly while minimizing the potential barrier height at the interface makes it possible to have electron transport not hampered by the interface. Using this strategy, Venkatasubramanian's group has reported cross-plane ZT values between 2 and 3 [13].

Figure 5 summarizes *ZT* values reported by various groups. It must be emphasized that the measurement of the thermoelectric properties of nanostructured materials has been extremely challenging and that is one reason that some of the reported *ZT* values have not been reproduced. The most direct measurements of *ZT* are from the net cooling effect but heat loss and contact resistance can greatly diminish the intrinsic *ZT* of the material. Thermal conductivity measurement techniques of superlattices and other thin films used in the exploration of high *ZT* materials have been significantly advanced by several representative techniques such as the 3ω method, the optical pump-and-probe technique, and other microfabricated sensor techniques [25–28]. Direct measurements of the Seebeck coefficient in the cross-plane direction were only reported recently [29]. Measurements of thermoelectric properties of nanowire systems await further developments.

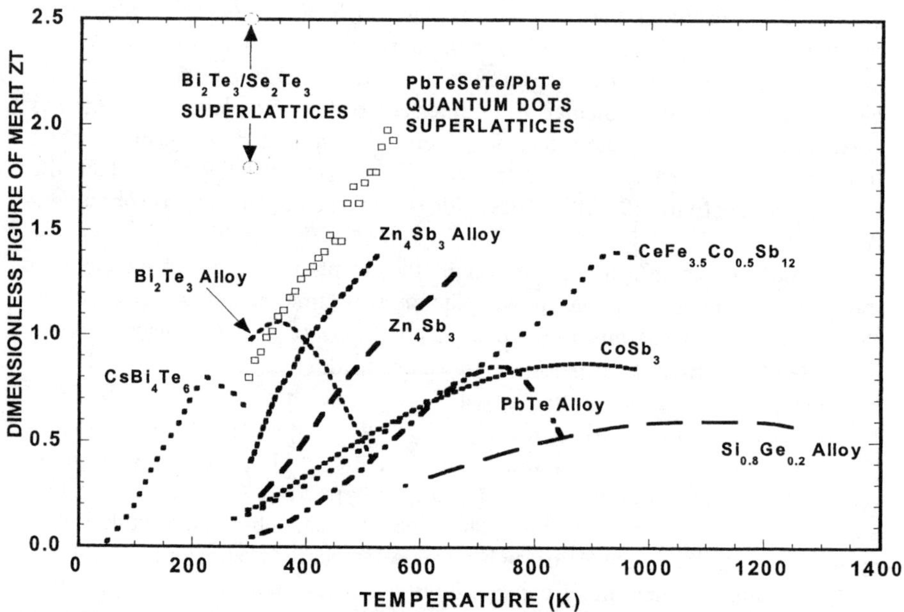

Figure 5: Summary of reported *ZT* values for various bulk and nanostructured materials.

It is also interesting to mention the bulk material work by Kanatzidis and co-workers [30]. They have reported high *ZT* values first in layered compounds similar to Bi_2Te_3 but with a much more complicated unit cell. More recently, they worked on a class of cubic crystals with large unit cells and found that there are microscopic inhomogeneities in these materials that are similar to superlattices and quantum dot superlattices [31]. This suggests that strategies learned in nanostructures can be potentially scaled up to bulk materials that are more useful for large-scale applications.

Thermal conductivity reduction in superlattices and quantum dot superlattices has thus played a central role in the high *ZT* reported in nanostructured materials. The first reported experiment on the thermal conductivity of superlattices was by Yao on GaAs/AlAs superlattices along the film plane direction [32]. The experiments showed that the thermal conductivity of superlattices along this direction is lower than predictions from the Fourier theory but higher than that of their equivalent alloy. Tien and Chen suggested that superlattices may be used to make super-thermal insulators [33]. Chen and co-workers reported the anisotropic thermal diffusivity of GaAs/AlGaAs superlattices and found that the cross-plane thermal conductivity is four times larger than along the in-plane direction [34]. The equivalent thermal conductivity of the structure is lower than the equivalent alloy. The temperature dependency behavior along the in-plane direction was first reported for a GaAs/AlAs superlattice [35] and in the cross-plane direction for Si/Ge superlattices [36] and GaAs/AlAs superlattices [28]. Extensive experimental data on the thermal conductivity of various superlattices emerged in recent years, including Bi_2Te_3/Sb_2Te_3 [37–39], GaAs/AlAs [32–35, 40], Si/Ge [36, 41–44], InAs/AlSb [45] InP/InGaAs [46], $CoSb_3/IrSb_3$ [47] and PbTe based superlattices [48]. Most of these measurements are in the cross-plane direction [28, 36–41, 45–47] using the 3ω method or the optical pump-and-probe method. Measurements along the film plane direction relied heavily on the removal of the substrate [32–35, 43, 48]. Very few studies have reported thermal conductivity in both the in-plane and cross-plane directions [34, 43]. All these experiments confirmed that the thermal conductivities of the superlattices in both directions are significantly lower than the predictions based on the Fourier law and the properties of their bulk parent materials. In the cross-plane direction, the thermal conductivity values can definitely be reduced below that of their corresponding alloys. In the in-plane direction, the reduction is generally above or comparable to that of their equivalent alloys, although a few experimental data indicate that k values lower than the corresponding alloys' values are possible [48].

Figure 6 gives typical thickness and temperature dependency behavior observed in several materials systems. Typically, the thermal conductivity increases with increasing period thickness. In the in-plane direction, the thermal conductivity usually reaches a peak at a much higher temperature than observed in bulk materials, suggesting the onset of size effects. In the cross-plane direction, the thermal conductivity can have a peak as in the in-plane direction but its magnitude more often decreases monotonically with decreasing temperature. In the very thin period limit, it was observed experimentally in Bi_2Te_3/Sb_2Te_3 superlattices [42] and less systematically in GaAs/AlAs superlattices [40], that the thermal conductivity can recover as the period thickness decreases, and thus a minimum exists in the thermal conductivity when plotted as a function of the period thickness.

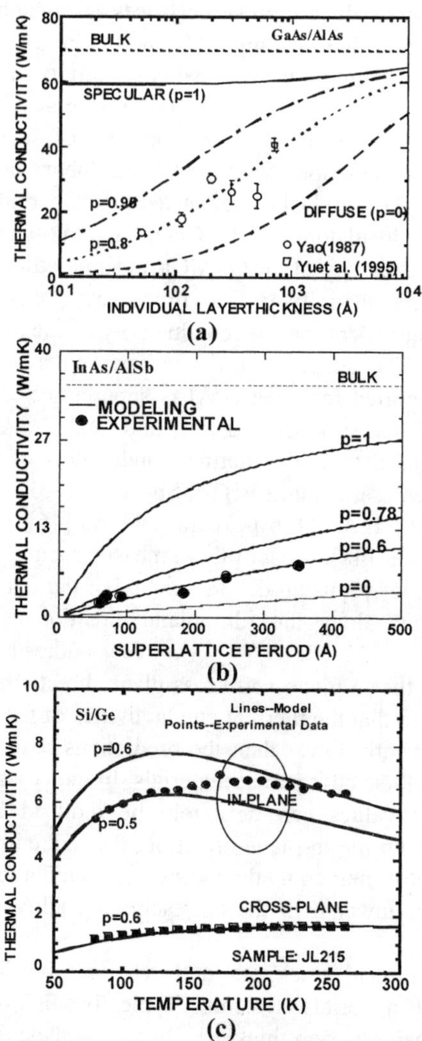

Figure 6: Examples of the thermal conductivity of superlattices: (a) GaAs/AlAs along the in-plane direction, (b) InAs/AlSb along the cross-plane direction, and (c) Si/Ge in both in-plane and cross-plane directions. Dots are experimental points and lines are Boltzmann equation-based modeling results in which p represents the fraction of specularly scattered phonons [43,78,80]. (Figure 6 (a), courtesy of ASME.)

3.2 General transport picture

The modeling of the thermoelectric properties of nanostructures is often conducted along a similar line as for bulk materials, i.e., using the relaxation time approximation and the solution of the linearized Boltzmann equation. The key difference lies in that the electron band structure and the phonon dispersion of nanostructures may differ from that of the bulk materials. The changes in the energy spectra of electrons and phonons have consequences in the density of states, the group velocity, and the relaxation time, which can be calculated from the new dispersion relations obtained for the nanostructures. An important but often ignored question is when one should use the energy spectra of electrons and phonons in bulk materials and when one must resort to the energy spectra of nanostructures. The calculation of the dispersion relations for electrons and phonons in nanostructures is usually done under the assumption of idealized potentials, such as the Kronig–Penny model for electrons and harmonic force interaction for phonons. Electron and phonon waves in such calculations are completely harmonic and coherent and do not interact with each other. Their interactions and effects of anharmonicity are included in the relaxation time term. This approach is identical to the modeling of transport properties in bulk materials. The basic requirement for such a picture to be valid is that the phase destroying scattering processes are much less frequent than the minimum periods necessary for the carriers to form new bands when going through multiple reflections at the interfaces of the nanostructures. How many periods are needed to form a new band in a superlattice? The answer to this question may vary depending on materials and structures but one can gain some idea by examining the quarter wavelength stack (Bragg reflector) used in optical coatings [49]. Taking a GaAs/AlAs quarter wavelength stack as an example, although the reflectivity at an individual interface between GaAs and AlAs is small, a reflectivity close to unity can be created with a small number of periodic quarter wavelength layers. As another example, we show in Figure 7 the average transmissivity for acoustic waves through a superlattice structure with different number of periods [50]. It shows that the transmissivity change is small after 10 periods. Thus, we can infer that if carriers (electrons and phonons) can maintain their phase coherence over a few to tens of periods of the unit cell, new energy bands will form. Under this condition, it is justifiable to use the energy bands calculated under the idealized coherence picture in combination with the relaxation time approximation. In bulk solids, 10 unit cells is often converted into around 50 Å, which is typically shorter than the mean free path and thus the traditional relaxation time approach to calculate the thermoelectric transport properties is easily justified.

The extension of the bulk approach to nanostructures, however, needs to be exercised with discretion and care. Taking again a superlattice as an example, if the thickness of each period is larger than the mean free path in bulk materials, the band structure calculated from the harmonic potential is hard to justify. In this case, the bulk band structure can still be established because the mean free path is much longer than the bulk material unit cell length but the carriers will not follow the dispersion relations as predicted by the simplified band theories due to the phase destroying scattering processes. This situation can be further extended to imperfect interfaces. If the scattering (reflection and transmission) at each interface is not phase preserving due to diffuse interface scattering, the band structures that are obtained from harmonic wave approximations cannot be justified even if the carrier mean free path in bulk materials is much longer than the periods of the superlattice. Clearly, it is very hard to treat these partially coherent regimes. One approximation that has often been taken is to assume that the carriers are completely incoherent and obey the bulk dispersion relations, except at interfaces where they experience reflections and transmissions.

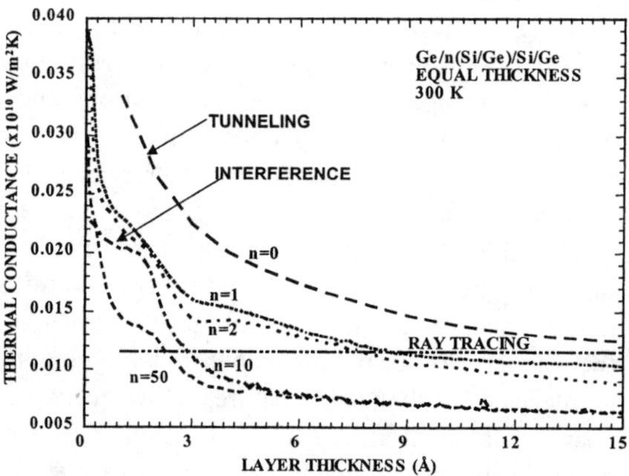

Figure 7: Thermal conductance of superlattices based on acoustic wave transmission calculations, showing that the conductance becomes independent of the number of periods when the total number of periods is larger than 10, except in the very thin period region where phonon tunneling becomes important [50]. (Courtesy of ASME.)

We summarize the above discussion in Figures 8(a) and 8(b). There are two characteristic lengths. One is the length of the minimum domain needed to form a band. It is typically a few to a few tens of unit cells. The other is the carrier mean free path (assuming that the mean free path limiting scattering processes are phase destroying). If the mean free path is larger than the minimum domain length, no bands will form and the general density-of-states approach described in section 2 will not be valid. For bulk materials, the mean free path is typically longer than the minimum domain length except in amorphous materials. For superlattices, because the minimum domain length is much larger, and because diffuse interface scattering can shorten the mean free path, the condition for band formation is not always easily satisfied. If the mean free path inside the superlattice is longer than the minimum domain size needed for superlattice band formation, the approach in section 2 can be easily adapted to superlattices by modifying the density-of-states and the relaxation time. This is the totally coherent transport regime. If this condition is not satisfied, one should treat each layer of the superlattice as a bulk medium and impose interfaces as boundary conditions for the Boltzmann equation. This is the totally incoherent regime. The real transport processes for electrons and phonons are probably in between the two extremes. In this intermediate region, it is more difficult to model the transport properties and not much has been done. It is even more complicated if one considers the transport of both electrons and phonons since they have different mean free paths and phase-destroying scattering events, and thus may fall into different transport regimes in the same structure. Typically, the electron wavelength is longer than that of the dominant phonons. Consequently, it is easier for electrons to be in the coherent regime than for phonons. In the following, we will divide our discussion mainly along two lines. One treats electrons and phonons as coherent carriers. This can be thought of as the quantum size regime. The other treats them as incoherent carriers, which can be thought of as the classical size effect regime. In section 3.5, we will discuss some recent progress in modeling the intermediate region when both quantum and classical size effects exist.

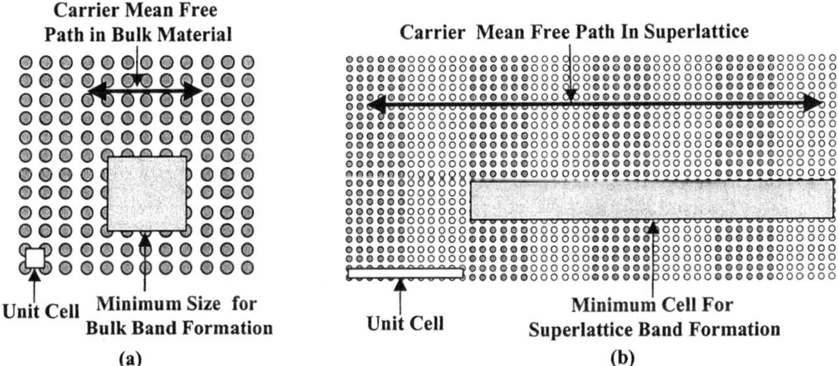

Figure 8: Characteristic lengths determining whether transport is in the coherent regime or incoherent regime in bulk materials (a) and in superlattices (b). One is the minimum domain size needed to form a band and the other is the mean free path limited by phase destroying scattering events. For the coherent regime, the mean free path should be longer than the minimum domain length in the direction of interest. This requirement is relatively easy to satisfy in bulk materials but more difficult in nanostructures due to increased unit cell size.

We should point out that the above picture is a departure from previous work on the coherence problem of photons [49, 51]. Classical partial coherence theory of light treats the integration effects of polychromatic light but does not consider the phase destroying scattering events. The optical coherence length is a measure of the width of the wave packets formed due to the superposition of the polychromatic waves. Each of the monochromatic components is completely coherent and their behavior can always be predicted by Maxwell's equations.

3.3 Coherent electron and phonon transport in nanostructures

Dresselhaus and co-workers pioneered the idea that quantum size effects on electrons can be used to improve the electron thermoelectric properties. They started with the modeling of quantum wells and later extended the work to quantum wires [5, 6]. In an infinite quantum well as shown in Figure 9(a), the eigenstates of electron energy for the nth quantized mode are,

$$E(k_x, k_y, n) = \frac{\hbar^2}{2m^*}\left(k_x^2 + k_y^2\right) + n^2 \frac{\hbar^2 \pi^2}{2m^* L^2} \tag{40}$$

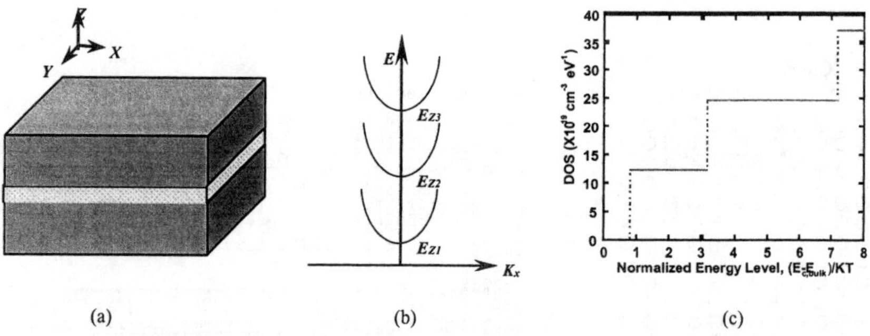

Figure 9 Illustration of (a) a single quantum well sandwiched between barriers, (b) electronic band structure inside a quantum well with infinite potential barrier, and (c) density of states in the quantum well.

Figure 9(b) shows the dispersion relation for electrons in a quantum well and Figure 9(c) the density of states. Because of energy quantization in the z-direction, electrons can only have quasi-continuum wavevectors in the k_x and k_y directions and the subbands can be thought of as the projection of the standing waves in the z-direction to the x-y plane. This energy quantization has several consequences on the expressions we used before for the Seebeck coefficient and electrical conductivity calculations for bulk materials. One is that the integration of the wavevector should be performed only in the k_x and k_y directions. In the k_z direction, the integration should be replaced by a summation. Equivalently, the electron density of states must be replaced by the new density of states inside the quantum wells. As nanostructures are progressively moved into 2D, 1D, and 0D, the density of states become more discrete, as Figure 4 indicates. It is possible to change the doping level such that the Fermi level is placed close to one subband edge and the power factor can be increased compared to that in bulk materials. Such an idea has been demonstrated in PbTe quantum wells and Si/SiGe quantum wells [6].

It is hard to use a single quantum well for thermoelectric energy conversion because the quantum well is too thin and also it has two barrier layers. Superlattices that include periodic quantum wells and barriers are thus a natural extension. When the barriers are very thin, electrons can tunnel between wells. For a single band, calculations show that such tunneling can reduce the benefit of quantum confinement [52, 53]. In addition, heat conduction in the barrier can also reduce the gain inside quantum wells. A possible solution for this problem is to use multiple carrier pockets in different layers such that every layer can be utilized for thermoelectric transport [54, 55]. As the exploration on superlattices moved on, it was discovered that PbTe-based superlattices form quantum dots with high *ZT* values [11]. Although the electron quantization effect discussed here may play a role, there is no quantitative theory at this stage to explain the experimental data.

The studies discussed in the previous paragraph are based on electron transport along the film plane direction. There are also a few investigations on the quantum confinement effect for electron transport perpendicular to the interface, based on calculations of the electronic band structures [56, 57]. So far, modeling does not suggest that one can gain much in the electron performance due to quantization effects along this direction, but phonon transport can benefit tremendously, as we will discuss later.

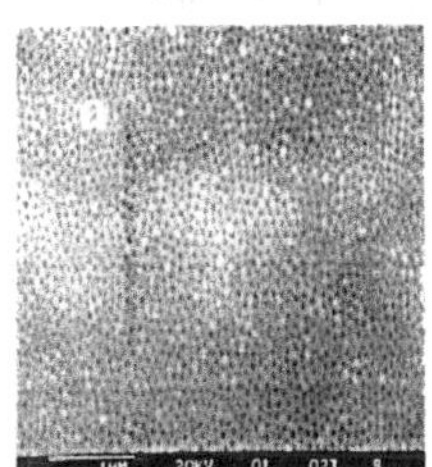
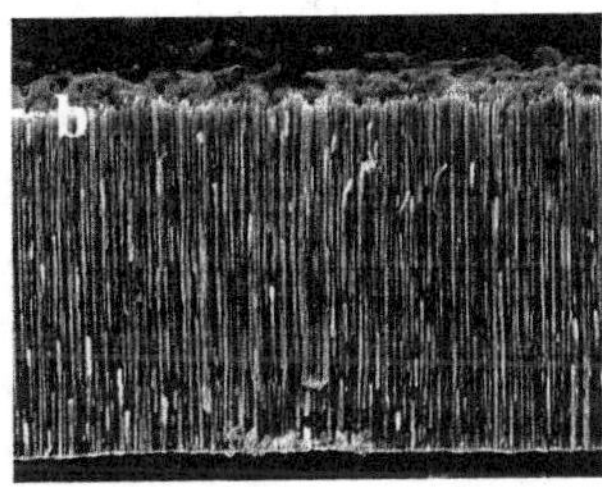
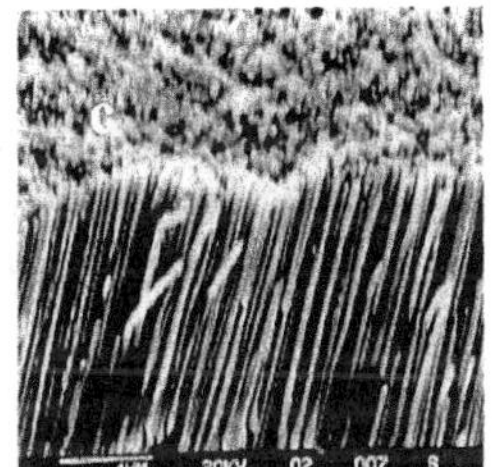

Figure 10: Nanowires can be fabricated by various template based methods: (a) anodization of aluminum leads to the formation of regular nanochannels under appropriate conditions, (b) electrodeposition of nanowires into template, and (c) nanowires after template removal.

Compared to quantum wells and superlattices, which are made using expensive equipment such as molecular beam epitaxy, nanowires can be made using more crude methods based on templates [58, 59]. Theories have predicted a much larger benefit for electron quantization effects in quantum wires and thus research on thermoelectric effects in quantum wires has been gaining increasing attention. By anodizing alumina, regular nanoscale channels can form under proper conditions as shown in Figure 10(a). Nanowires can be formed using such nanochannels as templates by pressure injection, vapor condensation, or electrodeposition [10(b) and 10(c)] [60–62]. A semimetal-to-semiconductor transition due to quantum size effects on electrons has been observed for bismuth nanowires [63] and bismuth–antimony nanowires [64]. Thermoelectric property measurements have been difficult and thus the predictions of enhanced *ZT* due to electron energy quantization have not been confirmed. Recent experiments from the Dresselhaus group shows that the Seebeck coefficient of bismuth and bismuth antimony nanowires increasing with decreasing diameter [65], consistent with the trend predicted by theories. Thermophysical properties of nanowire arrays have also been measured [66]. Single nanowire thermal conductivity measurements are also reported [67].

Similar to the potential well for electrons, phonon waves are also confined in quantum wells and superlattices. Studies on the wave confinement effects on phonons in quantum wells and superlattices have been carried out for phonon heat conduction both along and perpendicular to the film plane. There are two effects that have been discussed in some depth in the literature. One is the group velocity change and the other is the relaxation time change due to the group velocity change. We will discuss these effects below.

An example of the phonon dispersion relations in quantum wells is shown in Figure 11(a) [68]. Both lattice dynamics and continuum elastic theories lead to qualitatively similar results. Compared to the normal representation of the phonon dispersion in bulk materials, one can see that there are more phonon branches in a quantum well because of the increase in the size of the unit cell. This behavior is similar to Figure 9(b) for electrons because theoretically, traveling waves in the *z*-direction are not allowed. Different branches for the in-plane wavevectors can be considered as the superposition of opposing traveling waves with both *z*- and in-plane wave components [10]. The corresponding density of states of phonons and their group velocity in a quantum well are shown in Figures 11(b) and 11(c), in comparison to that in a bulk material. If the group velocity of the quantum well is substituted into the expression for the thermal conductivity, the reduction in the thermal conductivity is not as large as shown in Figure 11(d). Balandin [69] argued that since there is a significant drop in the phonon group velocity along the film plane direction compared to that in bulk materials, the scattering rates, particularly the impurity scattering, are significantly increased. Because a quantum well does not allow traveling waves in the *z*-direction, the group velocity in a quantum well can be considered as the projection of the bulk group velocity into the *x*-*y* plane and is thus always smaller than that of the bulk. Whether one should use this group velocity in predicting the behavior of thermal conductivity depends on whether or not the new dispersion relation can be established, a topic that we discussed at the beginning of this section.

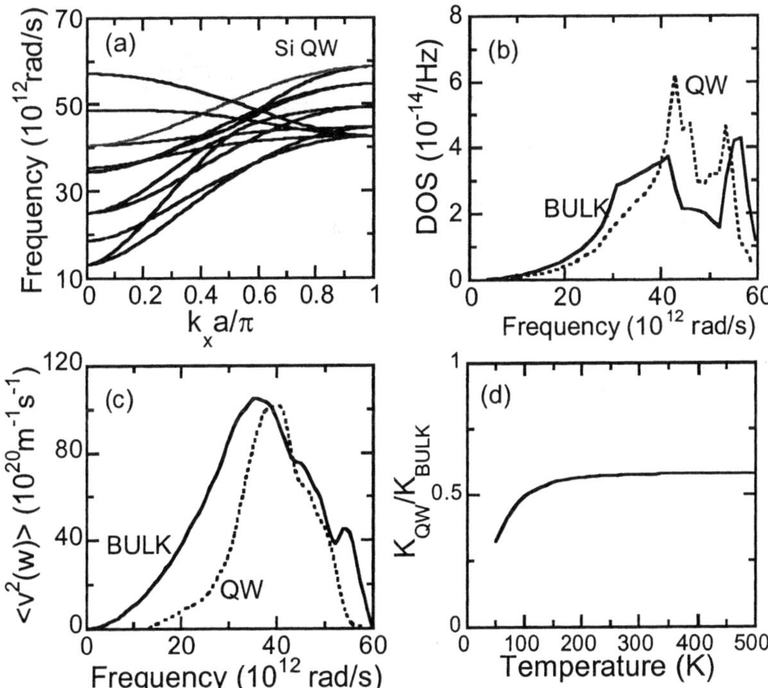

Figure 11: Phonon behavior in an ideal quantum well: (a) dispersion curves are plotted along in-plane directions, which allow traveling waves, (b) density of states in quantum wells show a few sharp features due to energy quantization but is similar overall to that in bulk materials, (c) group velocity averaged over density of states is slightly smaller than that in bulk materials, leading to (d) a slight reduction in the thermal conductivity. The reduction is not enough to explain experimental observations in single crystalline films [68].

For superlattices, waves can propagate in both the cross-plane and in-plane directions. One can again calculate the phonon dispersion in superlattices and evaluate its impact on the thermal conductivity reduction. A series of papers has been published examining this mechanism for explaining the cross-plane thermal conductivity reduction based on various lattice dynamics models of superlattices [70–74]. Figures 12(a) and 12(b) show the phonon dispersion in a Si/Ge like superlattice along the in-plane [12(a)] and cross-plane [12(b)] directions, considering the acoustic phonons in the bulk materials only [75]. The corresponding density of states and group velocity are shown in Figure 12(c) and 12(d), respectively. Clearly, the overall density of states change is relatively small compared to bulk materials, but the cross-plane group velocity has been significantly reduced.

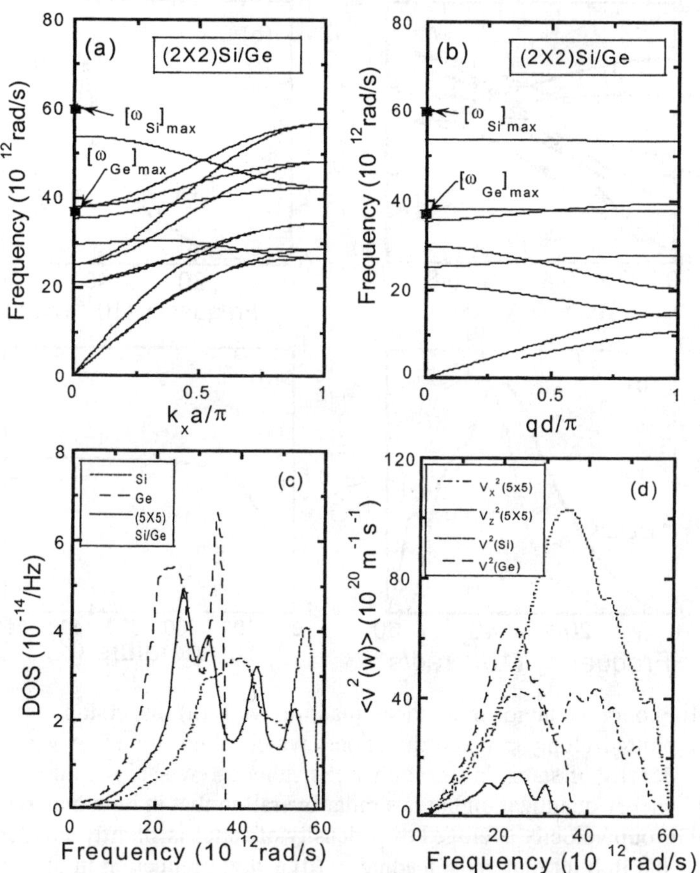

Figure 12: Phonon behavior in Si/Ge-like superlattices: (a) phonon dispersion along the film plane and (b) phonon dispersion perpendicular to the film plane, (c) density of states showing sharp features but are approximately an average of those of Si/Ge, and (d) group velocity along the in-plane and cross-plane directions. The cross-plane direction shows a large reduction while the in-plane direction is roughly an average of their bulk materials [75].

Using this group velocity, one can obtain the thermal conductivity as shown in Figures 13(a) and 13(b), which demonstrate a relatively large reduction compared to bulk materials overall in the cross-plane direction but only a small reduction in the in-plane direction. These lattice-dynamics-based results predict that in the very thin period limit, the thermal conductivity actually increases with decreasing period thickness while in the relatively thick period limit, the thermal conductivity approaches a constant. Such predictions are consistent with acoustic wave propagation based modeling as shown in Figure 7.

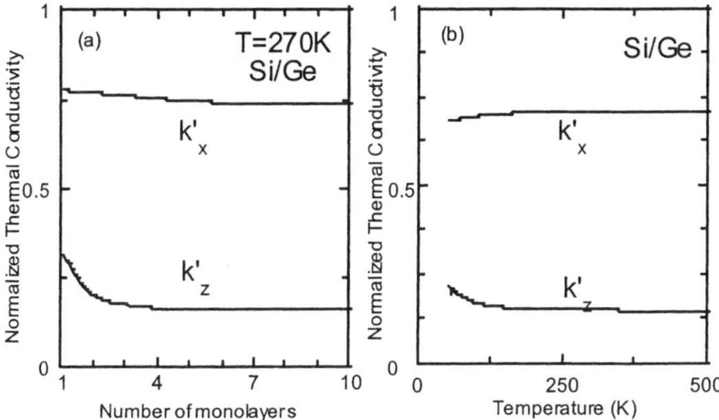

Figure 13: (a) Period thickness and (b) temperature dependency of the thermal conductivity of superlattices due to group velocity reduction [75]. The reduction along the in-plane direction is too small to explain the experimental data. The reduction along the cross-plane direction, although considerable, is still smaller than experimental data. Temperature dependency is not in agreement with experimental data either. These suggest that incoherent scattering processes play an important role in the thermal conductivity reduction exemplified in Fig. 6.

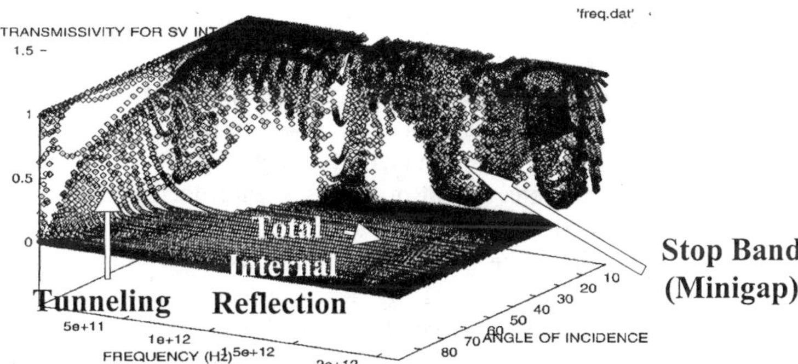

Figure 14: Phonon transmissivity through a superlattice. Long wavelength phonons can tunnel through the superlattice above the critical reflection angle [50]. This is the mechanism behind the recovery of the thermal conductivity as predicted in Fig. 7 and Fig. 13. In reality, thinner period superlattices may approach alloy structures and thus may create a recovery of the thermal conductivity. Which one of these two mechanisms is responsible for some experimental observations is not clear at this stage. (Courtesy of ASME.)

From the acoustic wave propagation model [50], it is clear that the cause of the increase in thermal conductivity with decreasing period thickness as predicted by the lattice dynamics model is due to phonon tunneling above the critical reflection angle, as shown in Figure 14. The acoustic wave propagation model treats a superlattice as an inhomogeneous medium while the lattice dynamics model treats a superlattice as a homogeneous crystal with a large unit cell. Although these are two very different approaches, both methods assume coherent phonon propagation and thus give similar trends.

The thickness and temperature dependencies of the predicted reduction by the lattice dynamics and the acoustic wave propagation models (coherent phonon models), however, are generally different from experimental observation in most superlattices. Experimentally, it has been observed that the cross-plane thermal conductivity increases with increasing period thickness and approaches bulk values when the period thickness is large, as shown in Figure 6. This trend contradicts the predictions of lattice dynamics and acoustic wave models. A few experimental data do show that the thermal conductivity recovers somewhat in the very thin period limit [38, 40, 42] that is similar to the predictions of the coherent transport models. In the cross-plane direction, although the group velocity reduction does lead to a significantly lower value than the bulk values calculated from the Fourier law, the absolute magnitude of the thermal conductivity reduction due to the group velocity reduction is still smaller compared to experimental values observed in Si/Ge and GaAs/AlAs superlattices. In addition, the trend of the temperature dependency of the relative thermal conductivity reduction is opposite to experimental observation. As the temperature decreases, the experimental film-to-bulk thermal conductivity ratio in the cross-plane direction decreases, while the lattice dynamics model predicts an opposite trend, because phonon tunneling is stronger at lower temperatures.

We should point out that a new group velocity model developed by Simkin and Mahan [74] is able to explain the correct anticipated trends in the cross-plane direction, i.e., the thermal conductivity increases with increasing period thickness for thick periods and also increases with decreasing periods for very thin periods. The model is based on assuming an extinction coefficient for the wavevector. This allows the group velocity to recover to that of the bulk in the thick film limit because exponentially decaying waves do not sample interfaces when the period thickness is large. This model, however, is unlikely to explain the experimental data in the in-plane direction and suffers from the same problem as other lattice dynamics models in that the magnitude of the predicted cross-plane thermal conductivity reduction is smaller than experimental data. Figure 13(a) shows that the in-plane thermal conductivity reduction due to the reduced group velocity is very small, while experimentally a much larger reduction was observed, as shown in Figure 6(a). Later on, we will show that diffuse interface scattering holds the key to explaining the experimental results. Diffuse interface scattering can be treated in two different ways: one is to incorporate it into the relaxation time but using the dispersion of nanostructures. The other is to use the bulk relaxation time and bulk dispersion, considering electrons and phonons as particles, and to incorporate interface scattering as the boundary conditions for the Boltzmann equation. Clearly, which approach is more appropriate depends on how strong the diffuse and internal scattering processes are in destroying the necessary phase coherence required in forming the new dispersion relations in nanostructures. So far, there have been a few attempts at the first approach, i.e., calculating the dispersion relations first based on the harmonic potential approximation and then imposing diffuse scattering mechanisms as either an internal scattering mechanism or a boundary scattering event [76, 77]. Most other treatments are based on the second approach [78–82], i.e., assuming bulk dispersion relations and a bulk relaxation time inside the nanostructures while imposing diffuse interface scattering as new boundary conditions through the Boltzmann equation. This latter approach treats electrons and phonons as completely incoherent particles and is the topic of the next section.

3.4 Incoherent electron and phonon transport in nanostructures

In the incoherent regime, we can assume that electrons and phonons in each region of a composite nanostructure have the same dispersion relations as in their bulk materials and the same relaxation time. Boundaries and interfaces in nanostructures are imposed onto the transport processes as an additional scattering mechanism. Boundary scattering is not new and classical size effects on electrons and phonons were studied a long time ago [83, 84]. There are two different approaches to deal with the boundary scattering. One is to add an extra boundary scattering term into the relaxation time through the Mathiessen rule. This approach is simple and had been widely used in dealing with low-temperature transport problems in bulk materials where size effects are important, but it is not accurate because it treats the boundary scattering processes equally with the internal scattering [84]. The other is to treat interfaces and boundaries through boundary conditions to the Boltzmann equation. We will focus on the latter approach.

3.4.1 Transport processes at a single interface

Both electrons and phonons are reflected at an interface. Electron reflection at an interface is caused by the electrical potential difference at the interface caused by the intrinsic electronic band structure mismatch or by dopants or dislocations near the interface region, while phonon reflection is due to the mismatch in acoustic properties such as the mass and the force constants of neighboring materials. Since both electrons and phonons are wave packets and the width of the wave packets is always, in theory, larger than an interface, the calculation of the interface reflectivity and transmissivity for an ideal interface should be based on the wave picture. Calculation of the reflection and transmission coefficients for electrons at an interface is a standard problem in quantum mechanics. The reflection and transmission of acoustic waves are often treated in acoustic wave theories [85]. In optics, the equivalent analogies are the Fresnel formula for the photon reflection and transmission coefficient. Despite these established theories, models based on these theories do not necessary lead to results that are in agreement with experiments because interfaces are normally not ideal for these theories to be applicable. Theories for nonideal interfaces are not well developed, as one can easily appreciate from the thermal radiative properties of rough interfaces. An often-used approximation is to treat an interface as partially specular and partially diffuse. Specular reflection and transmission of carriers can be calculated using standard theories established for ideal interfaces. The diffuse part is subject to various modeling approximations and assumptions. An example is the diffuse mismatch model for phonon reflection and transmission at a single interface [86].

Transport near the interfacial region is also a subject worthy of careful examination. Consider electron and phonon transport across an interface. A typical model for such an interfacial transport process is to assume that there is a group of carriers coming towards the interface from each side of the medium, part of it experiencing reflection and the rest transmission. It is normally assumed that the incoming carriers are at an equilibrium distribution, as determined by the local temperature and Fermi level on each side of the interface, as shown in Figure 15(a). The net flux of a microscopic quantity of interest, b (which can be energy per electron or phonon, for example), is

$$J_b = \frac{1}{(2\pi)^3} \sum_p \left[\int_{k_x>0} t_{12} b g_o(E, T_1, E_{f1}) v_{x1} d^3 k_1 + \int_{k_x<0} t_{21} b g_o(E, T_2, E_{f2}) v_{x2} d^3 k_2 \right] \tag{41}$$

where E is the energy and E_f the Fermi level or chemical potential, g_o is the equilibrium distribution function for particles coming towards the interface, t_{12} and t_{21} are the transmissivities of carriers from one side to the other as denoted by the order of the subscripts, v is the carrier velocity, the integration is over the allowable wavevector (**k**) space, and the summation is over all polarizations. Using the principle of detailed balance, eqn. (41) can be written as

$$J_b = \sum_p \int_{k_x>0} t_{12} b \left[g_o(E, T_1, E_{f1}) - g_o(E, T_2, E_{f2}) \right] v_{x1} d^3 k_1 / (2\pi)^3 \tag{42}$$

which is the familiar Landauer formalism [87]. From this formalism, one can derive many familiar expressions for interfacial and quantum transport problems. For example, we apply the above formula for electron transport at an interface. Taking b equal to the electron charge and $t_{12} = 1$ for $v_x > v_{xo}$, where v_{xo} is the minimum velocity needed to overcome the barrier height, we can derive the following Bethe formula for thermionic emission across a Schottky barrier [88]

$$J_e = AT^2 \exp\left(-\frac{q\phi}{kT}\right)\left[\exp\left(\frac{qV}{kT}\right) - 1\right]. \tag{43}$$

As another example, consider phonon transport across an interface and take $b = h\nu$. Equation (42) leads to an equation that determines the relationship between the heat flux and the temperature drop across the interface as [86],

$$J_q = (T_1 - T_2)\left[\int \tau_{1\to 2}\hbar\omega \frac{\partial g_o}{\partial T} d^3k/(2\pi)^3\right] = (T_1 - T_2)/R \qquad (44)$$

where R is the thermal boundary resistance. Although both eqns. (43) and (44) are often used in modeling electron and phonon flow across an interface, they cannot be applied to the limit when the interface disappears because they lead to an artificial drop in the Fermi level or temperature at such a fictitious interface. Thus, well-established interface transport formulae, such as thermionic emission theories and phonon thermal boundary resistance theory, all assume a distinct interface in existence.

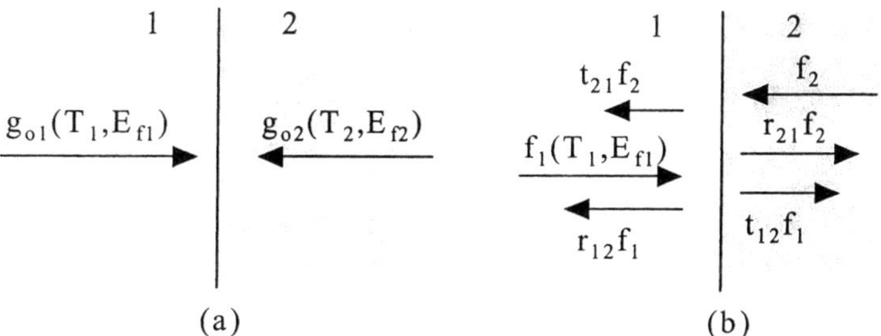

Figure 15: Modeling transport processes at a single interface. In (a), the equilibrium properties of incoming carriers are used. These equilibrium properties do not reflect the local particle or energy density and thus not the local Fermi level and temperature because the local particles are made of incoming, reflected, and transmitted carriers, as shown in (b). Consistent use of the local equilibrium quantities as based on (b) leads to diffusion-transmission boundary conditions that can be used as an approximation in combination with the use of drift-diffusion descriptions away from the interface.

One reason that these interfacial transport expressions do not degrade gracefully to the case of the existence of no interface is in the definition of the local temperature and Fermi level. In eqn. (42), g_o represents the distribution of only the carriers coming towards the interface. As shown in Figure 15(b), the local particle density on each side of the interface consists of three parts, one is those coming towards the interface as represented by f_1. The other two parts are (1) reflected from the interface, $r_{12}f_1$, and (2) transmitted from the other side of the interface $t_{21}f_2$, where r_{12} is the reflectivity. These carriers are at highly nonequilibrium states from each other and thus local equilibrium quantities such as temperature and Fermi level are hard to define in a strict thermodynamic sense. However, if these particles are allowed to scatter and to approach an equilibrium state (which we will denote as the equivalent equilibrium state), represented by f_o at the interface, this state will have a temperature and chemical potential different from those of the incoming particles that are represented by g_o. Thus, the Fermi level and temperature of the incoming electrons and/or phonons as represented by g_o are not the local (equivalent) equilibrium values that are used in classical transport theories. This inconsistency was treated before for phonon transport near interfaces. Chen [80] found that a consistent treatment is very important for explaining the experimental data on superlattices and later extended the consideration to include the nonequilibrium distribution as determined from the diffusion theory on both sides of the interface [89, 90]. With such an approach, a new general interface transport formula, called the diffusion-transmission boundary condition, was derived,

$$\left[1-\left(t_{12}''+t_{21}''\right)/2\right]J_b=\frac{t_{12}'}{(2\pi)^3}\int_{k_x>0} b\left[f_{o1}\left(E,T_1,E_{f1}\right)-f_{o1}\left(E,T_2,E_{f2}\right)\right]v_{x1}d^3k_1 \tag{45}$$

where

$$t_{12}'=\frac{\sum_p \int_{k_x>0} t_{12}bv_{x1}f_o\left(E,T_1,E_{f1}\right)d^3k_1}{\sum_p \int_{k_x>0} bv_{x1}f_o\left(E,T,E_{f1}\right)d^3k_1} \tag{46}$$

$$t_{12}''=-(2\pi)^{-3}\sum_p \int_{k_x>0} t_{12}\tau_1 bv_{x1}{}^2\left(\frac{\partial f_{o1}}{\partial x}+F_x\frac{\partial f_{o1}}{\partial E}\right)d^3k_1/(J_{b1}/2) \tag{47}$$

with a similar definition for t_{21}''. Equation (45) applies to the limit where no interface is in existence. In this case, $t_{12}=t_{21}=t_{12}'=t_{21}'=t_{12}''=t_{21}''$, leading to the solutions $T_1=T_2$, $E_{f1}=E_{f2}$ and leaving the flux J_b completely determined by drift-diffusion equations in the bulk regions. The diffusion-transmission boundary condition can be used as an approximation that is consistent with a drift-diffusion or diffusion type of treatment inside the bulk region. See refs. [89] and [90] for their applications to phonon and electron transport.

3.4.2 Classical size effects on thermoelectric transport in nanostructures

Classical size effects on electrical conductivity in single layer and multilayer structures for transport along the film plane direction have been dealt with extensively in the past, as summarized in the book by Tellier and Tosser [83].

These studies are based on solving the Boltzmann equation with interface reflection as the boundary conditions. Electron reflection and transmission are typically modeled as a diffuse or a specular process or a mixture of diffuse and specular processes. Similarly, the thermoelectric transport properties under the classical size effect can be calculated. Figure 16 shows the modeled power factor subject to quantum confinement only [Figure 16(a)] and combined quantum confinement and diffuse boundary scattering [Figure 16(b)] [91]. Compared to the quantum size effects as discussed in the previous section, the power factor decreases, rather than increases, with increasing diffuse boundary scattering. This does suggest that experimentally observed thermoelectric enhancement may come from either quantum size effects as exemplified in the work of Dresselhaus and co-workers [6].

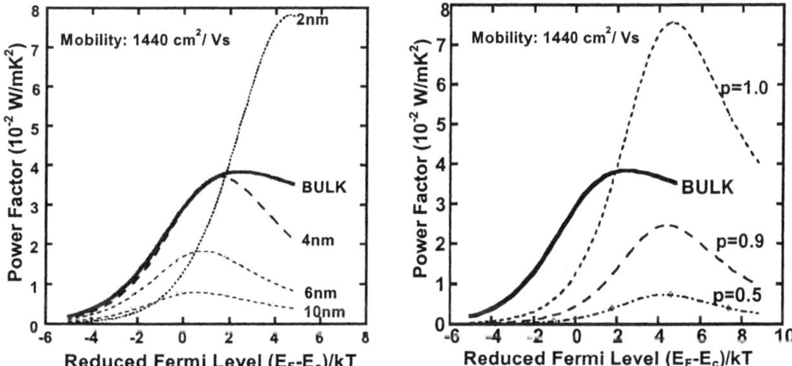

Figure 16: Power factor for a silicon quantum well calculated (a) based on quantum size effects at different quantum well widths, and (b) based on combined quantum confinement and diffuse boundary scattering effects for a 2 nm wide quantum well. It indicates that classical size effects decrease the optimum power factor [91].

The classical size effect for transport in the cross-plane direction is a topic considerably more interesting than along the in-plane direction. This is due to the thermionic emission of electrons and holes when they go across a potential barrier [88]. It has been known that vacuum thermionic emission can be used for power generation since the 1950s [92]. Mahan studied the reverse process for cooling applications and concluded that vacuum thermionic emission will have difficulty for cooling because of the large work function of available materials [93]. Solid-solid interfaces, on the other hand, can have arbitrary barrier height. For example, the barrier height between GaAs/$Al_{1-x}Ga_xAs$ depends on the concentration x and by controlling x, one can in theory change the barrier height to any desired value. Shakouri and Bower [6] proposed the use of solid-state heterojunctions as a way to get around the large work function of vacuum thermionics. Such thermionic emission coolers share the same disadvantage that the reverse heat conduction cannot be eliminated. Mahan and co-workers [8, 93, 94] introduced superlattices as cascade heterojunction structures but later concluded that such thermionic emission processes will not be more efficient than thermoelectric processes [95]. Shakouri and co-workers carried out a series of experiments exploring the thermionic cooler in a single layer and in superlattices and observed a small cooling effect in some of the superlattice coolers [96–99]. At this stage, however, it is difficult to claim whether the cooling is due to thermionic emission or normal thermoelectric processes. In addition, there are also attempts to combine resonant tunneling with field emission to create refrigeration effects, although such efforts have not so far led to any net cooling [100, 101]. Recently, there has been a new claim of a thermal diode structure that seems to have a high energy-conversion efficiency [102]. Although these claims have not been independently confirmed and thus must be viewed with care, it is possible that there are additional mechanisms that one can explore for increasing the energy conversion efficiency using solid-state structures.

One topic that is worthy of more discussion is the hot and cold electron effects for carrier transport across interfaces. A Monte Carlo simulation showed that the heat generation in heterojunction structures is not uniform as one would assume in a normal Ohmic heating process in thermoelectric devices [103]. This can be attributed to hot electron effects and has been studied quite extensively in the Russian literature [104]. However, how to include both ballistic and hot electron effects in the modeling of energy conversion process is still an open question. Zeng and Chen [105] modeled the ballistic and nonequilibrium transport of electrons and phonons, starting from the Boltzmann equation for electrons and phonons but used approximations similar to those used for photon thermal radiation heat transfer. There is much to be learned about the hot electron effects on the energy conversion processes.

Studies on the classical size effects on phonon transport have a history that is as old as that for electron transport, starting with the work of Casimir in the 1920s [106]. Size effects on thermal conductivity are normally significant at low temperature when the phonon mean free path is long. Size effects on thin films at room temperature began to draw attention in the 1970s [107]. Polycrystalline thin films have been studied by various groups [108, 109]. Goodson's group [81, 82] carried out studies on the thermal conductivity of single crystal thin silicon films and demonstrated that size effects begin to kick in at relatively high temperatures, and they also showed that the observed phonon mean free path is much longer than estimations based on simple kinetic theory, a point that was discussed in detail by Hyldgaard and Mahan [110], and Chen [78] in their modeling of the in-plane thermal conductivity in superlattices.

The use of an incoherent particle transport model based on the Boltzmann equation was developed by Chen and co-workers in a series of papers for transport in both in-plane and cross-plane directions [78–80], and by Hyldgaard and Mahan for the in-plane direction [110]. Chen's models assume partially diffuse and partially specular interfaces, with the fraction of specular interface scattering left as a fitting parameter. For the in-plane direction, the model seems to capture both the trends in the experimentally observed thickness and temperature dependency experimental data on GaAs/AlAs superlattices and more recently on Si/Ge superlattices, as exemplified by the curves in Figure 6. These successes seem to suggest that diffuse interface scattering holds the key for the observed thermal conductivity reduction in the in-plane direction. In the cross-plane direction, the model can again lead to a satisfactory explanation for a wide range of superlattices, demonstrating the reflection at a single interface, rather than the superlattice band formation, hold the key to explain the experimentally observed thermal conductivity reduction. It is worth emphasizing, however, that unlike transport along the in-plane direction, whether diffuse or specular reflection will create a larger thermal conductivity reduction depends on which type of process gives a larger interface reflectivity. It is conceivable that for interfaces with a large mismatch in acoustic properties, a diffuse interface will actually reduce the phonon reflectivity and thus increase the thermal conductivity, while for interfaces with a small mismatch, diffuse scattering may increase the phonon reflectivity and thus reduce thermal conductivity [86]. This is a topic that needs further experimental investigation. Borca-Tasciuc *et al.* [45] studied the effects of the growth and annealing temperatures on the thermal conductivity of InAs/AlSb superlattices and observed that the thermal conductivity of this superlattice system decreases with increasing growth annealing and growth temperature [Figure 17(a)]. On the other hand, a Si/Ge superlattice system seems to show an opposite trend, as shown in Figure 17(c), which suggests that quantum dot superlattices grown at a lower temperature have a smaller thermal conductivity. We have also observed similar trends for pure Si/Ge superlattices [Figure 17(b)].

The thermal conductivity of quantum-dot superlattice systems is also of great practical interest due to the experimental discovery by Harman and co-workers that PbTe-based superlattices lead to regular quantum dot arrays. Liu *et al.* [43] measured the thermal conductivity of Si/Ge quantum dot superlattices and found that the in-plane thermal conductivity is larger than that in the cross-plane direction. Recent experiments seem to suggest that the in-plane thermal conductivity of PbTe superlattices is much lower than their equivalent alloys [24, 48]. A diffuse-interface scattering-based model does suggest that values lower than that of the alloy thermal conductivity are possible along the in-plane direction of a superlattice [78] due to the different frequency dependency of the interface scattering in the superlattice versus mass difference scattering in an alloy. However, more systematic studies are needed to understand this phenomenon.

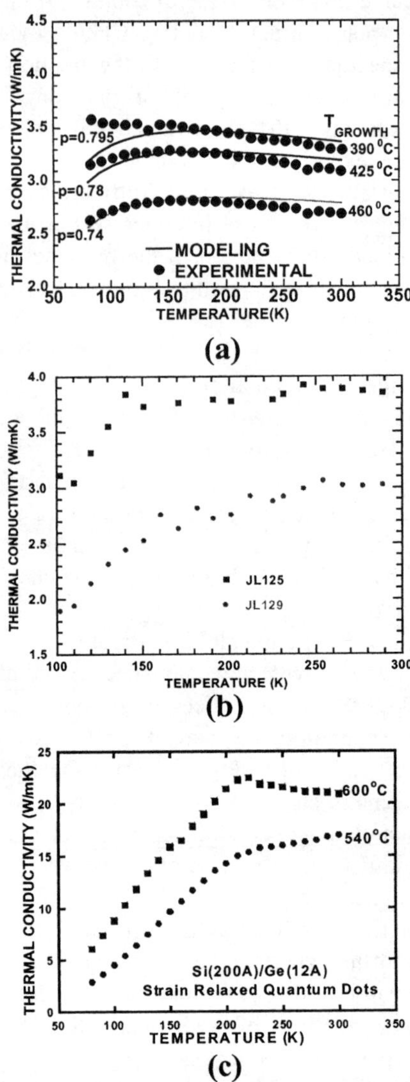

Figure 17: Effects of growth temperature on the cross-plane thermal conductivity of (a) InAs/AlSb superlattices, (b) Si/Ge superlattices, and (c) Si/Ge quantum dots superlattices [45].

3.5 Transport in the partially coherent regime

The regime where both coherent and incoherent transport processes exist is much more difficult to model. Models for such mixed transport regimes are beginning to emerge for phonons [111, 112] and for electrons [91]. As illustrated in Figure 8, the formation of a new coherent state requires that a wave maintains its phase coherence over a minimum size. For electrons inside an infinite potential quantum well, one perfect reflection at each interface is needed to form standing waves as predicted by the simple quantum-well model. If the interfaces are not perfect and scatter electrons into other directions, the ideal quantum-well model will no longer be valid. In theory, one can use the real interface potentials and solve the Schrödinger equation to obtain the true energy states of electrons. This approach is clearly quite difficult. One reasonable approximation is that if the scattering is random enough, the bulk electron energy state is still appropriate. Thus, depending on the fraction of phase destroying scattering events at the interface, the fraction of quantum and bulk states can be estimated. Electrons in the bulk states can be described by classical size effects, while electrons in the quantum states can be modeled based on the approach developed by Dresselhaus and co-workers [6]. Figure 16(b) shows results from such modeling for the electron power factor of a quantum well [91].

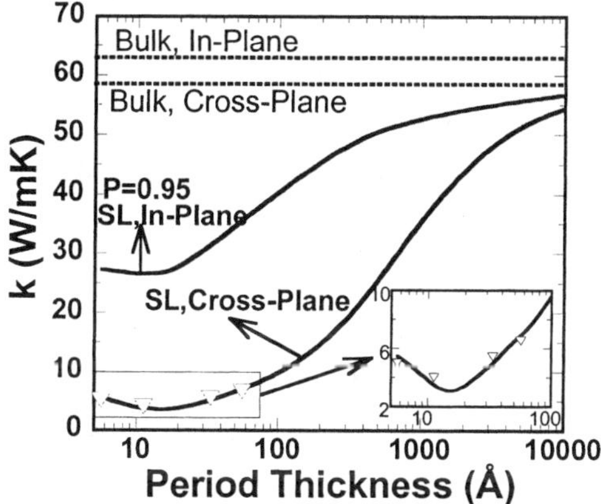

Figure 18: Phonon thermal conductivity as a function of period thickness in both in-plane and cross-plane directions of GaAs/AlAs superlattices, based on the combined wave-particle model [113]. The triangles are experimental data reported in Ref. [40]. (Courtesy of APS.)

Phonon transport in a superlattice has been treated based on a similar strategy. Figure 7 shows that the integrated phonon transmissivity through a superlattice does not change much after 10 periods, indicating that a coherent superlattice state is formed if phonons maintain their phase coherence over a distance of a few periods. Using this clue, Yang and Chen [111, 113] established transport model for phonon heat conduction in superlattices. Figure 18 shows the obtained thermal conductivity for GaAs/AlAs superlattices along the in-plane and the cross-plane directions. In the cross-plane direction, a minimum appears in the very thin period limit due to phonon tunneling. Similar behavior was observed experimentally in Bi_2Te_3/Sb_2Te_3 superlattices [42]. Lattice dynamics modeling with an extinction coefficient in the wavevector can also lead to a similar trend [74] but is not able to explain the thermal conductivity reduction in the in-plane direction. This is because such a lattice dynamics approach can predict the transition from superlattice phonon states to bulk phonon states because the extinction coefficient effectively limits the coherence length of phonons. However, the approach does not include the diffuse interface mechanism, which is the main cause of the short phonon coherence length since the bulk phonon mean free path is much longer than period thickness. It is this diffuse interface scattering mechanism that creates a large drop in the in-plane thermal conductivity.

With the above discussion, it is also instructive to compare the magnitude of the thermal conductivity reduction predicted by different models and observed experimentally. The lattice dynamics model predicts a thermal conductivity of superlattices that is consistently larger than that observed experimentally [68, 70–74]. On the other hand, consideration of diffuse interface scattering and particle transport can lead to a thermal conductivity that is comparable to experimental data and even lower values are possible [10, 80, 111]. This is because in a period structure, coherent waves form extended states that can propagate through the superlattice and thus contribute to heat conduction. By destroying these states, as implied in the particle models, lower thermal conductivity can be reached. This strongly suggests that it is not necessary to have periodic structures for thermal conductivity reduction. Nanocomposites with more interfaces can actually have a lower thermal conductivity by exploring the increased interface areas and randomness. By choosing the composite materials to minimize the disturbance to electron transport, one can potentially make high performance thermoelectric materials at low cost and scale up the size effects observed in expensive superlattices to macroscale materials for large scale applications. We believe that this will be a fruitful strategy to pursue in future research.

4 Summary

In this paper, the principles of thermoelectric energy conversion in bulk materials and devices are briefly presented, followed by a more detailed discussion of why nanostructures can be used to improve the energy conversion efficiency. The discussion on bulk materials and devices clearly shows that transport properties of electrons and phonons hold the key to energy conversion efficiency. Inspired by theoretical work on electron and phonon thermoelectric transport in nanostructures, experimental studies in the past few years have led to encouraging results that promise high energy conversion efficiency using nanostructures. One major objective of this chapter is to summarize our understanding on the electron and phonon thermoelectric transport in nanostructures, resulting from extensive studies carried out in this field.

Nanostructures can be employed to change the transport processes through interface and size effects, some of which can potentially be used to improve the energy conversion efficiency. The size and interfaces of nanostructures impact their transport properties mainly through two different mechanisms: coherent transport and incoherent transport. In the coherent transport regime, electrons and/or phonon wavefunctions are changed, causing a change in the corresponding density of states, group velocity, and scattering rates. Some of the mechanisms can be used for increasing the electron performance while decreasing the reverse phonon heat conduction. Examples are that changes in the density of states in quantum wells can be used to enhance the electron energy carrying capability while the phonon group velocity reduction can be used for thermal conductivity reduction. Although the coherent transport is attractive in physics and gives more control in design, phase destroying scattering processes inside the materials and at the interfaces can make the coherent transport regime hard to realize. In this incoherent transport regime, single interface scattering processes can be used to steer the electron and phonon transport towards desired directions. The actual transport processes in real materials may include both coherent and incoherent mechanisms. It is possible to explore the coherent effect for one type of carrier, for example, electrons, while capitalizing the incoherent effect for phonons, based on their difference in their wavelength and scattering mechanisms. Much work remains to be done in the understanding and modeling of the mixed transport regimes.

Based on the current understanding of phonon thermal conductivity reduction mechanisms in superlattices, we conclude that incoherent transport is not only dominant, but also favorable for thermal conductivity reduction. We further conclude that it is unnecessary to have periodic structures for thermal conductivity reduction. With this conclusion, we suggest that nanocomposites of thermoelectric materials can be used as a new way to realize high performance thermoelectric materials for large-scale applications. The effectiveness of this approach remains to be demonstrated.

Although the discussion in this chapter has focused on the fundamental understanding, we should emphasize that experimental work in materials synthesis and characterization has been very challenging. Continued improvements in materials and in characterization techniques are essential for future success in nanostructured-based thermoelectrics.

Acknowledgments

We would like to acknowledge support from DoD/ONR MURI on Thermoelectrics, the DARPA HERETIC project, DOE and NSF for the work reported here. We would like to thank Professors M.S. Dresselhaus, K.L. Wang, T. Sands, and R. Gronsky, and their students for their contributions to the discussed work. We also would like to thank contributions of current and former members in G. Chen's group working on nanoscale thermoelectrics and heat transfer, and to thank D. Borca-Tasciuc, C. Dames and R.G. Yang for critically reading the manuscript.

References

[1] Goldsmid, H.J., *Thermoelectric Refrigeration*, Plenum Press: New York, 1964.

[2] Ioffe, A.F., *Semiconductor Thermoelements and Thermoelectric Cooling*, Infosearch Limited: London, 1957.

[3] Rowe, D.M., (ed). *CRC Handbook of Thermoelectrics*, CRC Press: Boca Raton, Florida, 1995.

[4] Tritt, T.M., (ed). Recent trends in thermoelectric materials research. *Semiconductors and Semimetals*, **69–71**, Academic Press: San Diego, 2001.

[5] Hicks, L.D. & Dresselhaus, M.S., Effect of quantum-well structures on the thermoelectric figure of merit. *Physical Review B*, **47**, pp. 12727–12731, 1993.

[6] Dresselhaus, M.S., Lin, Y.M., Cronin, S.B., Rabin, O., Black, M.R., Dresselhaus, G. & Koga, T., Quantum wells and quantum wires for potential thermoelectric applications. *Semiconductors and Semimetals*, **71**, pp. 1–121, 2001.

[7] Shakouri, A. & Bowers, J.E., Heterostructure integrated thermionic coolers. *Applied Physics Letters*, **71**, pp. 1234–1236, 1997.

[8] Mahan, G.D., Thermionic refrigeration. *Semiconductors and Semimetals*, **71**, pp. 157–174, 2001.

[9] Venkatasubramanian, R., Phonon blocking electron transmitting superlattice structures as advanced thin film thermoelectric materials. *Semiconductors and Semimetals*, **71**, pp. 175–201, 2001.

[10] Chen, G., Phonon transport in low-dimensional structures. *Semiconductors and Semimetals*, **71**, pp. 203–259, 2001.

[11] Harman, T.C., Taylor, P.J., Spears, D.L. & Walsh, M.P., Thermoelectric quantum-dot superlattices with high *ZT*. *Journal of Electronic Materials*, **29**, pp. L1–L4, 2000.

[12] Dubios, L., An introduction to the DARPA program in advanced thermoelectric materials and devices. *Proceedings of 18th International Conference of Thermoelectrics*, ICT'99, pp. 1–4, 1999.

[13] Venkatasubramanian, R., Thin-film thermoelectric devices with high room-temperature figures of merit. *Nature*, **413**, pp. 597–602, 2001.

[14] DiSalvo, F.J., Thermoelectric cooling and power generation. *Science*, **285**, pp. 703–706, 1999.

[15] Chen, G. & Shakouri, A., Heat transfer in nanostructures for solid-state energy conversion. *Journal of Heat Transfer*, **124**, pp. 242–252, 2002.

[16] Chen, G., Dresselhaus, M.S., Dresselhaus, G., Fleurial, J.P. & Caillat, T., Some recent developments in thermoelectric materials. *International Materials Review*, in press.

[17] Mahan, G.D., Good thermoelectrics. *Solid State Physics*, **51**, pp. 81–157, 1998.

[18] Aschroft, N.W. & Mermin, N.D., *Solid State Physics*, Saunders College Publishing: Fort Worth, 1976.

[19] Callaway, J., Model for lattice thermal conductivity at low temperatures. *Physical Review*, **113**, pp. 1046–1051, 1959.
[20] Slack, G.A., The thermal conductivity of nonmetallic crystals. *Solid State Physics*, **34**, pp. 1–71, 1979.
[21] Moyzhes, B.Y. & Nemchinsky, V., Thermoelectric figure of merit of metal–semiconductor barrier structure based on energy relaxation length. *Applied Physics Letters*, **73**, pp. 1895–1897, 1998.
[22] Hicks, L.D., Harman, T.C. & Dresselhaus, M.S., Experimental study of the effect of quantum-well structures on the thermoelectric figure of merit. *Physical Review B*, **53**, pp. 10493–10496, 1996.
[23] Sun, X., Cronin, S.B., Liu, J.L., Wang, K.L., Koga, T., Dresselhaus, M.S. & Chen, G., Experimental study of the effect of the quantum well structures on the thermoelectric figure of merit in $Si/Si_{1-x}Ge_x$ system. *Proceedings of 18th International Conference on Thermoelectrics*, ICT'99, pp. 652–655, 1999.
[24] Harman, T, Integrated thermoelectric materials and device research. Presented at *DARPA/ONR Program Review and DOE High Efficiency Thermoelectric Workshop*, Coronado, CA, March 24–27, 2002.
[25] Volklein, F. & Kesseler, E., A method for the measurement of thermal conductivity, thermal diffusivity and other transport coefficients of thin films. *Physica Status Solidi A*, **81**, pp. 585–596, 1984.
[26] Hatta, I., Thermal diffusivity measurement of thin films and multilayered composites. *International Journal of Thermophysics*, **11**, pp. 293–304, 1990.
[27] Lee, S.M. & Cahill, D.G., Heat transport in thin dielectric films. *Journal of Applied Physics*, **81**, pp. 2590–2595, 1997.
[28] Capinski, W.S. & Maris, H.J., Thermal conductivity of GaAs/AlAs superlattices. *Physica B*, **219** and **220**, pp. 699–701, 1996.
[29] Yang, B., Liu, J.L., Wang, K.L. & Chen, G., Simultaneous measurements of seebeck coefficient and thermal conductivity across superlattice. *Applied Physics Letters*, **80**, pp. 1758–1760, 2002.
[30] Kanatzidis, M.G., The role of solid-state chemistry in the discovery of new thermoelectric materials. *Semiconductors and Semimetals*, **69**, pp. 51–100, 2001.
[31] Kanatzidis, M.G., New chacolgenide materials and devices with superior thermoelectric performance. Presented at *DARPA/ONR Program Review and DOE High Efficiency Thermoelectric Workshop*, March 24–27, 2002, Coronado, CA, 2002.
[32] Yao, T., Thermal properties of AlAs/GaAs superlattices. *Applied Physics Letters*, **51**, pp. 1798–1800, 1987.
[33] Tien, C.L. & Chen, G., Challenges in microscale conductive and radiative heat transfer, *Proceedings of the American Society of Mechanical Engineers*, **227**, 1–12, 1992; also in *Journal of Heat Transfer*, **116**, 799–807, 1994.

[34] Chen, G., Tien, C.L., Wu, X. & Smith, J.S., Measurement of thermal diffusivity of GaAs/AlGaAs thin-film structures. *Journal of Heat Transfer*, **116**, pp. 325–331, 1994.
[35] Yu, X.Y., Chen, G., Verma, A. & Smith, J.S., Temperature dependence of thermophysical properties of GaAs/AlAs periodic structure. *Applied Physics Letters*, **67**, pp. 3554–3556, 1995.
[36] Lee, S.M., Cahill, D.G. & Venkatasubramanian, R., Thermal conductivity of Si-Ge superlattices. *Applied Physics Letters*, **70**, pp. 2957–2959, 1997.
[37] Venkatasubramanian, R., Thin-film superlattice and quantum-well structures – a new approach to high-performance thermoelectric materials. *Naval Research Review*, **58**, pp. 44–54, 1996.
[38] Yamasaki, I., Yamanaka, R., Mikami, M., Sonobe, H., Mori, Y. & Sasaki, T., Thermoelectric properties of Bi_2Te_3/Sb_2Te_3 superlattice structure. *Proceedings of 17th International Conference on Thermoelectrics*, ICT'98, pp. 210–213, 1998.
[39] Touzelbaev, M.N., Zhou, P., Venkatasubramanian, R. & Goodson, K.E., Thermal characterization of Bi_2Te_3/Sb_2Te_3 superlattices. *Journal of Applied Physics*, **90**, pp. 763–767, 2001.
[40] Capinski, W.S., Maris, H.J., Ruf, T., Cardona, M., Ploog, K. & Katzer, D.S., Thermal-conductivity measurements of GaAs/AlAs superlattices using a picosecond optical pump-and-probe technique. *Physical Review B*, **59**, pp. 8105–8113, 1999.
[41] Borca-Tasciuc, T., Liu, W.L., Zeng, T., Song, D.W., Moore, C.D., Chen, G., Wang, K.L., Goorsky, M.S., Radetic, T., Gronsky, R., Koga, T. & Dresselhaus, M.S., Thermal conductivity of symmetrically strained Si/Ge superlattices. *Superlattices and Microstructures*, **28**, pp. 119–206, 2000.
[42] Venkatasubramanian, R., Lattice thermal conductivity reduction and phonon localization like behavior in superlattice structures. *Physical Review B*, **61**, pp. 3091–3097, 2000.
[43] Liu, W.L., Borca-Tasciuc, T., Chen, G., Liu, J.L. & Wang, K.L., Anisotropy thermal conductivity of ge-quantum dot and symmetrically strained Si/Ge superlattice. *Journal of Nanoscience and Nanotechnology*, **1**, pp. 39–42, 2001.
[44] Huxtable, S.T., Abramson, A.R., Tien, C.L., Majumdar, A., LaBounty, C., Fan, X., Zeng, G., Bowers, J. & Croke, E.T., Thermal conductivity of Si/SiGe and SiGe/SiGe superlattices. *Applied Physics Letters*, **80**, pp. 1737–1739, 2002.
[45] Borca-Tasciuc, T., Achimov, D., Liu, W.L., Chen, G., Ren, H.-W., Lin, C.-H. & Pei, S.S., Thermal conductivity of InAs/AlSb superlattices. *Microscale Thermophysical Engineering*, **5**, pp. 225–231, 2001.
[46] Huxtable, S.T., Shakouri, A., LaBounty, C., Fan, X., Abraham, P., Chiu, Y.J., Bowers, J.E. & Majumdar, A., Thermal conductivity of Indium Phosphide based superlattices. *Microscale Thermophysical Engineering*, **4**, pp. 197–203, 2000.

[47] Song, D.W., Liu, W.L., Zeng, T., Borca-Tasciuc, T., Chen, G., Caylor, C. & Sands, T.D., Thermal conductivity of skutterudite thin films and superlattices. *Applied Physics Letters*, **77**, pp. 3854–3856, 2000.

[48] Beyer, H., Nurnus, J., Bottner, H., Lambrecht, Roch, T. & Bauer, G., PbTe based superlattice structure with high thermoelectric efficiency. *Applied Physics Letters*, **80**, pp. 1216–1218, 2002.

[49] Born, M. & Wolf, E., *Principles of Optics*, 6th edn. Pergamon Press: London, 1993.

[50] Chen, G., Phonon wave effects on heat conduction in thin films and superlattices. *Journal of Heat Transfer*, **121**, pp. 945–953, 1999.

[51] Chen, G. & Tien, C.L., Partial coherence theory of thin film radiative properties. *Journal of Heat Transfer*, **114**, pp. 636–643, 1992.

[52] Sofo, J.O. & Mahan, G.D., Thermoelectric figure of merit of superlattices. *Applied Physics Letters*, **65**, pp. 2690–2692, 1994.

[53] Broido, D.A. & Reinecke, T.L., Effect of superlattice structure on the thermoelectric figure of merit. *Physical Review B*, **51**, pp. 13797–13800, 1995.

[54] Koga, T., Concept and applications of carrier pocket engineering to design useful thermoelectric materials using superlattice structures, Ph.D. thesis, Harvard University, 2000.

[55] Koga, T., Sun, X., Cronin, S.B. & Dresselhaus, M.S., Carrier pocket engineering to design superior thermoelectric materials using GaAs/AlAs superlattices. *Applied Physics Letters*, **73**, pp. 2950–2952, 1998.

[56] Whitlow, L.W. & Hirano, T., Superlattice application to thermoelectricity. *Journal of Applied Physics*, **78**, pp. 5460–5466, 1995.

[57] Radtke, R.J., Ehrenreich, H. & Grein, C.H., Multilayer thermoelectric refrigeration in $Hg_{1-x}Cd_xTe$ superlattices. *Journal of Applied Physics*, **86**, pp. 3195–3198, 1999.

[58] Hicks, L.D. & Dresselhaus, M.S., Thermoelectric figure of merit of a one-dimensional conductor. *Physical Review B*, **47**, pp. 16631–16634, 1993.

[59] Sun, X., Zhang, Z. & Dresselhaus, M.S., Theoretical modeling of thermoelectricity in Bi nanowires. *Applied Physics Letters*, **74**, pp. 4005–4007, 1999.

[60] Zhang, Z., Sun, X., Dresselhaus, M.S., Ying, J.Y. & Heremans, J., Electronic transport properties of single-crystal bismuth nanowire arrays. *Physical Review B*, **61**, pp. 4850–4861, 2000.

[61] Heremans, J., Trush, C.M., Lin, Y.-M., Cronin, S., Zhang, Z., Dresselhaus, M.S. & Mansfield, J.F., Bismuth nanowire arrays: Synthesis and galvanomagnetic properties. *Physical Review B*, **61**, pp. 2921–2930, 2000.

[62] Prieto, A.L., Sander, M.S., Stacy, A.M., Gronsky R. & Sands, T., Electrodeposition of Bi_2Te_3 nanowire composites, Thermoelectric materials 2000 – the next generation materials for small-scale refrigeration and power generation applications. *Materials Research Society Symposium Proceedings*, **626**, pp. Z14.1.1–Z14.1.5, 2000.

[63] Lin, Y.M., Sun, X. & Dresselhaus, M.S., Theoretical investigation of thermoelectric transport properties of cylindrical Bi nanowires. *Physical Review B*, **62**, pp. 4610–4623, 2000.

[64] Lin, Y.-M., Rabin, O., Cronin, S.B., Ying, J.Y. & Dresselhaus, M.S., Semimetal-semiconductor transition in $Bi_{1-x}Sb_x$ alloy nanowires and their thermoelectric properties. *Applied Phyics Lett*ers **81**, pp. 2403–2405, 2002.

[65] Lin, Y.-M. Cronin, S.B., Rabin, O., Ying, J.Y. & Dresselhaus, M.S., Thermoelectric properties of $Bi_{1-x}Sb_x$ nanowire arrays, in *Thermoelectric Materials 2001 – Research and Applications: MRS Symposium Proceedings*, **691**, pp. G10.6.1–10.6.6, 2001.

[66] Borca-Tasciuc, D.A., Chen, G., Gonzales, M.S.M., Prieto, A.L., Stacy, A., Sands, T., Borshchevesky, A., Fleurial, J.-P. & Ryan, A., Thermal diffusivity characterization of Bi_2Te_3 nanowire arrays in amorphous anodized alumina template. Presented at 2002 *International Mechanical Engineering Congress and Exhibition,* November 17–22, New Orleans, LA.

[67] Li, D., Abramson, A., Huxtable, S., Shi, L. & Majumdar, A., Thermal Transport in Single Nanowires and Nanowire Arrays. Presented at *International Symposium on Micro/Nanoscale Energy Conversion and Transport Phenomena*, April 14–19, 2002, Antalya, Turkey.

[68] Yang, B. & Chen, G., Lattice dynamics study of phonon heat conduction in quantum wells. *Physics of Low-Dimensional Structures*, **5–6**, pp. 37–48, 2000.

[69] Balandin, A. & Wang, K.L., Significant decrease of the lattice thermal conductivity due to phonon confinement in a free-standing semiconductor quantum well. *Physical Review B*, **58**, 1544–1549, 1998.

[70] Hyldgaard, P. & Mahan, G.D., Phonon superlattice transport. *Physical Review B*, **56**, pp. 10754–10757, 1997.

[71] Tamura, S., Tanaka, Y. & Maris, H.J., Phonon group velocity and thermal conduction in superlattices. *Physical Review B*, **60**, pp. 2627–2630, 1999.

[72] Kiselev, A.A., Kim, K.W. & Stroscio, M.A., Thermal conductivity of Si/Ge superlattices: a realistic model with diatomic unit cell. *Physical Review B*, **62**, pp. 6896–6899, 2000.

[73] Bies, W.E., Radtke, R.J. & Ehrenreich, H., Phonon dispersion effects and the thermal conductivity reduction in GaAs/AlAs superlattices. *Journal of Applied Physics*, **88**, pp. 1498–1503, 2000.

[74] Simkin, M.V. & Mahan, G.D., Minimum of thermal conductivity of superlattices. *Physical Review Letters*, **84**, pp. 927–930, 2000.

[75] Yang B. & Chen, G., Anisotropy of heat conduction in superlattices. *Microscale Thermophysical Engineering*, **5**, pp. 107–116, 2001.

[76] Volz, S. & Lemonnier, D., Confined phonon and size effects on nanowire thermal conductivity: The radiative transfer approach. *Physics of low-dimensional structures*, **5–6**, pp. 91–107, 2000.

[77] Zou, J. & Balandin, A., Phonon heat conduction in a semiconductor nanowire. *Journal of Applied Physics*, **89**, pp. 2932–2938, 2001.
[78] Chen, G., Size and interface effects on thermal conductivity of superlattices and periodic thin-film structures. *ASME Journal of Heat Transfer*, **119**, pp. 220–229, 1997.
[79] Chen, G. & Neagu, M., Thermal conductivity and heat transfer in superlattices. *Applied Physics Letters*, **71**, pp. 2761–2763, 1997.
[80] Chen, G., Thermal conductivity and ballistic phonon transport in cross-plane direction of superlattices. *Physical Review B*, **57**, pp. 14958–14973, 1998.
[81] Ju, Y.S. & Goodson, K.E., Phonon scattering in silicon films of thickness below 100 nm. *Applied Physics Letters*, **74**, pp. 3005–3007, 1999.
[82] Asheghi, M., Touzelbaev, M.N., Goodson, K.E., Leung, Y.K. & Wong, S., Temperature-dependent thermal conductivity of single-crystal silicon layers in SOI substrates. *ASME Journal of Heat Transfer*, **120**, pp. 30–36, 1999.
[83] Tellier, C.R. & Tosser, A.J., *Size Effects in Thin Films*, Elsevier: Amsterdam, 1982.
[84] Ziman, J.M., *Electrons and Phonons*, Clarendon Press: Oxford, 1960.
[85] Auld, B.A, *Acoustic Fields and Waves in Solids*, 2nd edn. Kruger, Malabar, 1990.
[86] Swartz, E.T. & Pohl, R.O., Thermal boundary resistance. *Review of Modern Physics*, **61**, pp. 605–668, 1989.
[87] Imry, Y. & Landauer, R., Conductance viewed as transmission. *Reviews of Modern Physics*, **71**, pp. S306–S312, 1999.
[88] Sze, S.M., *Physics of Semiconductor Devices*, 2nd edn. Wiley, 1981.
[89] Chen, G. & Zeng, T., Nonequilibrium phonon and electron transport in thin films and superlattices. *Microscale Thermophysical Engineering*, **5**, pp. 71–88, 2001.
[90] Chen, G., Diffusion–transmission interface condition for electron and phonon transport, *Applied Physics Letters*, **82**, pp. 991–993, 2003.
[91] Liu, W.L. & Chen, G, Classical size effects on in-plane thermoelectric properties in superlattices. Presented at MRS Fall Meeting, Nov. 26–30, 2001, *Symposium G: Thermoelectric Materials – Research and Applications*, pp. G8.28.1–8.28.7, 2001.
[92] Hatsopoulos, G.N. & Kaye, J., Measured thermal efficiencies of a diode configuration of a thermo electron engine. *Journal of Applied Physics*, **29**, pp. 1124–1125, 1958.
[93] Mahan, G.D., Thermionic refrigeration. *Journal of Applied Physics*, **76**, pp. 4362–4366, 1994.
[94] Mahan, G.D. & Woods, L.M., Multilayer thermionic refrigeration. *Physical Review Letters*, **80**, pp. 4016–4019, 1998.
[95] Vining, C.B. & Mahan, G.D., The B factor in multilayer thermionic refrigeration. *Journal of Applied Physics*, **86**, pp. 6852–6853, 1999.

[96] Shakouri, A., LaBounty, C., Piprek, J., Abraham, P. & Bowers, J.E., Thermionic emission cooling in single barrier heterostructures. *Applied Physics Letters*, **74**, pp. 88–89, 1999.

[97] Zeng, G.H., Shakouri, A., La Bounty, C., Robinson, G., Croke, E., Abraham, P., Fan, X.F., Reese, H. & Bowers, J.E., SiGe micro-cooler. *Electronics Letters*, **35**, pp. 2146–2147, 1999.

[98] Fan, X.F., Zeng, G.H., LaBounty, C., Bowers, J.E., Croke, E., Ahn, C.C., Huxtable, S., Majumdar, A. & Shakouri, A., SiGeC/Si superlattice microcoolers. *Applied Physics Letters*, **78**, pp. 1580–1582, 2001.

[99] Fan, X.F., Zeng, G., Croke, E., LaBounty, C., Ahn, C.C., Vashaee, D., Shakouri, A. & Bowers, J.E., High cooling power density SiGe/Si micro-coolers. *Electronics Letters*, **37**, pp. 126–127, 2001.

[100] Korotkov, A.N. & Likharev, K.K. Possible cooling by resonant Fowler-Nordheim emission. *Applied Physics Letters*, **75**, pp. 2491–2493, 1999.

[101] Miskovsky, N.M. & Cutler, P.H., Microelectronic cooling using the Nottingham effect and internal field emission in a diamond (wide-band gap material) thin-film device. *Applied Physics Letters*, **75**, pp. 2147–2149, 1999.

[102] Haglestein, P.L. & Kucherov, Y., Enhancement of thermal to electric energy conversion with thermal diode. *MRS Proceedings,* 2001 MRS Fall Meeting, *Symposium G*, **691**, 2001.

[103] Shakouri, A., Lee, E.Y., Smith, D.L., Narayanamurti, V. & Bowers, J.E., Thermoelectric effects in submicron heterostructure barriers. *Microscale Thermophysical Engineering*, **2**, pp. 37–42, 1998.

[104] Anisimov, S.I., Kapeliovich, B.L. & Perelman, T.L., Emission of electrons from the surface of metals induced by ultrashort laser pulses. *Teoreticheskoi Fiziki (Soviet Physics – JETP)* , **66**, pp. 776–781, 1974.

[105] Zeng, T.F. & Chen, G., Nonequilibrium electron and phonon transport in heterostructures for energy conversion. *Proceedings of International Mechanical Engineering Congress and Exhibition (IMECE2000)*, ASME HTD-366-2, pp. 361–372, 2000.

[106] Casimir, H.B.G., Note on the conduction of heat in crystals. *Physica*, **5**, pp. 495–500, 1938.

[107] Nath, P. & Chopra, K.L., Thermal conductivity of copper films. *Thin Solid Films*, **20**, pp. 53–62, 1974.

[108] Volklein, F. & Kessler, E., Analysis of the lattice thermal conductivity of thin films by means of a modified mayadas-shatzkes model: the case of bismuth films. *Thin Solid Films*, **142**, pp. 169–181, 1986.

[109] Tai, Y.C., Mastrangelo, C.H. & Muller, R.S., Thermal conductivity of heavily doped low-pressure chemical vapor deposited polycrystalline silicon films. *Journal of Applied Physics*, **63**, pp. 1442–1447, 1987.

[110] Hyldgaard, P. & Mahan, G.D., Phonon knudson flow in superlattices. *Thermal Conductivity*, Technomic: Lancaster, **23**, pp. 172–182, 1996.

[111] Yang, B. & Chen, G., A unified wave-particle model for phonon transport in superlattices. Presented at *4th US–Japan Nanotherm Seminar:*

Nanoscale Thermal Science and Engineering, June 24–26, 2002, Berkeley, CA.

[112] Chen, G., Yang, B, Liu, W.L., Borca-Tasciuc, D. & Jacquot, A., Energy conversion and transport in nanostructures. Presented at *International Symposium on Micro/Nanoscale Energy Conversion and Transport*, April 14–19, 2002, Antalya, Turkey, extended abstract book, pp. 42–43.

[113] Yang, B. & Chen, G., Partially coherent phonon heat conduction in superlattices, *Physical Review B*, **67**, pp. 195311–195314, 2003. Also selected for the May 19, 2003 issue of the *Virtual Journal of Nanoscale Science & Technology*.

CHAPTER 3

Heat transport in superlattices and nanowires

S.T. Huxtable, A.R. Abramson & A. Majumdar
Department of Mechanical Engineering
University of California – Berkeley, USA

Abstract

Nanostructures such as superlattices, nanowires, and nanotubes are currently being examined for applications in a variety of fields including microelectronics, optoelectronics, and thermoelectrics. Regardless of the application, understanding the basic physics of thermal transport in these structures is crucial for engineering new devices. As the size of a structure decreases, the surface area-to-volume ratio increases, thereby increasing the importance of boundaries and interfaces. Furthermore, as the critical lengthscale of a structure approaches the phonon wavelength or mean free path, other interesting phenomena become significant. For these reasons, heat transport in micro and nanostructures can be much different than in bulk solids. This chapter discusses thermal transport in superlattices, nanowires, and nanotubes. Heat transport in bulk solids by phonons is first reviewed, followed by a discussion of effects that become important for structures at the micro and nanoscale, such as mismatches in the phonon spectra and acoustic impedance, phonon tunneling, interface and boundary scattering, and phonon wave interference and bandgap formation. Recent studies on different structures and materials systems are reviewed and summarized with an emphasis on experimental results.

1 Introduction

In recent years, the scaling down of feature sizes in microelectronic devices has renewed interest in the unusual physics underlying thermal transport in micro and nanoscale structures. Research has shown that thermal phenomena in micro and nanoscale components can vary significantly from the thermal behavior of bulk materials. For example, multilayered structures of thin films, or superlattices, made of semiconductor materials have often been shown to be relatively poor conductors of heat, even when they are composed of materials that individually possess high bulk thermal conductivities. This fact causes problems when these multilayered structures are used as vertical cavity surface emitting lasers (VCSELs), where a high thermal conductivity is desired in order to dissipate the high density of heat that is generated within the device. On the other hand, efficient thermoelectric devices, used for the conversion between thermal and electrical energy, require a low thermal conductivity to prevent the backflow of heat between the hot and cold junctions. In addition to the two-dimensional superlattice structures, recent advances in the growth techniques of nanotubes and nanowires have led to the prospect of fabricating similar devices using one-dimensional components. The ability to understand, predict, and manipulate micro/nanoscale thermal phenomena within these structures will be required in the development and optimization of new nanostructured devices.

In this chapter, heat transport in low-dimensional structures such as the aforementioned superlattices, nanowires, and nanotubes is reviewed. A brief and qualitative review of heat transport in bulk materials is presented followed by an extension for heat conduction at the micro and nanoscale. Finally, recent studies on heat transport within superlattices, nanowires, and nanotubes are reviewed.

2 Superlattices

Superlattices are periodic structures that are composed of alternating layers of different materials as shown in Figure 1. The artificial periodicity created in superlattices can lead to the tailoring of certain material properties for applications in fields such as optoelectronics [1, 2] and thermoelectrics [3, 4]. In both cases the thermal conductivity is an important parameter for device performance.

Superlattices are typically made from semiconducting materials and are grown by either molecular beam epitaxy (MBE) [5, 6] or metal organic chemical vapor deposition (MOCVD) [7]. Both techniques allow for the growth of atomically abrupt interfaces between different materials and control of layer thickness at the atomic level. In general, the thickness of the individual layers may range from several to hundreds of ångstroms. A superlattice may contain up to several hundred of these layers for a total thickness on the order of 1 μm. Growing superlattices thicker than a few μm is difficult due to the build up of strain within the layers.

3 Nanowires and nanotubes

In addition to the interest in two-dimensional nanostructures, one-dimensional structures such as nanowires and nanotubes have also received attention recently. Although the definitions vary throughout the literature, nanowires generally refer to solid cylinders with diameters less than ~100 nm, while nanotubes are hollow cylinders with diameters typically much less than 100 nm.

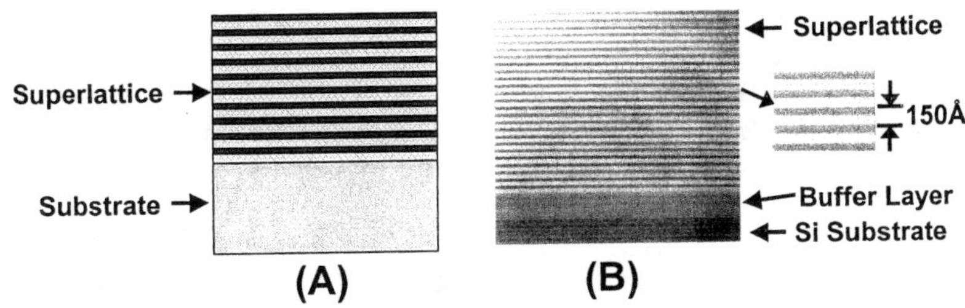

Figure 1: (A) Schematic diagram of a generic superlattice. Superlattices are periodic structures made up of alternating layers of different materials grown by molecular beam epitaxy (MBE) or metalorganic chemical vapor deposition (MOCVD) on substrates. Layers range from tens to hundreds of ångstroms in thickness and superlattices may have up to several hundred periods. (B) Transmission electron microscope (TEM) images of a $Si_{0.76}Ge_{0.24}/Si_{0.84}Ge_{0.16}$ superlattice on a Si substrate. The inset at the right is a high magnification image showing each layer (TEM images courtesy of E. Croke, HRL Labs, and C. Ahn, CalTech).

3.1 Nanowires

Theoretical predictions have shown that nanowires have the potential to make for excellent thermoelectric elements [8]. However, techniques for actually fabricating nanowires have only recently been developed.

The majority of the growth methods for semiconductor nanowires may be loosely categorized as either vapor [9–15] or solution-based techniques [16-19]. At the present time, no technique has been shown to allow for the reliable and predictable synthesis of nanowires with diameters less than ~10 nm where strong quantum confinement effects are expected to be found. Arrays of silicon and ZnO nanowires grown by a vapor technique known as the vapor–liquid–solid (VLS) method are shown in Figure 2. This technique is particularly attractive because a wide range of materials can be used to fabricate nanowires that are single crystalline and have varying diameters. This type of nanowire growth process involves dissolving gas reactants in nanoparticles of a catalytic metal. The metal nanoclusters may be pre-formed and then distributed on a substrate. Alternatively, metal nanobeads of a particular diameter that naturally form upon melting a thin metal film on a substrate under specific growth conditions could be utilized. Laser ablation of a thin metal film has also been used to produce nanometer diameter metal clusters [20]. Equilibrium phase diagrams can be used to choose a suitable liquid alloy for the specific nanowire material and the appropriate temperature range to enable alloy formation. The metal nanocluster serves as a preferential site for absorption of a reactant gas containing the desired nanowire material. As the volume of the alloy droplet increases due to the absorption of the gas and becomes increasingly saturated, nucleation begins. As additional material precipitates, the liquid/solid interface rises and a solid nanowire is grown. One-dimensional growth will proceed as long as the catalyst remains liquid and there is ample reactant gas available [20]. Each wire forms with an alloy droplet that has solidified on the nanowire tip. The VLS growth process is summarized in Figure 3.

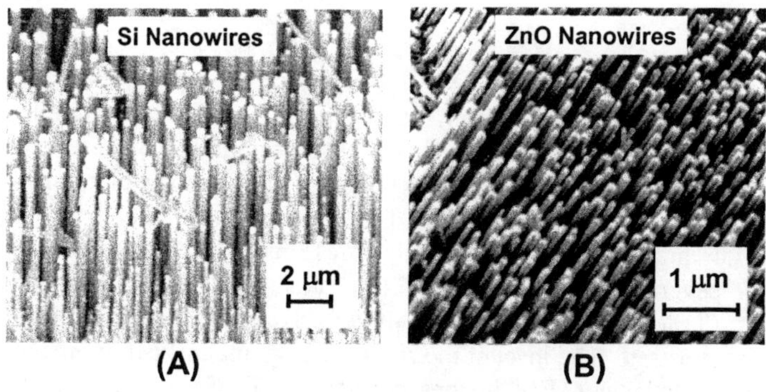

Figure 2: (A) Scanning electron micrograph (SEM) of Si nanowires grown by the vapor–liquid–solid (VLS) method. (B) SEM of ZnO nanowires grown by the VLS method (image courtesy of P. Yang group, Chemistry, UCB).

Dense nanowire arrays have also been fabricated through the use of templates with nanoscale pores or channels. These porous templates may be filled with pressurized liquid [21, 22] or vapor [23] injection or through growth by electrodeposition [24, 25]. The size of the pores may be adjusted down to ~10 nm in diameter, although consistently growing wires in pores of this size remains a challenge. Figure 4 shows a transmission electron micrograph (TEM) of a Bi_2Te_3 nanowire array grown by electrodeposition in an alumina template.

3.2 Nanotubes

There has been extensive interest in nanotubes over the last decade. Nearly all of the attention has focused on carbon nanotubes [26], although nanotubes have been successfully synthesized out of other materials such as boron-nitride [27].

Carbon nanotubes can be thought of as a sheet of carbon atoms that have been rolled up to form a seamless hollow cylinder. These nanotubes are called single-walled nanotubes (SWNTs) and are approximately one nanometer in diameter and can be up to tens of microns long. Additional structures exist where several concentric nanotube shells form what are called multi-walled nanotubes (MWNTs). Depending on the chirality of the nanotube – that is the angle with which the graphene sheet is twisted as it is rolled up – it may be semiconducting or metallic.

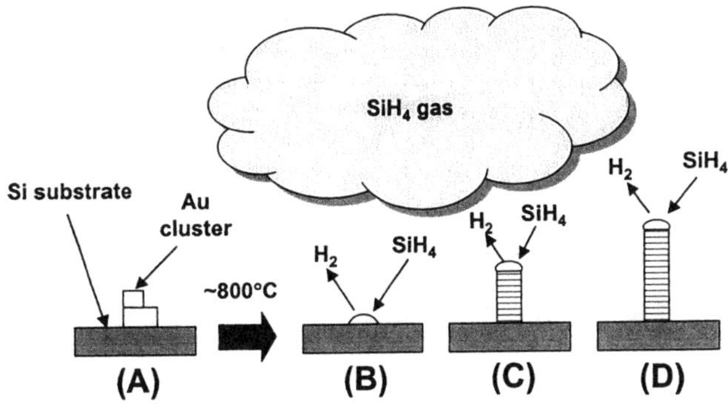

Figure 3: Schematic diagram of Si nanowire growth by the vapor–liquid–solid (VLS) method. (A) Gold clusters on a Si substrate. (B) After heating to ~800°C in a silane (SiH_4) environment, the gold liquefies and forms an alloy with Si. (C) When the liquid is supersaturated with Si, the Si precipitates and crystallizes as nanowires. (D) Further condensation and dissolution of Si increases the length of the nanowire.

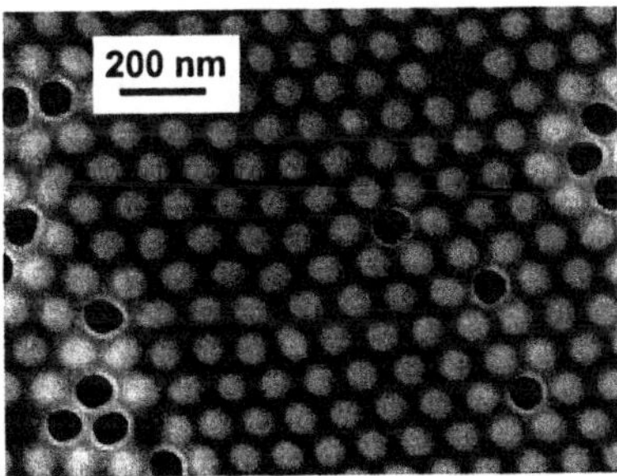

Figure 4: SEM of Bi_2Te_3 nanowires grown in a porous anodic alumina template. The black spots are unfilled pores and the white spots are pores filled with Bi_2Te_3 (image courtesy A. Prieto, Chemistry, UCB).

4 Heat transport in bulk materials by phonons

Before discussing heat transport in these micro and nanostructures, a brief and largely qualitative review of heat transport theory for bulk materials is presented. More rigorous treatments may be found elsewhere [28–30].

In a solid, heat is transported by atomic lattice vibrations called phonons and by charge carriers such as electrons and holes. This "electronic" contribution to thermal conductivity is the reason that materials that are good conductors of electricity, such as metals, also have a high thermal conductivity. While the electronic component of the thermal conductivity may be significant for moderate to heavily doped semiconductors, the lattice contribution to heat transport typically dominates. Therefore, the focus of this work is on heat transport by the lattice.

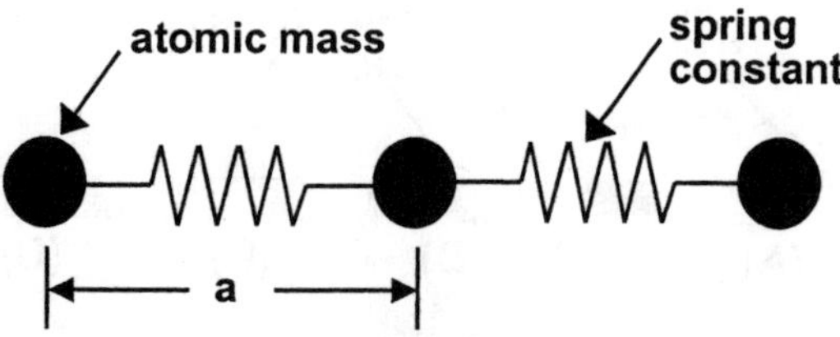

Figure 5: One-dimensional representation of atoms and chemical bonds as a spring–mass system. The chemical bond acts as a spring, and the atom as a mass. The distance between adjacent atoms is called the lattice constant or lattice parameter and is represented by the letter *a*.

The atoms in a solid crystal are held together in the form of a lattice by the chemical bonds between the atoms. These bonds are not rigid, but act like springs which connect the atoms, creating a spring–mass system as shown, in one dimension, in Figure 5. When an atom or plane of atoms is displaced, this displacement can travel as a wave through the crystal, transporting energy as it propagates. These waves may either be longitudinal where the displacement of the atom is in the same direction as the propagation of the wave, or they may be transverse where the atomic displacement is perpendicular to the direction of propagation as shown in Figure 6.

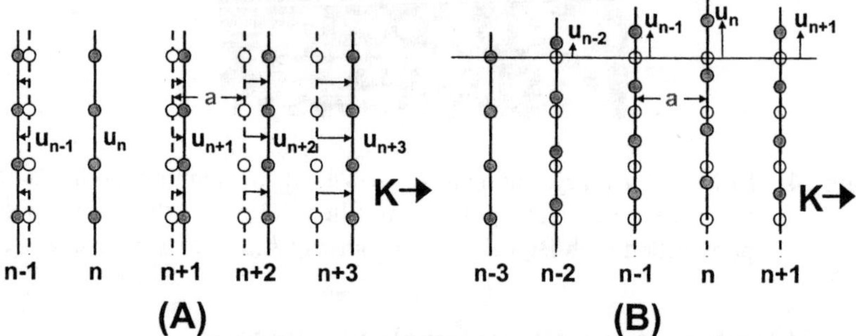

Figure 6: Schematic diagram of atomic lattice waves. (A) Longitudinal wave, where *n* is the lattice plane, *u* is the displacement of each plane, *K* is the wave vector, and *a* is the lattice spacing. (B) Transverse wave. The wave in (A) is a longitudinal wave since the displacement is in the direction of propagation, whereas the wave in (B) is a transverse wave because the displacement is perpendicular to the wave propagation direction.

These lattice vibrations are quantized and are known as phonons. Phonon transport in a crystal may be treated in a manner similar to that of gas molecules in a container, and therefore the lattice thermal conductivity may be approximately determined by applying kinetic theory to phonons. The formulation for the thermal conductivity (k) may then be expressed as

$$k = \frac{1}{3} c v_g l \tag{1}$$

where c is the specific heat, v_g is the phonon group velocity and l is the phonon mean free path. These variables are discussed in more detail later in this section.

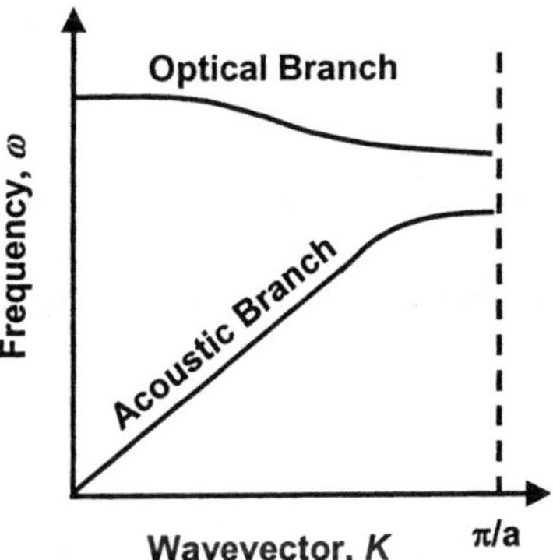

Figure 7: Generic dispersion relation for phonons. The slope of each branch gives the group velocity, or speed of sound, of the phonons.

By solving the equations of motion for these waves, one can determine the angular frequency (ω) of the waves as a function of their wavelength (λ) or their wave number (also called wave vector) (K), where $K = 2\pi/\lambda$. The relationship between ω and K may then be plotted in the form of a dispersion relation as shown in Figure 7. One important parameter that can be found from the dispersion relation is the group velocity (v_g), or speed of sound, of the phonons. The group velocity is defined as

$$v_g = \frac{d\omega}{dK} \tag{2}$$

which is simply the slope of the phonon branch on the dispersion relation.

There are two distinct phonon branches in the dispersion relation. The lower branch is known as the acoustic branch since its group velocity is linear over a large range of wave numbers, which is similar to sound waves. The upper branch is called the optical branch since this branch interacts with electromagnetic radiation and is responsible for the infrared properties of the crystal [28, 29]. Since the group velocity in the acoustic branch is considerably larger than in the optical branch, the acoustic phonons contribute to the thermal conductivity to a much greater extent [30].

4.1 Phonon scattering

When discussing heat transport in solids by phonons, an important factor is the degree to which these lattice waves are disrupted, or scattered. Phonons can be scattered by defects or dislocations in the crystal, crystal boundaries, impurities such as dopants or alloying species, or by interactions with other phonons. These scattering mechanisms can be grouped into two categories: elastic scattering between a phonon and an imperfection where the frequency of the incident phonon does not change, or inelastic scattering between interacting phonons where the frequency does change.

An important metric in the discussion of phonon scattering mechanisms is the phonon mean free path (l), which is the average distance a phonon travels between collisions or scattering events. The mean free path is defined as

$$l = v\tau \tag{3}$$

where v is the phonon velocity and τ is the average time between scattering events (also called the mean free time). If the thermal conductivity of a material is known, then the mean free path may be estimated using kinetic theory as

$$l = \frac{3k}{cv} \tag{4}$$

where c is the heat capacity per unit volume and k is the thermal conductivity. However, this method often underestimates the mean free path for several of the following reasons [31]. From Figure 7 it is clear that phonons are dispersive and that the group velocity is not constant for all frequencies. So the actual average group velocity is less than the group velocity at low frequency, which is typically used in eqn. (4). Furthermore, optical phonons contribute to the specific heat, but do not contribute appreciably to the thermal conductivity [30]. Finally, phonon scattering is highly frequency dependent and high frequency phonons are typically scattered more than low frequency phonons. Although this is not the most accurate way to determine the mean free path, it is a simple expression that is useful to quickly obtain an approximate value. Each scattering mechanism has its own mean free path, which depends on the material and may also depend on the temperature. In the following subsections several of these different phonon scattering mechanisms are described.

4.1.1 Normal and umklapp scattering

Inelastic scattering processes arise due to the fact that the forces between atoms are not purely harmonic. For a harmonic oscillator, the spring constant is independent of the spring deformation. However, this is not the case for the bonds between atoms, as the spring constant can change if the bond is strained. So as one lattice wave propagates across a plane of atoms, the atoms will be displaced slightly from their equilibrium positions and the spring constant between those atoms will be modified. If another lattice wave is incident on these same atoms it will come across this different spring constant and it may scatter.

There are two types of phonon–phonon scattering processes called "normal" and "umklapp" processes, which are sometimes referred to as N or U processes. The normal scattering process is the simpler of the two to understand and it is shown schematically in Figures 8A and 8B. Two phonons with wave vectors K_1 and K_2 can combine to produce a third phonon with wave vector K_3 (Figure 8A), or one phonon can scatter into two phonons (Figure 8B). For these processes, phonon momentum, $\hbar K$, is conserved, thus normal processes do not produce any direct resistance to heat flow.

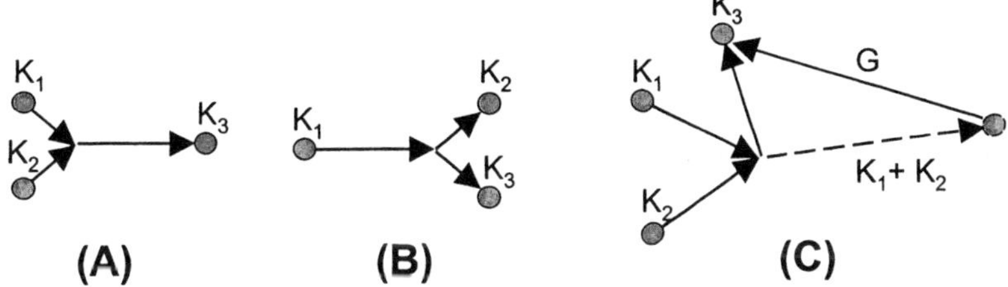

Figure 8: (A) Vectorial representation of a normal phonon scattering process where two phonons combine to create a third. (B) Normal process where one phonon scatters into two phonons. (C) Umklapp process where two phonons combine to create a third. Due to the discrete nature of the atomic lattice there is a minimum phonon wavelength, which corresponds to a maximum allowable wave vector. If two phonons combine to create a third phonon which has a wave vector greater than this maximum, the direction of the phonon will be reversed or "flipped over" with a reciprocal lattice vector G, such that its wave vector is allowed.

The important phonon scattering process for thermal conductivity is the umklapp process shown in Figure 8C. Similar to the N process, two phonons with wave vectors K_1 and K_2 combine and attempt to produce a third phonon with wave vector $K_3 = K_1 + K_2$. However, due to the discrete nature of the atomic lattice there is a minimum phonon wavelength which can propagate through the crystal of $2a$, where a is the spacing between atoms. Since $K = 2\pi/\lambda$, this places an upper limit of π/a on the wave vector. If the sum of K_1 and K_2 is greater than this maximum, the process can only be completed with the addition of what is known as the reciprocal lattice vector, G [28–30]. This reciprocal lattice vector "flips" the phonon over to a lower frequency and wave vector, and reverses the direction of the phonon. Phonon momentum is not conserved during the umklapp process, therefore producing a resistance to heat flow. Umklapp processes are generally the dominant scattering mechanism in semiconductors at room temperature.

4.1.2 Scattering by defects, dislocations, impurities, and boundaries

In addition to phonon–phonon interactions, phonons may scatter as a result of imperfections in the crystal lattice. These imperfections may be in the form of defects in the location of atoms in the lattice, or by impurity atoms placed within the lattice.

Defects or dislocations in the atomic lattice have the effect of acting as a different mass and/or spring constant to the incident phonons. This is perhaps best explained in terms of the acoustic impedance (Z) of the lattice waves, which is analogous to the index of refraction for electromagnetic waves. The acoustic impedance is defined as

$$Z = \rho v \tag{5}$$

where ρ is the mass density and v is the speed of sound in the material. The speed of sound is related to the elastic stiffness (C) of the chemical bonds by

$$v = \sqrt{\frac{C}{\rho}}\,. \tag{6}$$

Therefore, a change in either the mass (density) or the stiffness of the bond will alter the acoustic impedance of the material. When a phonon encounters a change in acoustic impedance, it may scatter just as a photon scatters when it encounters a change in the optical index of refraction.

Impurity atoms will have a different mass and spring constant from the host atoms, thus disrupting phonon transport in a similar manner as described above. These impurity atoms can be in the form of isotopes [32], dopant atoms, or they can be species introduced to form an alloy. Alloying is a particularly effective way to reduce thermal conductivity. For example, at room temperature silicon [33] has a thermal conductivity of approximately 140 W/m-K and germanium [34] has a thermal conductivity of roughly 60 W/m-K. However, Si_xGe_{1-x} alloys [35] have a room temperature thermal conductivity of approximately 5–10 W/m-K for Ge concentrations of 10–90%.

Dopant atoms can affect heat transport in a similar fashion, just to a smaller degree. As an example, one cubic centimeter of silicon contains approximately 5×10^{22} atoms. For thermoelectric applications, the optimum doping level in many semiconductors [36] is on the order of 1×10^{19} cm^{-3} (which is considered heavily doped), meaning that approximately one out of every 5,000 atoms is a dopant atom. However, doping levels on the order of 10^{19}–10^{20} cm^{-3} may still be significant to heat transport in certain materials. Experimental data [33] show that the thermal conductivity of Si drops from 140 W/m-K to 120, 47, and 40 W/m-K for doping levels of 2×10^{19} (phosphorus), 3×10^{19} (boron) and 5×10^{19} (boron) cm^{-3}, respectively. So, while doping a semiconductor leads to additional charge carriers that increase the electronic contribution to the thermal conductivity the dopant atoms also act as scattering sites, which serve to reduce the lattice component of the thermal conductivity. Depending on the doping level and the material, the overall thermal conductivity may increase *or* decrease with an increase in dopant atoms.

4.1.3 Temperature dependence of phonon scattering

While different phonon scattering mechanisms dominate for different materials, different mechanisms may also dominate in the same material depending on the temperature. This is a result of the fact that the phonon wavelength that is dominant in energy depends on temperature according to Wien's displacement law for phonons. The relationship between phonon wavelength and temperature can be approximated as [30, 37, 38]

$$\lambda_{dom} \approx \frac{hv}{3k_B T} \tag{7}$$

where v is the phonon velocity, k_B is Boltzmann's constant, and h is Planck's constant. Therefore, as temperature increases, the wavelength of the dominant phonons decreases. This effect on thermal conductivity is best explained through the use of an example. Figure 9 is a plot of the thermal conductivity of silicon as a function of temperature. At low temperatures the phonon wavelength is long. If the phonon wavelength is larger than the size of the defects, there will be little scattering from the defects. Additionally, long wavelengths correspond to small wave vectors. Since large wave vectors are required for umklapp scattering, umklapp scattering is unimportant at low temperatures and it is said to be "frozen out." Therefore, at low temperatures boundary scattering is dominant, and the measured thermal conductivity may depend on the size of the sample. As the temperature increases, the phonon wavelength decreases and becomes comparable to the size of the defects and defect scattering becomes the dominant scattering mechanism. For further increases in temperature the phonon wavelength continues to decrease and eventually the wave vectors become large enough for umklapp processes to dominate. This type of temperature dependence is typical for a single crystal semiconductor. For alloys, or materials with a large number of defects, defect scattering may dominate over much of the temperature range and umklapp processes may be relatively unimportant.

By measuring thermal conductivity over a range of temperature, the dominant phonon wavelength changes and different scattering mechanisms become significant and lead to different trends in thermal conductivity. Therefore, even if a material is to be used in a device at room temperature, it is often useful to measure the thermal conductivity of the material as a function of temperature to determine which scattering mechanism is dominant.

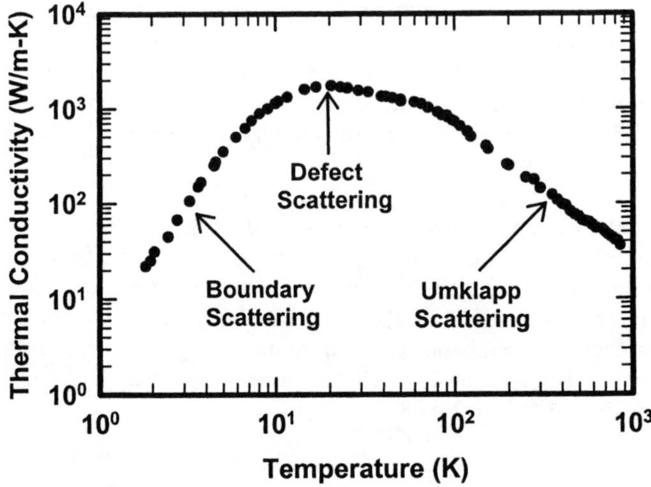

Figure 9: Thermal conductivity of bulk silicon as a function of temperature. At low temperatures when the dominant phonon wavelength is long, umklapp scattering is frozen out and boundary scattering dominates. As the temperature increases, the dominant phonon wavelength decreases and becomes comparable to the size of the defects, and defect scattering dominates. At high temperatures, the phonon wavelengths become shorter (larger wave vectors) and umklapp scattering becomes dominant (data from Touloukian [33], curves 4, 25, and 28).

5 Heat transport in low-dimensional structures

Heat transport in low-dimensional structures may be different from transport in bulk materials for several reasons. The main reason for this is the fact that as the size of a structure decreases, its surface area-to-volume ratio increases, thereby increasing the importance of surface effects such as boundary and interface scattering in relation to volumetric effects such as defect and umklapp scattering.

In addition to the scattering mechanisms covered in the previous section, other mechanisms that are either not present or not significant in bulk materials may play a role in heat conduction in nanostructured materials. For example, the artificial periodicity of a superlattice may lead to interference of the lattice waves. Phonon tunneling also becomes possible as the size of the structure becomes comparable to the phonon wavelength. These, and other important effects for nanostructures are discussed in the following subsections.

5.1 Acoustic impedance mismatch at a single interface

Acoustic impedance was discussed previously in terms of defects or impurity atoms where the local acoustic impedance was different from the host material and phonons were scattered. A similar effect may take place at the interface between different materials. Instead of a wave encountering just one atom with a different mass and spring constant, it may encounter an entire plane of atoms with a different mass and spring constant. At this plane, the incident wave can either be reflected or transmitted.

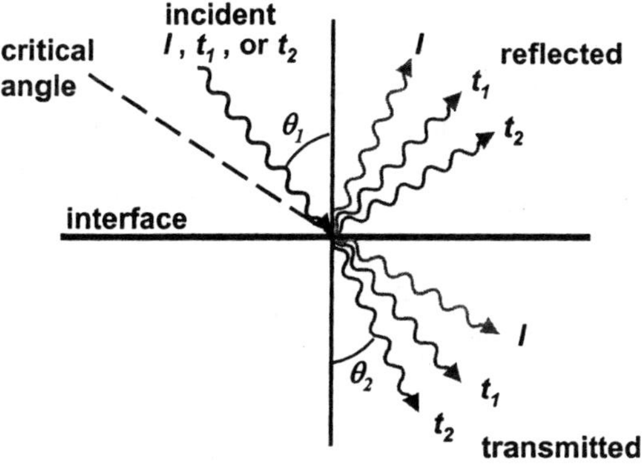

Figure 10: Schematic diagram of a phonon incident on an interface between two different materials. The incident phonon may be either a longitudinal mode (l) or one of two transverse modes (t_1 or t_2), and it may: (i) reflect specularly, (ii) reflect and mode convert, (iii) refract, or (iv) refract and mode convert.

Determining the phonon transmission probabilities through an interface is not a straightforward problem. However, the effect of acoustic impedance mismatch between materials can be isolated and addressed through the use of the acoustic mismatch model (AMM) [39], which incorporates some simplifying assumptions. First, if the phonon wavelength is much larger than the interatomic spacing, the phonons can be treated as plane waves in a continuous medium. Second, the interface is treated as a perfectly flat plane. Under these assumptions, when a phonon is incident on an interface there may be only four general outcomes as shown in Figure 10. The phonon may: (i) reflect specularly, (ii) reflect and mode convert (i.e., convert from a longitudinal mode to one of the two transverse modes or vice versa), (iii) refract, or (iv) refract and mode convert. The angle of a refracted phonon may be calculated using an acoustic analog of the familiar Snell law [40] for electromagnetic waves.

For example, the angle of the transmitted phonon in Figure 10 can be calculated from

$$\sin\theta_2 = \frac{v_2}{v_1}\sin\theta_1 \tag{8}$$

where the subscripts 1 and 2 refer to the incident and transmitted phonons, respectively. This relationship holds regardless of whether or not mode conversion takes place. An important consequence of the Snell law is that when $\sin\theta_1$ equals v_1/v_2, a critical angle, beyond which the probability of transmission is zero, is reached. Therefore, in three-dimensional space, this angle creates a critical cone. All phonons incident from outside of this cone will be reflected, leading to a thermal boundary resistance and a temperature jump at the interface.

However, even phonons within the critical cone may be reflected due to differences in the acoustic impedance of the two materials. For example, the probability of a phonon being transmitted from material A to material B ($T_{A\to B}$) at normal incidence is [39]

$$T_{A\to B} = \frac{4Z_A Z_B}{(Z_A + Z_B)^2} \tag{9}$$

where Z again is the product of the mass density and sound velocity. Similar relations can be derived for other polarizations and angles of incidence [41].

The acoustic mismatch model generally predicts thermal boundary resistance values reasonably well at very low temperatures [39, 41], but is less successful at higher temperatures. One possible reason for this discrepancy is that at low temperatures the phonon wavelength can be much larger than the average roughness and the interface may "appear" to be perfectly flat to the phonons, which is one of the assumptions in the model. However, at high temperatures the dominant phonon wavelength is short and may be comparable to the interfacial roughness of the sample, which leads to diffuse scattering. This effect will be discussed in more detail later.

The acoustic impedance for a material may change depending on the phonon polarization and propagation direction, as the phonon velocity may depend on both factors. Therefore, defining a single acoustic impedance value for one material is a difficult task.

5.2 Phonon spectra mismatch

In addition to the mismatch in acoustic impedance, a mismatch in the phonon spectra of the two materials may also play a role in phonon transport across an interface [42, 43]. For example, the number of states in material A for phonons of frequency ω may be much greater than the number of states at that frequency that are available in material B. If that is the case, then the transmission percentage of those phonons from A to B will be small. To further complicate matters, instead of being confined to the incident medium, a high frequency phonon may split into two phonons of lower frequency and still be transmitted through the interface.

While it can be difficult to quantify the phonon spectra mismatch between two materials, a simple qualitative method to estimate the extent of the mismatch in phonon spectra is to compare the Debye temperature between materials. The Debye temperature can be regarded as measure of the elastic stiffness of a crystal [28]. In general, if there is a large difference in the Debye temperature, there will be a large mismatch in the phonon spectra.

5.3 Phonon tunneling

When the phonon wavelength becomes comparable to the size of the layer, it becomes possible for the phonon to tunnel through the barrier. This makes it possible for a phonon to be transmitted when it is incident from outside of the critical cone.

However, assuming that the speed of sound for phonons in a semiconductor is on the order of several thousand meters per second, the dominant phonon wavelength at room temperature may be calculated from eqn. (7) to be less than one nanometer. Since the layer thickness in a superlattice is typically greater than a few nanometers, phonon tunneling can largely be ignored. Using a more rigorous analysis, Chen [44] estimated that at room temperature phonon tunneling occurs only when the constituent layers are less than three monolayers thick and is not important for most practical superlattices.

If phonon tunneling were important for heat transport in superlattices, the thermal conductivity should increase for decreasing layer thickness for ultra-short period samples. At the present time, few superlattices have exhibited this trend experimentally [45, 46]. In one of those studies, Venkatasubramanian [46] found that the thermal conductivity of Bi_2Te_3/Sb_2Te_3 superlattices increased as the period decreased from 60 Å to 20 Å. To identify if this behavior could be attributed to tunneling, he measured three different superlattices with a 60 Å period that had thickness ratios of Bi_2Te_3 to Sb_2Te_3 of 30 Å/30 Å, 20 Å/40 Å, and 10 Å/50 Å. If phonon tunneling were dominant for the 10 Å layers in the sample with the shortest (20 Å) period to explain its higher thermal conductivity, one would expect that the 10/50 sample would have a higher conductivity than the 30/30 sample. However, this was not the case as the 10/50, 20/40, and 30/30 samples all had essentially the same conductivity. Thus phonon tunneling was ruled out in this case, and the reason attributed to this rise in thermal conductivity will be discussed later.

5.4 Phonon wave interference and mini-bandgap formation

Due to the periodic nature of superlattices, it is possible for "phonon bandgaps" to be created just as electronic band gaps arise due to the periodic potential of ion cores in a lattice [28, 29]. In both cases the bandgaps are created by interference of the waves known as Bragg reflection. This is shown schematically for lattice waves in Figure 11.

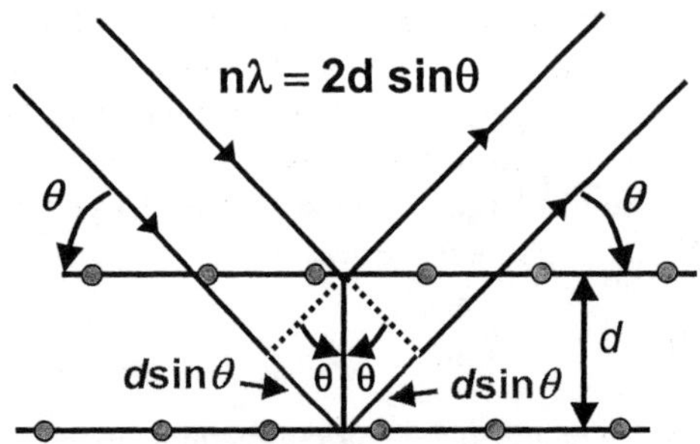

Figure 11: Schematic diagram of constructive phonon interference by Bragg reflection. When two phonons reflect from adjacent interfaces they may interfere constructively if the path difference between the two waves is an integral number (*n*) of wavelengths (λ).

When two phonons reflect from adjacent interfaces they will interfere constructively if the path difference between the two waves is an integral number (*n*) of wavelengths (λ) such that they satisfy the Bragg condition of

$$n\lambda = 2d \sin\theta \tag{10}$$

where d is the superlattice period and θ is the incident angle of the phonon. In order for reflection to take place, the phonons must be coherent and they must reflect specularly from the interfaces.

This selective phonon filtering effect was first shown experimentally by Narayanamurti *et al.* [47] in the late 1970s. However, their experiments used the somewhat optimized conditions of externally generated monochromatic phonons at extremely low temperature (1 K). At room temperature, a wide range of phonon frequencies are thermally generated and various scattering mechanisms limit the coherence length of the phonons, thus restricting wave interference. Furthermore, it is not clear if the selective filtering of a rather narrow band of phonons would have a significant effect on thermal conductivity, since a broad range of phonon frequencies transport heat.

Another important consequence of phonon wave interference is the reduction of the average phonon group velocity for frequencies near the band gaps [48–50]. This effect is perhaps best explained qualitatively through Figure 12. Figure 12A compares two superlattices, with superlattice (i) having a period of 50 (arbitrary units) and superlattice (ii) a period of 100. For the sake of keeping the math simple, let us assume that the minimum wavelength due to the discrete nature of the lattice is also 50. For phonons normal to the layers in superlattice (i) there are only two wavelengths (50 and 100) which fulfill the Bragg condition in Eqn. (10) and correspond to bandgaps in the dispersion relation shown in Figure 12B. However, for superlattice (ii), which has a period twice as thick as superlattice (i), there are four wavelengths (50, 66, 100, and 200) that satisfy the Bragg condition and lead to four bandgaps. Since there are standing waves associated with each gap, the group velocity must go to zero for wave vectors approaching each gap.

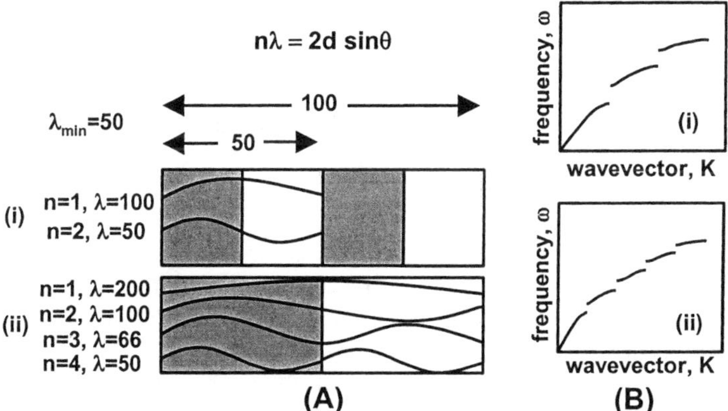

Figure 12: Example of phonon group velocity reduction in superlattices. (A) The period of superlattice (i) is 50 (arbitrary units), while the period of superlattice (ii) is 100. The minimum phonon wavelength is also 50 in both superlattices. For phonons normal to the layers in superlattice (i) there are only two wavelengths which fulfill the Bragg condition and correspond to bandgaps in the dispersion relation shown in Figure 12B. However, in superlattice (ii), which has a period twice as thick as superlattice (i), there are four wavelengths that satisfy the Bragg condition and lead to four bandgaps. Since there are standing waves associated with each gap, the group velocity must go to zero for wave vectors approaching each gap. Also note that the superlattice with thicker layers has more gaps, a lower average velocity and, therefore, a lower thermal conductivity.

An interesting aspect of mini-band formation that may be inferred from Figure 12B is that the thermal conductivity decreases for increasing period thickness, if only wave effects are considered [50]. In general, an increase in the thickness of each superlattice period leads to an increase in the number of band gaps. An increase in the number of band gaps leads to a lower average phonon velocity, which decreases the thermal conductivity of the superlattice. This trend is exactly the opposite of what one would expect if the thermal resistance of the superlattice were dominated by scattering at the interfaces. If the thermal boundary resistance at the interfaces is dominant, thinner superlattice periods create more interfaces per unit length, and therefore a greater thermal resistance and lower conductivity.

5.5 Interface scattering

Although MBE and MOCVD are capable of growing nearly atomically perfect superlattices, there are instances where defects may arise at interfaces and cause significant phonon scattering. Specifically, when the phonon wavelength becomes comparable to, or smaller than, the average interfacial roughness, the phonons will scatter as particles rather than interfere as waves.

Defects and dislocations are of major concern for superlattices where there is a large mismatch between the lattice constants of the two layers. For example, the lattice constants of silicon and germanium differ by approximately 4%. For the sake of argument, assume that a superlattice of equal layer thickness of Si and Ge is grown on top of a layer of $Si_{0.5}Ge_{0.5}$ that has a lattice constant exactly between that of the two elements. In this case, the Si layer would be under a tensile strain of 2% and the Ge under 2% compressive strain. The total strain energy within the layer depends on several factors including the material properties, the thickness of the layer, the growth temperature, and the lattice parameter of the underlying layer. There is a critical layer thickness (h_c), beyond which the strain energy becomes greater than the maximum the layer can withstand, and defects and dislocations will be created in order for the layer to partially relax [51–54]. This is particularly important for the growth of superlattices with many periods, where the strain builds up and can create undulations or dislocations in the layers.

The number of defects or dislocations due to lattice-mismatched growth is difficult to quantify. One method is known as “defect reveal etching.” In this technique the sample is etched with a specific solution in which the etch rate is known to vary depending on strains in the material. Large strains close to a defect enhance the etch rate and an etch pit develops. These etch pits then may be tallied using a microscope that is calibrated to detect the phase difference between light reflected from a plane and from the pit. This is often done using a Schimmel [55] etch solution and Nomarski microscopy.

There are also several techniques that may give a qualitative indication of the amount of defects or dislocations in a sample. High-resolution x-ray diffraction (XRD) may be used to accurately determine the thickness of each layer, which can also lend some insight into the quality of the interfaces [6]. Atomically abrupt interfaces will give sharp peaks in the x-ray data, while rougher interfaces will have broader peaks. By comparing the full width at half maximum (FWHM) of the peaks of similarly composed superlattices one can get a relative idea of the amount of defects at the interfaces. Transmission electron microscopy (TEM) is another method by which the roughness of superlattice interfaces may be qualitatively examined. Figure 13 shows TEM images for a superlattice with high quality interfaces (A) and a superlattice with many defects (B). The two layers in (A) are $Si_{0.76}Ge_{0.24}$ and $Si_{0.84}Ge_{0.16}$ which have a mismatch in lattice parameters of only ~0.3%. Superlattice (B) is made of layers of $Si_{0.9}Ge_{0.1}$ and

$Si_{0.1}Ge_{0.9}$, which have a mismatch of ~3.1%. The effect of the strain in superlattice (B) is evident by the wavy nature of the interfaces and the defects. TEM is useful in this case to get an idea of the quality of the periodic structure, but it is not possible to quantify the defect density with this technique.

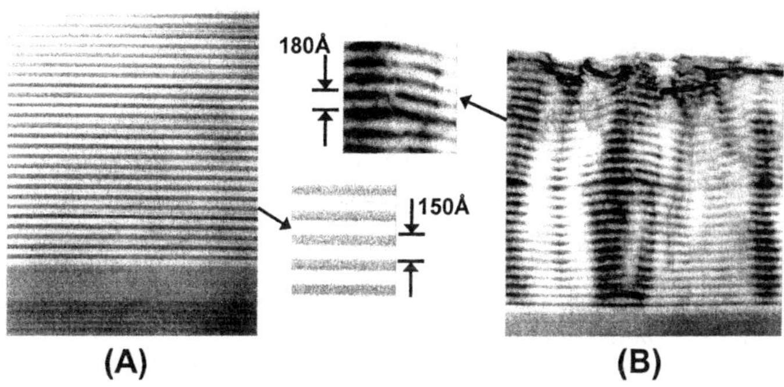

Figure 13: Transmission electron microscopy (TEM) images of two superlattice cross-sections. (A) $Si_{0.76}Ge_{0.24}/Si_{0.84}Ge_{0.16}$ superlattice with high quality interfaces. (B) $Si_{0.9}Ge_{01}/Si_{0.1}Ge_{0.9}$ superlattice with defects. The mismatch in lattice parameters of the two layers in (A) is only ~0.3%, while in (B) it is ~3.1%, which leads to higher strain and, ultimately, many defects. (images courtesy of E. Croke, HRL Labs and C. Ahn, CalTech).

Just as the acoustic mismatch model may be used to treat the case of purely specular reflection and refraction, a diffuse mismatch model (DMM) [39] exists for the limiting case where all scattering at an interface is diffuse. This model puts an upper limit on the effect that diffuse scattering might have at an interface. At a boundary between materials with identical acoustic properties, the transmission according to the acoustic mismatch model would be 100%. However, if all of the phonons scatter diffusely at the interface, the transmission would be, by definition, 50%. Therefore, in this case, diffuse scattering would increase the thermal boundary resistance. However, if the acoustic impedances of the two materials are considerably different, diffuse scattering can decrease the interfacial resistance. A quantitative analysis [39] shows that for typical solid-solid interfaces, the thermal boundary resistance as calculated by the acoustic mismatch model and diffuse mismatch model differ by less than 30%.

6 Survey of previous work

Over the past 15 years, there have been a considerable number of studies that investigated the thermal conductivity of a variety of superlattices both experimentally and theoretically. More recently, the thermal properties of nanowires have also received some theoretical attention, however, little experimental work has been reported in the literature to date. In this section, previous work on heat transport in both superlattices and nanowires is reviewed with an emphasis on experimental data.

6.1 Superlattices

The first measurements of the thermal properties of superlattices were done in 1987 by Yao [56] for samples of AlAs/GaAs at room temperature with periods ranging from 100 Å to 1000 Å. He measured the thermal diffusivity and conductivity in the direction parallel to the layers using an ac calorimetric method and found that they were higher than that of an $Al_{0.5}Ga_{0.5}As$ alloy. The thermal conductivity values increased from a low of ~15 W/m-K at 100 Å to ~32 W/m-K at 400 Å, then dropped to ~25 W/m-K at 1000 Å. The thermal conductivity of the $Al_{0.5}Ga_{0.5}As$ alloy was only 9 W/m-K and the fact that the superlattice conductivities were all greater than that of the alloy was attributed to the lack of alloy scattering in the superlattice.

The thermal conductivity of an AlAs/GaAs superlattice may be estimated by simply calculating a weighted average of the two materials as conductors in parallel. Using bulk values for the thermal conductivity of GaAs (~46 W/m-K, [57]) and AlAs (~84 W/m-K, [58]), the superlattice thermal conductivity was estimated to be ~65 W/m-K, which is greater than that of any of the superlattices. The fact that the measured thermal conductivity of the superlattices was always less than the estimated value indicated that there was some additional scattering mechanism involved that presumably was related to the interfaces. While no structural characterization of the interfaces was presented in this work, the lattice parameters of GaAs (5.654 Å, [59]) and AlAs (5.639 Å, [59]) differ by only ~0.25%, so it is expected that superlattices made of these materials would have high quality interfaces with few defects.

Chen and coworkers [60] later measured the thermal diffusivity of a GaAs/AlGaAs multilayer structure in directions both normal and parallel to the layers. The structure was identical to that used for a vertical-cavity surface emitting laser (VCSEL) and contained 50 periods (each 1400 Å thick) of GaAs and $Al_{0.67}Ga_{0.33}As$ along with three InGaAs quantum wells. The $Al_{0.67}Ga_{0.33}As$ layers were grown by a digital alloy method and were actually made up of short period AlAs/GaAs superlattices. The measurement was done using a modulated laser as a heat source and a microfabricated metal line as a temperature sensor. The authors found that the measured diffusivity in both directions was ~5–7 times smaller than a corresponding bulk value, and that there was a stronger reduction in the normal direction. The fact that the thermal diffusivity showed anisotropy further indicated that the phonons were being scattered at the interface. However, as with Yao's work, it was not possible to clarify which interface scattering mechanisms might be responsible for the decrease in thermal diffusivity.

Both of the previous studies by Chen *et al.* [60] and Yao [56] were done at room temperature. However, as was discussed earlier, measuring the thermal properties as a function of temperature can often shed some light on the phonon scattering mechanisms in the material. Yu *et al.* [61, 62] continued the work on the GaAs/AlAs system by measuring the thermal diffusivity and conductivity of a GaAs/AlAs superlattice with a period of 1400 Å in the direction parallel to the film plane over a temperature range of 190 to 450 K using an ac calorimetric method. Their results were consistent with Yao's work in that they found the room temperature thermal conductivity of 41 W/m-K was lower than the value calculated for a corresponding structure using bulk properties.

They postulated that the decrease in thermal conductivity could be the result of (a) scattering from the dopant atoms, (b) quantum size effects, or (c) interface scattering of the phonons. The doping effect was not considered to be important because the bulk values used for the calculation were from samples that contained a similar doping level and other work [63] had shown that the thermal conductivity of GaAs did not depend strongly on doping. Quantum size effects were also ruled out, as the phonon mean free paths in GaAs and AlAs were calculated from kinetic theory to never be greater than 450 Å over the measured temperature range, while the thickness of each layer in the structure was 700 Å. Therefore, they concluded that the reduction in thermal conductivity was due to interface scattering.

The interface scattering effects were also evident from the temperature dependence of the thermal conductivity. The superlattice exhibited slightly weaker temperature dependence than bulk GaAs, which the authors attributed to the increased importance of boundary scattering, since the thermal resistance of boundary scattering should remain relatively constant over the temperature range, while the thermal resistance due to internal (umklapp) scattering should decrease. This work provided further evidence that interface scattering was an important mechanism for the reduction of thermal conductivity in these samples, although the specific type of interface scattering was still not clear.

Capinski and Maris [64, 65] also measured the cross-plane thermal conductivity of two GaAs/AlAs superlattices with periods of ~35 Å and ~150 Å over a temperature range of 100 to 330 K using a picosecond optical technique. The room temperature thermal conductivities for the 35 and 150 Å samples were 3.1 and 5.8 W/m-K, respectively, which is approximately an order of magnitude lower than the in-plane value of 41 W/m-K that Yu *et al.* [61] measured for a GaAs/AlAs superlattice with a period of 1400 Å. The large difference between the two studies could be anticipated since the cross-plane thermal conductivity is expected to be reduced much more strongly due to interface scattering, particularly for these short period samples.

Capinski *et al.* [66] later continued their work on the cross-plane thermal conductivity of GaAs/AlAs superlattices by measuring eight different samples with periods ranging from ~10 Å (one monolayer of each material) to ~450 Å. They found a general decrease in thermal conductivity with a decrease in period thickness, where the shortest period samples had a room temperature thermal conductivity of ~5 W/m-K, which is below the value of 12 W/m-K reported [67] for an $Al_{0.5}Ga_{0.5}As$ alloy. Additionally, the thermal conductivity of all of the superlattices decreased with increasing temperature. This is particularly interesting as it indicates that umklapp processes still played a significant role even for the samples where the period thickness was only a few monolayers.

Superlattices made from silicon, germanium, or SiGe alloys are very different from the GaAs/AlAs samples for several reasons. The first difference is there is a much larger lattice mismatch between Si and Ge (~4%) than between GaAs and AlAs (~0.25%). Therefore, while GaAs/AlAs superlattices of nearly any period can be grown with excellent crystal quality, the strain between the layers of Si and Ge in a Si/Ge SL will lead to defects and dislocations when the thickness of an individual layer exceeds a critical value [51]. Along with the large mismatch in lattice constants, there is a considerably larger mismatch between the acoustic impedances of Si and Ge than of GaAs and AlAs.

Lee *et al.* [68] measured the cross-plane thermal conductivity of a series of Si/Ge superlattices with periods ranging from 30 Å to 275 Å over a temperature range of 80 K to 400 K using a technique known as the 3ω method [69–71], which has become a standard technique for thin-film thermal conductivity measurements. The thermal conductivities of all of the superlattices were below that of SiGe alloys, but above the conductivities of amorphous Si and amorphous Ge. For periods less than 70 Å thick, the thermal conductivity decreased with decreasing period thickness. However, superlattices with periods of 140 Å or greater had thermal conductivities that were smaller than the short period samples. The authors reasoned that the samples with longer periods likely contained a high density of defects due to the build up of strain in the lattice-mismatched layers. The thermal conductivity of all of the samples increased gradually with temperature up to ~200 K above which it remained approximately constant. This temperature dependence is further evidence that defect scattering was the dominant scattering mechanism.

Borca-Tasciuc *et al.* [72] measured the cross-plane thermal conductivity of a series of Si/Ge superlattices over a temperature range of 80 K to 300 K using the 3ω method. Unlike the superlattices studied by Lee *et al.* [68] in which the ratio of Si to Ge was different for each sample, all of these SLs were grown by MBE with Si and Ge layers of equal thickness such that the superlattice was symmetrically strained. They measured three undoped samples with periods of 44, 90, and 140 Å, and three samples with the same period (40 Å), but different doping levels. For the samples with different periods, they found that the thermal conductivity values were slightly higher than what Lee *et al.* [68] measured, but that they had nearly identical temperature dependences. The lower thermal conductivity values reported by Lee and co-workers for superlattices of similar period thickness may be the result of additional scattering by dopant atoms, as their samples had carrier concentrations between 3×10^{18} cm^{-3} and 2×10^{19} cm^{-3}. The thermal conductivity of Borca-Tasciuc's samples also decreased for increasing period thickness. However, the authors believed that this trend was not a result of dislocations as suggested by Lee. They cited work by Chen and Neagu [73] which indicated that dislocation densities on the order of 10^{11} to 10^{12} cm^{-2} were required to account for the strong reduction in thermal conductivity found in Si/Ge superlattices, while they only measured dislocation densities of ~1.5×10^4 cm^{-2} in their samples. However, the models used by Chen and Neagu

[73] are highly idealized and accurately modeling the effects of defects and dislocations is a difficult task. No alternate explanation for the decrease in thermal conductivity for the long period samples was given, but the authors suggested that one possible reason could be increased residual stress in the thicker period layers.

The effects of carrier concentration and carrier type were also examined in this study, as four superlattices with identical periods of 40 Å, but varying doping levels, were examined. Three samples were n-doped with antimony to 1×10^{18} cm^{-3}, 4×10^{18} cm^{-3}, and 2×10^{19} cm^{-3} respectively, and one was p-doped with boron to 1×10^{19} cm^{-3}. All four superlattices had lower thermal conductivities than the undoped samples, and the n-type SLs all exhibited thermal conductivities that were 30 to 65% less than the p-type sample. The n-type sample with the lowest dopant concentration also had the lowest thermal conductivity. Since the thermal conductivities of all four doped samples were less than the undoped samples it appears as though a certain amount of doping will decrease the thermal conductivity, as it does in bulk Si and Ge [33]. However, unlike bulk samples, the thermal conductivity of the superlattices increased for doping above 1×10^{18} cm^{-3}. In bulk samples of Si and Ge the increase in scattering due to the dopant ions outweighs the increase in the electronic contribution to the thermal conductivity and the total thermal conductivity decreases with doping. Since the lattice thermal conductivity of the superlattices is already quite low to begin with, apparently above a certain carrier concentration the increased electronic contribution of the dopant ions offsets the additional scattering and the thermal conductivity increases. However, as interesting as these results are, they should probably be considered to be somewhat preliminary as there were two important differences between the doped and undoped samples that make a fair comparison between the two sets of samples difficult. The undoped samples were grown at 510 °C, while the doped samples were grown at 380 °C and later annealed at 580°C for 10 minutes to activate the dopants. The difference in growth temperatures may influence the interface conditions as will be discussed later. The doped samples were also grown on different buffer layers and had threading dislocation densities of $\sim 1.5 \times 10^8$ cm^{-2}, compared to $\sim 1.5 \times 10^4$ cm^{-2} for the undoped samples.

In addition to the work done on Si/Ge superlattices, Chen *et al.* [74] examined the cross-plane thermal conductivity of a Si/$Si_{0.79}Ge_{0.21}$ (50 Å/10 Å) superlattice using the 3ω technique. The cross-plane thermal conductivity was reported to be 13 W/m-K which is about 2 to 3 times smaller than a weighted average calculated from bulk values for Si and Ge, but also is 3 to 4 times larger than values reported for Si/Ge superlattices of similar period thickness. The larger reduction of thermal conductivity in the Si/Ge samples is likely due to the greater mismatch in material properties such as the acoustic impedance, phonon spectra, and lattice parameter, although it is not possible to pinpoint which one is the most significant. These data are interesting in that they provide further evidence that the thermal conductivity of the individual layers within the superlattice is relatively unimportant. One might guess that replacing a layer of pure Ge, which has a room temperature thermal conductivity of ~46 W/m-K [57], with a lower conductivity $Si_{0.79}Ge_{0.21}$ alloy (~5–10 W/m-K, depending on doping) [35] would decrease the thermal conductivity of the overall structure, but clearly this is not the case.

One deficiency of much of the work on superlattices is that many studies are done on only one or two superlattices with somewhat arbitrary periods. By measuring similarly composed superlattices over a range of periods, it is sometimes possible to determine the relative importance of certain scattering mechanisms. Venkatasubramanian [46] used the 3ω technique to systematically measure the cross-plane lattice thermal conductivity of a series of MOCVD grown Bi_2Te_3/Sb_2Te_3 superlattices with periods ranging from 20 Å to 180 Å. The lattice thermal conductivity was extracted from the total thermal conductivity by subtracting the electronic contribution, which was calculated using the Wiedemann–Franz law [30]. He found that a minimum thermal conductivity of ~0.22 W/m-K existed for a period of ~50 Å. This value is a factor of two times smaller than that of a comparable solid solution alloy. For periods greater than 50 Å, the thermal conductivity increased above the alloy value and approached the value of a weighted average of Bi_2Te_3 and Sb_2Te_3. For periods less than 50 Å, the thermal conductivity also increased and began to approach the alloy value. The fact that a minimum thermal conductivity was observed is consistent with the phonon filtering by Bragg reflection that was experimentally observed by Narayanamurti *et al.* [47]. The overall trend was similar to that predicted theoretically by Simkin and Mahan [50]. They determined that there should be a transition between wave effects and particle effects at a certain period. When the period is smaller than the phonon mean free path, wave interference effects (mini-bandgap formation) should dominate and the thermal conductivity should decrease for increasing period thickness. When the period becomes larger than the mean free path, the phonons act as particles and interface scattering dominates. The mean free path for phonons in these superlattices as calculated from kinetic theory by Venkatasubramanian is less than 10 Å, which is smaller than the periods where wave effects were assumed to be important. Then again, the kinetic theory often underestimates the phonon mean free path as discussed earlier, so it is possible that the mean free path was comparable to the period thickness. Venkatasubramanian proposed a phonon "localization-like" behavior that was similar to localization of photons in superlattices [75] and consistent with the observations of Narayanamurti *et al.* [47] and the predictions of Simkin and Mahan [50]. Also of note is that phonon tunneling was eliminated as a possibility for the increase in thermal conductivity for the short period samples.

These Bi_2Te_3/Sb_2Te_3 superlattices exhibited a very low thermal conductivity and were later used in thermoelectric devices where Venkatasubramanian and co-workers measured ZT values of ~2.4 at room temperature [76]. While the exact origin of the competing effects that lead to a minimum in the thermal conductivity at 50 Å is not entirely clear, the fact that a minimum thermal conductivity was observed indicates that some type of phonon wave effects were possibly involved for the samples with the shortest periods.

One puzzling aspect regarding the thermal conductivity of Bi_2Te_3/Sb_2Te_3 superlattices is that there was a large discrepancy between the data from the previous work by Venkatasubramanian [46] and results published by Touzelbaev *et al.* [77] the following year. Touzelbaev and co-workers measured the total thermal conductivity of several Bi_2Te_3/Sb_2Te_3 superlattices with periods ranging from 40 Å to 120 Å using noncontact pulsed laser heating and thermoreflectance thermometry [78]. These samples apparently were grown in the same laboratory and under the same conditions [79] as those measured by Venkatasubramanian. However, Touzelbaev and co-workers did not find a minimum thermal conductivity. Instead their data showed a weak dependence on period. Furthermore, for three of the four periods studied, the total thermal conductivity they measured was *below* the lattice thermal conductivity measured by Venkatasubramanian. The discrepancy between the two studies was not addressed by the authors, and the reason for it is unclear. Possible explanations include errors with the different measurement techniques, or, more likely, the growth conditions were not exactly the same leading to different structures for each set of samples.

Yamasaki *et al.* [45] also measured the cross-plane thermal conductivity of a series of Bi_2Te_3/Sb_2Te_3 superlattices using a modified ac calorimetric method at room temperature. The superlattices were grown by pulsed laser deposition (PLD) with equal thickness of the Bi_2Te_3 and Sb_2Te_3 layers and with periods ranging from 30 Å to 1600 Å. They also found a trend similar to Venkatasubramanian [46] where the thermal conductivity decreased with decreasing period, reached a minimum, then generally increased for ultra-short periods. The minimum thermal conductivity was ~0.11 W/m-K at a period of ~120 Å. They attributed the rise in thermal conductivity for periods less than 120 Å to roughing of interfaces induced by island growth, rather than any wave effects.

Huxtable *et al.* [80] used the 3ω method to measure the cross-plane thermal conductivity of a series of four $Si/Si_{0.7}Ge_{0.3}$ superlattices and three $Si_{0.84}Ge_{0.16}/Si_{0.76}Ge_{0.24}$ superlattices with periods ranging from 45 to 300 Å and 100 to 200 Å, respectively. For the $Si/Si_{0.7}Ge_{0.3}$ samples, the thermal conductivity was found to decrease with decreasing period thickness, indicating that the additional interfaces associated with the thinner layers likely added a corresponding thermal boundary resistance. One possible reason for the thermal boundary resistance could be the mismatch in acoustic impedance between the Si and $Si_{0.7}Ge_{0.3}$ layers. However, the thermal conductivity of the $Si_{0.84}Ge_{0.16}/Si_{0.76}Ge_{0.24}$ superlattices was essentially independent of the period thickness. Since the mismatch in acoustic impedance (and other properties) between the two SiGe alloy layers is small, one could expect that the interfaces would have little impact on heat conduction through the structure.

While the previous experimental results have focused mainly on the differences in the structure of the superlattices, Borca-Tasciuc *et al.* [81] investigated the effects of growth temperature and post annealing on the thermal conductivity of InAs/AlSb superlattices. First, they grew three sets of samples by MBE that were identical in periodicity and composition, but were grown at temperatures of 390 °C, 425 °C, and 460 °C. They found that as the growth temperature increased, the measured thermal conductivity decreased, and there was a difference of approximately 20% between the room temperature conductivities of the samples grown at 390 °C and 460 °C. One explanation for this behavior that was given by the authors is that the samples grown at higher temperatures could have rougher interfaces, which would lead to more scattering. This explanation seems plausible, as elevated growth temperatures have been shown to produce films with greater strain and more defects [82].

In addition to examining the growth temperature dependence, Borca-Tasciuc and co-workers also examined the effect of annealing as two of the aforementioned samples that were grown at 390 °C and 425 °C were annealed at 490°C for 5 minutes. After the annealing step, both samples exhibited a reduction in thermal conductivity. Interestingly, the thermal conductivity of the sample grown at 425 °C decreased by ~10% while the conductivity of the sample grown at 390 °C dropped by ~25 % and had the lowest value of all the samples measured. No physical reason was given for this drop, but it seems possible that the annealing may have caused some diffusion of species at the interface, which could lead to a rougher surface and increased scattering.

This paper was the first to report on the relationship between growth temperature and annealing on thermal conductivity of superlattices. While there was no structural characterization included in this work that could have conclusively determined what specific effect the growth temperature and annealing had on the interfaces, the thermal conductivity results were promising and this area is deserving of further research.

Additional thermal conductivity measurements have been conducted on several other material systems. Song *et al.* [83] measured the cross-plane thermal conductivity of several $IrSb_3/CoSb_3$ superlattices that were grown by pulsed-laser deposition (PLD). They found thermal conductivity values that were lower than their bulk constituents, but comparable with similar alloys. Liu *et al.* [84] examined the thermal conductivity of a Ge quantum dot superlattice where the Si spacer layer was 200 Å and the dot base diameter and height were 750 Å and 70 Å, respectively. They reported a cross-plane thermal conductivity of 6 W/m-K at room temperature, which is half of the value for a SiGe alloy with the same Ge fraction.

The general consensus from the literature is that the thermal conductivity of superlattices is still not well understood. Theoretical models using lattice dynamics [48, 49, 85, 86] approaches that include wave effects such as phonon bandgaps and confinement cannot explain the observed reduction in thermal conductivity in many of the previous experimental studies. Furthermore, these models are only applicable for idealized interfaces, while many experimental studies indicate that the interfaces may be far from ideal.

Theories based on the Boltzmann transport equation (BTE) [73, 87, 88], in which phonons are treated as particles, are able to account for surface roughness and interface disorder. However, this approach requires estimates or assumptions for several inputs including phonon dispersions, whether phonons scatter from an interface in a specular or diffuse manner, and whether this scattering is elastic or inelastic. The BTE theory is more applicable for superlattices where the period thickness is greater than the phonon coherence length and wave effects are negligible.

From the experimental data and the varying degrees of success different models have had in predicting the thermal conductivity of superlattices, it seems apparent that there is still much to be learned about heat transport in superlattices. It is difficult to say which transport mechanisms are dominant since much of the research is on different materials under a variety of growth conditions and methods, with different periods, different property mismatches, different levels of defects, and so on. Furthermore, many of these differences, such as the defect density and mismatches in phonon spectra and acoustic impedance are difficult to quantify and may generally be unknown. As a result of these differences, the dominant scattering mechanism for one system may be entirely irrelevant for another.

One deficiency in much of the previous work is the lack of systematic measurements on well-characterized superlattices in an attempt to isolate different transport mechanisms. This is difficult because there are so many parameters involved and it is challenging to prepare samples that are similar enough such that one particular mechanism may be isolated. Further complicating matters is the fact that some degree of collaboration between groups is often necessary as these studies require expertise in epitaxial growth techniques, material characterization methods and tools, techniques for measuring thermal conductivity of thin films, along with knowledge of analytical or computational predictive methods.

6.2 Nanowires and nanotubes

Research on heat transport in nanowires and nanotubes is somewhat limited, and current results and corresponding explanations for a reduction in thermal conductivity are mainly speculative. Nonetheless, these explanations include a variation in the phonon spectrum and related properties [89–92], the influence of boundary scattering [91–99] and quantum conductance effects [100–104].

From eqn. (1), the lattice thermal conductivity is a function of heat capacity, group velocity, and phonon mean free path. If the phonon spectrum is affected by the one-dimensionality of the system, the group velocity will be changed, thereby influencing the thermal conductivity. Nonetheless, most studies account only for the effect of boundary scattering by constraining the phonon mean free path to be equal to the diameter of the nanowire and applying simple kinetic theory, or eqn. (1). Using a more complex approach by solving the elasticity equation in cylindrical coordinates for a one-dimensional system [105, 106], Khitun *et al.* [91] and Zou and Balandin [92] found the dispersion relation for particular nanowire parameters and dimensions, and the corresponding group velocities. The dispersion relation for a Si nanowire is considerably different from its bulk counterpart due to spatial confinement, and there is a significant reduction in phonon group velocity. By taking an average phonon group velocity over all phonon modes as a function of frequency, and accounting for isotope, umklapp and boundary-phonon scattering, the lattice thermal conductivity can be determined using eqn. (1). Studies [91, 92] showed that the lattice thermal conductivity of a typical Si nanowire is less than 10% of the bulk Si value.

Quantum thermal conductance effects may also influence thermal transport in particular nanowire systems. The phonon energy spectrum for all objects is actually discrete such that the allowed wave vectors can only occur at integer multiples of π/L, where L is the characteristic length of the object in the given direction. However, if L is large, then the finite spacing becomes small, and the spectrum may be treated as continuous. In a nanowire, continuous ω vs. K lines may no longer be a valid approximation for the discrete spectrum in the transverse direction. When phonons travel through the nanowire from a hot to a cold reservoir, the effects of the quantized thermal states restrict transport at specific frequencies resulting in ballistic phonon transport, and consequently, quantized thermal conductance. In other words, the phonon with the longest wavelength that just fits within the wire dimensions is allowed, and other longer wavelengths are not permitted. Therefore, the minimum allowed phonon energy is larger than in a macroscopic system. If this minimum phonon energy exceeds the thermal energy, k_BT, then quantum thermal transport is anticipated. The first conclusive observation of a quantized limiting value of thermal conductance at very low temperatures was reported by Schwab *et al.* [104]. The Landauer formula [103, 107] has been used to calculate the value of a quantum of thermal conductance in the ballistic, one-dimensional limit. Both the theoretical and experimental studies demonstrated that in these systems a quantum of thermal conductance is equal to $\pi^2 k_B^2 T/3h$. Consequently, the thermal conductance saturates at low temperatures (in the range 1 K down to 80 mK) to a finite value as opposed to approaching zero in a manner predicted by classical theory. Whether or not quantized thermal conductance will influence future advances in nanotechnology depends on the thermal operating range and the dimensions of specific components.

While there have been several theoretical studies on the thermal conductivity of nanowires, there have been very few experimental investigations. The lack of experimental data is a result of both the difficulty in growing the nanowires as well as in actually performing the thermal conductivity measurements.

Techniques for measuring the thermal conductivity of nanotubes and nanowires are still being developed. Hone *et al.* [96, 108] measured the thermal conductivity of single-walled carbon nanotubes using a comparative technique. However, this technique measures the conductivity of a bundle of nanotubes, rather than a single wire. Shi [109] and Kim *et al.* [110] developed a technique to measure the thermal conductivity of a single nanowire or nanotube using a suspended structure micromachined on a silicon wafer. In this technique, a wire is placed across two suspended islands as shown in Figure 14. On both islands metal lines act as electrical resistance thermometers, while on one island the metal line is also utilized as a resistive heater. If the temperature difference across the nanowire is measured, and the heat input is known, the thermal conductivity of the nanowire may be calculated. The major difficulty with this technique is placing the nanowire such that it bridges the two islands. Also, for wires of certain materials and diameters, contact resistance between the wire and the islands may be a large source of uncertainty. Using this technique, Kim *et al.* [110] measured a thermal conductivity of over 3000 W/m-K for a single-walled carbon nanotube at room temperature. This value was two orders of magnitude greater than the value reported for a mat of multi-walled carbon nanotubes [111] and an order of magnitude greater than that of an aligned single-walled nanotube film [108]. However, Kim and co-workers measured value was comparable to recently predicted theoretical values of 3000–6000 W/m-K [112–114]. The unusually high thermal conductivities are likely a result of the incredibly high atomic bond strength present in carbon nanotubes. Additionally, in contrast to their bulk counterparts, phonon–phonon (umklapp) scattering is suppressed in carbon nanotubes because their size limits the allowed wave vectors in the dispersion relation. In some semiconductor and semimetal nanowires, however, the thermal conductivities are expected to be significantly lower than their bulk counterparts [91, 97]. Li [115] addressed the contact resistance problem and used the technique described above to measure the thermal conductivity of Si nanowires with diameters less than 50 nm. Other researchers have also conducted low temperature experimental measurements of the thermal conductivity of GaAs nanowires [100] and silicon-nitride nanowires [101].

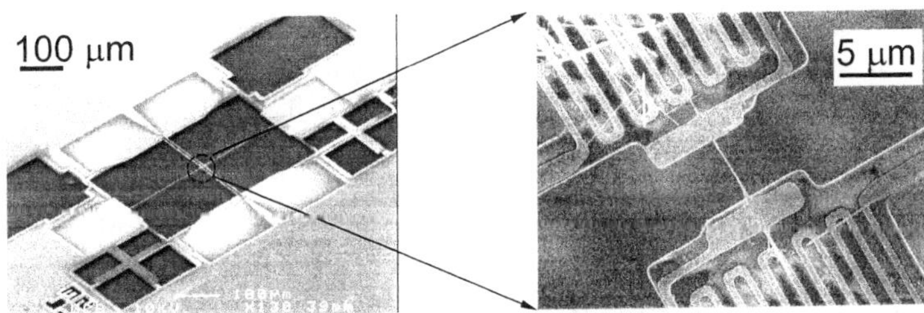

Figure 14: Left: SEM image of the microfabricated structure developed by Shi [109] and Kim *et al.* [110] to measure the thermal conductivity of nanotubes and nanowires. Right: Enlarged image of the suspended structure. The two islands are thermally isolated except for the nanowire that bridges between them. The metal lines on both islands act as thermometers and one also acts as a heater. If the thermal conductivities of the legs are known, the thermal conductivity of the wire may be calculated once the temperature difference between the two islands is measured. The nanowire bridging the two islands is made of Si and has a diameter of ~80 nm (images courtesy of D. Li, Mechanical Engineering, UCB, and L. Shi, Mechanical Engineering, University of Texas – Austin).

In the study of the lattice thermal conductivity of these nanowire systems, a variety of approaches and assumptions have been proposed to account for the reduced thermal conductivity. Factors such as modification of the phonon dispersion relation and group velocity due to confinement effects may play an important role [89, 91, 92]. However, boundary scattering should also influence thermal transport [92, 94–97]. Moreover, quantized thermal conductance may further contribute interesting effects, particularly at low temperature [100–104].

While measurements on individual wires are required in order to fully understand the nature of heat transport in these one-dimensional structures, experimental techniques are rather difficult and are still being refined. An alternative is to examine the thermal conductivity of nanowire arrays by adapting techniques originally developed for measurements on thin films. Although experiments on arrays do not directly reveal the properties of individual wires, they will measure overall properties for the arrays that will be useful in the development of devices. At the present time there has been even less work on measurements of nanowire arrays. Recently, Abramson [116] used an extension of the 3ω method to measure the thermal conductivity of an array of Si nanowires filled with parylene.

7 Summary

In this chapter, heat transport mechanisms in nanostructures such as superlattices, nanowires, and nanotubes were discussed and contrasted with transport mechanisms in bulk materials. Due to the increase in the ratio of surface area-to-volume, interfaces and boundaries play a more significant role for heat conduction in nanostructures. Additionally, when the critical lengthscale of a structure is on the order of, or less than, the phonon mean free path or wavelength, phonon confinement may also become an issue.

Experimental studies have shown that many semiconductor superlattices exhibit a reduced thermal conductivity when compared to bulk materials and corresponding alloys, however, the specific reasons for this reduction remain unclear. This is due partially to the lack of systematic studies carried out on carefully chosen structures and the difficulty in isolating the effects of the various mechanisms involved in thermal transport within these structures. However, it seems quite likely that different transport mechanisms dominate for different materials systems, growth techniques, and period thicknesses.

In recent years there have been numerous theoretical studies on heat transport in nanowires and nanotubes, however, experimental results have lagged behind. The lack of experimental results on these structures is due to the difficulty in synthesizing the materials along with the challenge of actually measuring the thermal conductivity. Reliable growth techniques and experimental methods are still being developed and refined.

References

[1] Weisbuch, C. & Vinter, B., *Quantum Semiconductor Structures*, Academic Press: San Diego, CA, 1991.

[2] Yariv, A., *Quantum Electronics*, 3rd edn, John Wiley and Sons: New York, 1989.

[3] Hicks, L.D. & Dresselhaus, M.S., Effect of quantum-well structures on the thermoelectric figure of merit. *Physical Review B*, **47**, pp. 12727–12731, 1993.

[4] Hicks, L.D., Harman, T.C. & Dresselhaus, M.S., Use of quantum-well superlattices to obtain a high figure of merit from nonconventional thermoelectric materials. *Applied Physics Letters*, **63**, pp. 3230–3232, 1993.

[5] Cho, A.Y. & Arthur, J.R., Molecular beam epitaxy. *Progress in Solid State Chemistry*, **10**, pp. 157–191, 1975.

[6] Herman, M.A. & Sitter, H., *Molecular Beam Epitaxy: Fundamentals and Current Status*, Springer-Verlag: Berlin, 1989.

[7] Dapkus, P.D., Manasevit, H.M., Hess, K.L., Low, T.S. & Stillman, G.E., High purity GaAs prepared from trimethylgallium and arsine. *Journal of Crystal Growth*, **55**, pp. 10–23, 1981.

[8] Hicks, L.D. & Dresselhaus, M.S., Thermoelectric figure of merit of a one-dimensional conductor. *Physical Review B*, **47**, pp. 16631–16634, 1993.

[9] Morales, A.M. & Lieber, C.M., A Laser ablation method for the synthesis of crystalline semiconductor nanowires. *Science*, **279**, pp. 208-211, 1998.

[10] Zhang, Y.F., Tang, Y.H., Wang, N., Yu, D.P., Lee, C.S., Bello, I. & Lee, S.T., Silicon nanowires prepared by laser ablation at high temperature. *Applied Physics Letters*, **72**, pp. 1835–1837, 1998.

[11] Duan, X. & Lieber, C.M., General synthesis of compound semiconductor nanowires. *Advanced Materials*, **12**, pp. 298–302, 2000.

[12] Gudiksen, M.S. & Lieber, C.M., Diameter-selective synthesis of semiconductor nanowires. *Journal of the American Chemical Society*, **122**, pp. 8801–8802, 2000.

[13] Tang, C.C., Fan, S.S., Dang, H.Y., Li, P. & Liu, Y.M., Simple and high-yield method for synthesizing single-crystal GaN nanowires. *Applied Physics Letters*, **77**, pp. 1961–1963, 2000.

[14] Wu, Y. & Yang, P., Germanium nanowire growth via simple vapor transport. *Chemistry of Materials*, **12**, pp. 605–607, 2000.

[15] Wu, Y. & Yang, P., Direct observation of vapor–liquid–solid nanowire growth. *Journal of the American Chemical Society*, **123**, pp. 3165–3166, 2001.

[16] Trentler, T.J., Hickman, K.M., Goel, S.C., Viano, A.M., Gibbons, P.C. & Buhro, W. E., Solution-liquid-solid growth of crystalline III-V semiconductors – an analogy to vapor-liquid-solid growth. *Science*, **270**, pp. 1791–1794, 1995.

[17] Holmes, J.D., Johnston, K.P., Doty, R.C. & Korgel, B.A., Control of thickness and orientation of solution-grown silicon nanowires. *Science*, **287**, pp. 1471–1473, 2000.

[18] Chang, S.S., Shih, C.W., Chen, C.D., Lai, W.C. & Wang, C.R.C., The shape transition of gold nanorods. *Langmuir*, **15**, pp. 701–709, 1999.

[19] Li, Y.D., Ding, Y. & Wang, Z.Y., A novel chemical route to ZnTe semiconductor nanorods. *Advanced Materials*, **11**, pp. 847–850, 1999.

[20] Hu, J., Odom, T.W. & Lieber, C.M., Chemistry and physics in one dimension: synthesis and properties of nanowires and nanotubes. *Accounts of Chemical Research*, **32**, pp. 435–445, 1999.

[21] Zhang, Z., Ying, J.Y. & Dresselhaus, M.S., Bismuth quantum-wire arrays fabricated by a vacuum melting and pressure injection process. *Journal of Materials Research*, **13**, pp. 1745–1748, 1998.

[22] Huber, T.E., Graf, M.J., Foss, C.A. & Constant, P., Processing and characterization of high-conductance bismuth wire array composites. *Journal of Materials Research*, **15**, pp. 1816–1821, 2000.

[23] Heremans, J., Thrush, C.M., Zhang, Z., Sun, X., Dresselhaus, M.S., Ying, J.Y. & Morelli, D.T., Magnetoresistance of bismuth nanowire arrays: A possible transition from one-dimensional to three-dimensional localization. *Physical Review B*, **58**, pp. R10091–R10095, 1998.

[24] Sapp, S.A., Lakshmi, B.B. & Martin, C.R., Template synthesis of bismuth telluride nanowires. *Advanced Materials*, **11**, pp. 402–404, 1999.

[25] Prieto, A.L., Sander, M.S., Martín-González, M.S., Gronsky, R., Sands, T. & Stacy, A.M., Electrodeposition of ordered Bi_2Te_3 nanowire arrays. *Journal of the American Chemical Society*, **123**, pp. 7160–7161, 2001.

[26] Dresselhaus, M.S., Dresselhaus, G. & Eklund, P., *Science of Fullerenes and Carbon Nanotubes*, Academic Press: New York, 1996.

[27] Chopra, N.G., Luyken, R.J., Cherrey, K., Crespi, V.H., Cohen, M.L., Louie, S.G. & Zettl, A., Boron-nitride nanotubes. *Science*, **269**, pp. 966–967, 1995.

[28] Ashcroft, N.W. & Mermin, D.N., *Solid State Physics*, Saunders College Publishing: Fort Worth, 1976.

[29] Kittel, C., *Introduction to Solid State Physics*, 7th edn, John Wiley and Sons: New York, 1996.

[30] Tien, C.-L., Majumdar, A. & Gerner, F.M., *Microscale Energy Transport*, Taylor and Francis: Bristol, 1998.

[31] Chen, G., Phonon transport in low-dimensional structures. *Semiconductors and Semimetals*, **71**, pp. 203–258, 2001.

[32] Haller, E.E., Isotopically engineered semiconductors. *Applied Physics Reviews*, **77**, pp. 2857–2878, 1995.

[33] Touloukian, Y.S., *Thermophysical Properties of Matter*, **2**, pp. 326–338, 1970.

[34] Touloukian, Y.S., *Thermophysical Properties of Matter*, **2**, pp. 108–111, 1970.

[35] Dismukes, J.P., Ekstrom, L., Steigmeier, E.F., Kudman, I. & Beers, D.S., Thermal and electrical properties of heavily doped Ge–Si alloys up to 1300°K. *Journal of Applied Physics*, **35**, pp. 2899–2907, 1964.

[36] Rowe, D.M., *CRC Handbook of Thermoelectrics*, CRC Press: Boca Raton, 1995.

[37] Klitsner, T. & Pohl, R.O., Phonon scattering at silicon crystal interfaces. *Physical Review B*, **36**, pp. 6551–6565, 1987.

[38] Narayanamurti, V., Phonon optics and phonon propagation in semiconductors. *Science*, **213**, pp. 717–723, 1981.
[39] Swartz, E.T. & Pohl, R.O., Thermal boundary resistance. *Reviews of Modern Physics*, **61**, pp. 605–668, 1989.
[40] Modest, M.F., *Radiative Heat Transfer*, McGraw-Hill: New York, 1993.
[41] Little, W.A., The transport of heat between dissimilar solids at low temperatures. *Canadian Journal of Physics*, **37**, pp. 334–349, 1959.
[42] Chen, G., Zeng, T., Borca-Tasciuc, T. & Song, D., Phonon engineering in nanostructures for solid-state energy conversion. *Materials Science and Engineering A*, **292**, pp. 155–161, 2000.
[43] Stoner, R.J. & Maris, H.J., Kapitza conductance and heat flow between solids at temperatures from 50 to 300 K. *Physical Review B*, **48**, pp. 16373–16387, 1993.
[44] Chen, G., Phonon wave heat conduction in thin films and superlattices. *Journal of Heat Transfer*, **121**, pp. 945–953, 1999.
[45] Yamasaki, I., Yamanaka, R., Mikami, M., Sonobe, H., Mori, Y. & Sasaki, T., Thermoelectric properties of Bi_2Te_3/Sb_2Te_3 superlattice structure. *Proceedings of the 17th International Conference on Thermoelectrics*, pp. 210–213, 1998.
[46] Venkatasubramanian, R., Lattice thermal conductivity reduction and phonon localization like behavior in superlattice structures. *Physical Review B*, **61**, pp. 3091–3097, 2000.
[47] Narayanamurti, V. Störmer, H.L., Chin, M.A., Gossard, A.C. & Wiegmann, W., Selective transmission of high-frequency phonons by a superlattice: the "dielectric" phonon filter. *Physical Review Letters*, **43**, pp. 2012–2016, 1979.
[48] Hyldgaard, P. & Mahan, G.D., Phonon superlattice transport. *Physical Review B*, **56**, pp. 10754–10757, 1997.
[49] Tamura, S., Tanaka, Y. & Maris, H.J., Phonon group velocity and thermal conduction in superlattices. *Physical Review B*, **60**, pp. 2627–2630, 1999.
[50] Simkin, M.V. & Mahan, G.D., Minimum thermal conductivity of superlattices. *Physical Review Letters*, **84**, pp. 927–930, 2000.
[51] Frank, F.C. & van der Merwe, J.H., One-dimensional dislocations. II. Misfiting monolayers and oriented overgrowth. *Proceedings of the Royal Society of London. Series A*, **198**, pp. 216–225, 1949.
[52] Matthews, J.W. & Blakeslee, A.E., Defects in epitaxial multilayers. *Journal of Crystal Growth*, **27**, pp. 118–125, 1974.
[53] Freund, L.B., The stability of a dislocation threading a strained layer on a substrate. *Journal of Applied Mechanics*, **54**, pp. 553–557, 1987.
[54] Dunstan, D.J., Strain and strain relaxation in semiconductors. *Journal of Materials Science: Materials in Electronics*, **8**, pp. 337–375, 1997.
[55] Schimmel, D.G., A comparison of chemical etchers for revealing <100> silicon crystal defects. *Journal of the Electrochemical Society*, **123**, pp. 734–741, 1976.

[56] Yao, T., Thermal properties of AlAs/GaAs superlattices. *Applied Physics Letters*, **51**, pp. 1798–1800, 1987.

[57] Holland, M.G., Phonon scattering in semiconductors from thermal conductivity studies. *Physical Review*, **134**, pp. A471–A480, 1964.

[58] Lide, D., *CRC Handbook of Chemistry and Physics*, 73rd edn, CRC Press: Boca Raton, 1993.

[59] Chen, A.-B. & Sher, A., *Semiconductor Alloys*, Plenum Press: New York, 1995.

[60] Chen, G., Tien, C.-L., Wu, X. & Smith, J.S., Thermal diffusivity measurement of GaAs/AlGaAs thin-film structures. *Journal of Heat Transfer*, **116**, pp. 325–331, 1994.

[61] Yu, X.Y., Chen, G., Verma, A. & Smith, J.S., Temperature dependence of thermophysical properties of GaAs/AlAs periodic structures. *Applied Physics Letters*, **67**, pp. 3554–3556, 1995.

[62] Yu, X.Y., Zhang, L. & Chen, G., Thermal-wave measurements of thin-film thermal diffusivity with different laser beam configurations. *Review of Scientific Instruments*, **67**, pp. 2312–2316, 1996.

[63] Amith, A., Kudman, I. & Steigmeier, E.F., Electron and phonon scattering in GaAs at high temperatures, *Physical Review*, **138**, pp. A1270–A1276, 1965.

[64] Capinski, W.S. & Maris, H.J., Thermal conductivity of GaAs/AlAs superlattices. *Physica B*, **219** and **220**, pp. 699–701, 1996.

[65] Capinski, W.S. & Maris, H.J., Improved apparatus for picosecond pump-and-probe optical measurements. *Review of Scientific Instruments*, **67**, pp. 2720–2726, 1996.

[66] Capinski, W.S., Maris, H.J., Ruf, T., Cardona, M., Ploog, K. & Katzer, D.S., Thermal-conductivity measurements of GaAs/AlAs superlattices using a picosecond optical pump-and-probe technique. *Physical Review B*, **59**, pp. 8105–8113, 1999.

[67] Adachi, S., *GaAs and Related Materials*, World Scientific Publishing: Singapore, 1994.

[68] Lee, S.-M., Cahill, D.G. & Venkatasubramanian, R., Thermal conductivity of Si-Ge superlattices. *Applied Physics Letters*, **70**, pp. 2957–2959, 1997.

[69] Cahill, D.G. & Pohl, R.O., Thermal conductivity of amorphous solids above the plateau. *Physical Review B*, **35**, pp. 4067–4073, 1987.

[70] Cahill, D.G., Thermal conductivity measurement from 30 to 750 K: the 3 omega method. *Review of Scientific Instruments*, **61**, pp. 802–808, 1990.

[71] Cahill, D.G., Katiyar, M. & Abelson, J.R., Thermal conductivity of a-Si:H thin films. *Physical Review B*, **50**, pp. 6077–6081, 1994.

[72] Borca-Tasciuc, T., Liu, W., Liu, J., Zeng, T., Song, D.W., Moore, C.D., Chen, G., Wang, K.L., Goorsky, M.S., Radetic, T., Gronsky, R., Koga, T. & Dresselhaus, M.S., Thermal conductivity of symmetrically strained Si/Ge superlattices. *Superlattices and Microstructures*, **28**, pp. 199–206, 2000.

[73] Chen, G. & Neagu, M., Thermal conductivity and heat transfer in superlattices. *Applied Physics Letters*, **71**, pp. 2761–2763, 1997.

[74] Chen, G., Zhou, S.Q., Yao, D.-Y., Kim, C.J., Zheng, X.Y., Liu, Z.L. & Wang, K.L., Heat conduction in alloy-based superlattices. *17th International Conference on Thermoelectrics*, pp. 202–205, 1998.

[75] John, S., Strong localization of photons in certain disordered dielectric superlattices. *Physical Review Letters*, **58**, pp. 2486–2489, 1987.

[76] Venkatasubramanian, R., Siivola, E., Colpitts, T. & O'Quinn, B., Thin-film thermoelectric devices with high room-temperature figures of merit. *Nature*, **413**, pp. 597–602, 2001.

[77] Touzelbaev, M.N., Zhou, P., Venkatasubramanian, R. & Goodson, K., Thermal characterization of Bi_2Te_3/Sb_2Te_3 superlattices. *Journal of Applied Physics*, **90**, pp. 763–767, 2001.

[78] Goodson, K.E., Käding, O.W., Rösler, M. & Zachai, R., Experimental investigation of thermal conduction normal to diamond-silicon boundaries. *Journal of Applied Physics*, **77**, pp. 1385–1392, 1995.

[79] Venkatasubramanian, R., Colpitts, T., Watko, E., Lamvik, M. & El-Masry, N., MOCVD of Bi_2Te_3, Sb_2Te_3 and their superlattice structures for thin-film thermoelectric applications. *Journal of Crystal Growth*, **170**, pp. 817–821, 1997.

[80] Huxtable, S.T., Abramson, A.R., Tien, C.-L., Majumdar, A., LaBounty, C., Fan, X., Zeng, G., Bowers, J.E., Shakouri, A. & Croke, E.T., Thermal conductivity of Si/SiGe and SiGe/SiGe superlattices. *Applied Physics Letters*, **80**, pp. 1737–1739, 2002.

[81] Borca-Tasciuc, T., Achimov, D., Liu, W.L. & Chen, G., Thermal conductivity of InAs/AlSb superlattices. *Microscale Thermophysical Engineering*, **5**, pp. 225–231, 2001.

[82] Van Nostrand, J.E., Cahill, D.G., Petrov, I. & Greene, J.E., Morphology and microstructure of tensile-strained SiGe (001) thin epitaxial films. *Journal of Applied Physics*, **83**, pp. 1096–1102, 1998.

[83] Song, D.W., Liu, W.L., Zeng, T., Borca-Tasciuc, T., Chen, G., Caylor, J.C. & Sands, T.D., Thermal conductivity of skutterudite thin films and superlattices. *Applied Physics Letters*, **77**, pp. 3854–3856, 2000.

[84] Liu, J.L., Khitun, A., Wang, K.L., Borca-Tasciuc, T., Liu, W.L. Chen, G. & Yu, D.P., Growth of Ge quantum dot superlattices for thermoelectric applications. *Journal of Crystal Growth*, **227–228**, pp. 1111–1115, 2001.

[85] Kiselev, A.A., Kim, K.W. & Stroscio, M.A., Thermal conductivity of Si/Ge superlattices: A realistic model with a diatomic unit cell. *Physical Review B*, **62**, pp. 6896–6899, 2000.

[86] Bies, W.E., Radtke, R.J. & Ehrenreich, H., Phonon dispersion effects and the thermal conductivity reduction in GaAs/AlAs superlattices. *Journal of Applied Physics*, **88**, pp. 1498–1503, 2000.

[87] Chen, G., Thermal conductivity and ballistic-phonon transport in the cross-plane direction of superlattices. *Physical Review B*, **57**, pp. 14958–14973, 1998.

[88] Hyldgaard, P. & Mahan, G.D., *Thermal Conductivity*, **23**, p. 172–182, Technomic: Lancaster, PA, 1995.

[89] Balandin, A. & Wang, K.L., Significant decrease of the lattice thermal conductivity due to phonon confinement in a free-standing semiconductor quantum well. *Physical Review B*, **58**, pp. 1544–1549, 1998.

[90] Prasher, R.S. & Phelan, P.E., Size effects on the thermodynamic properties of thin solid films. *Journal of Heat Transfer*, **120**, pp. 1078–1086, 1998.

[91] Khitun, A., Balandin, A. & Wang, K.L., Modification of the lattice thermal conductivity in silicon quantum wires due to spatial confinement of acoustic phonons. *Superlattices and Microstructures*, **26**, pp. 181–193, 1999.

[92] Zou, J. & Balandin, A., Phonon heat conduction in a semiconductor nanowire. *Journal of Applied Physics*, **89**, pp. 2932–2938, 2001.

[93] Chen, G. & Tien, C.-L., Thermal-conductivities of quantum-well structures. *Journal of Thermophysics and Heat Transfer*, **7**, pp. 311–318, 1993.

[94] Dresselhaus, M.S., Dresselhaus, G., Sugihara, K., Spain, I.L. & Goldberg, H.A., *Graphite Fibers and Filaments*, Springer Series in Materials Science, Vol. 5, Springer-Verlag: Berlin, 1988.

[95] Heremans, J., Rahim, I. & Dresselhaus, M.S., Thermal conductivity and raman spectra of carbon fibers. *Physical Review B*, **32**, pp. 6742–6747, 1985.

[96] Hone, J., Whitney, M., Piskoti, C. & Zettl, A., Thermal conductivity of single-walled carbon nanotubes. *Physical Review B*, **59**, pp. R2514–R2516, 1999.

[97] Volz, S.G. & Chen, G., Molecular dynamics simulation of thermal conductivity of silicon nanowires. *Applied Physics Letters*, **75**, pp. 2056–2058, 1999.

[98] Walkauskas, S.G., Broido, D.A., Kempa, K. & Reinecke, T.L., Lattice thermal conductivity of wires. *Journal of Applied Physics*, **85**, pp. 2579–2582, 1999.

[99] Volz, S. & Lemonnier, D., Confined phonon and size effects on nanowire thermal conductivity. The radiative transfer approach. *Physics of Low-Dimensional Structures*, **5–6**, pp. 91–107, 2000.

[100] Potts, A., Kelly, M.J., Hasko, D.G., Smith, C.G., Cleaver, J.R.A., Ahmed, H., Singleton, J. & Janssen, T.J.B., Thermal transport in free-standing semiconductor fine wires. *Superlattices and Microstructures*, **9**, pp. 315–318, 1991.

[101] Tighe, T.S., Worlock, J.W. & Roukes, M.L., Direct thermal conductance measurements on suspended monocrystalline nanostructures. *Applied Physics Letters*, **70**, pp. 2687–2689, 1997.

[102] Angelescu, D.E., Cross, M.C. & Roukes, M.L., 1998, Heat transport in mesoscopic systems. *Superlattices and Microstructures*, **23**, pp. 673–689, 1998.

[103] Rego, L.G.C. & Kirczenow, G., Quantized thermal conductance of dielectric quantum wires. *Physical Review Letters*, **81**, pp. 232–235, 1998.
[104] Schwab, K., Henriksen, E.A., Worlock, J.M. & Roukes, M.L., Measurement of the quantum of thermal conductance. *Nature*, **404**, pp. 974–977, 2000.
[105] Bannov, N., Aristov, V. & Mitin, V., Electron relaxation-times due to the deformation-potential interaction of electrons with confined acoustic phonons in a freestanding quantum-well. *Physical Review B*, **51**, pp. 9930–9942, 1995.
[106] Yu, S.G., Kim, K.W., Stroscio, M.A. & Iafrate, G.J., Electron acoustic-phonon scattering rates in cylindrical quantum wires. *Physical Review B*, **51**, pp. 4695–4698, 1995.
[107] Imry, Y. & Landauer, R., Conductance viewed as transmission. *Review of Modern Physics*, **71**, pp. S306–S312, 1999.
[108] Hone, J., Llaguno, M.C., Nemes, N.M., Johnson, A.T., Fischer, J.E., Walters, D.A., Casavant, M.J., Schmidt, J. & Smalley, R.S., Electrical and thermal transport properties of magnetically aligned single wall carbon nanotube films. *Applied Physics Letters*, **77**, pp. 666–668, 2000.
[109] Shi, L., *Mesoscopic Thermophysical Measurements of Microstructures and Carbon Nanotubes*, Ph.D. thesis, Dept. of Mechanical Engineering, University of California, Berkeley, 2001.
[110] Kim, P., Shi, L., Majumdar, A. & McEuen, P., Thermal transport measurements of individual multiwall carbon nanotubes. *Physical Review Letters*, **87**, pp. 215502-1–215502-4, 2001.
[111] Yi, W., Lu, L., Dian-lin, Z., Pan, Z.W. & Xie, S.S., Linear specific heat of carbon nanotubes. *Physical Review B*, **59**, pp. R9015–R9018, 1999.
[112] Berber, S., Kwon, Y.-K. & Tománek, D., Unusually High Thermal Conductivity of Carbon Nanotubes. *Physical Review Letters*, **84**, pp. 4613–4616, 2000.
[113] Che, J.W., Cagin, T. & Goddard, W.A., Thermal conductivity of carbon nanotubes. *Nanotechnology*, **11**, pp. 65–69, 2000.
[114] Osman, M.A. & Srivastava, D., Temperature dependence of the thermal conductivity of single-wall carbon nanotubes. *Nanotechnology*, **12**, pp. 21–24, 2001.
[115] Li, D., University of California – Berkeley, private communication, 2002.
[116] Abramson, A., *Thermal Energy Transport in Micro/Nanostructures*, Ph.D. thesis, Dept. of Mechanical Engineering, University of California, Berkeley, 2002.

CHAPTER 4

Thermomechanical formation and thermal detection of polymer nanostructures

W.P. King[1] & K.E. Goodson[2]
[1] *Woodruff School of Mechanical Engineering, Georgia Institute of Technology, Atlanta, GA, USA*
[2] *Department of Mechanical Engineering, Stanford University, Stanford, CA, USA*

Abstract

In thermomechanical data storage, a heated atomic force microscope (AFM) cantilever tip scans over and melts small data bit indentations as small as 20 nm into a thin polymer film. Small changes in the cantilever temperature, which correspond to changes in the thermal resistance between the cantilever and the data substrate, can detect the presence of previously written data bits. This chapter reviews work on modeling and measurement of heat and mass transfer during thermomechanical formation and thermal detection of polymer nanostructures for data storage. A study explores the impact of molecular confinement on viscous polymer flow in nanostructured thin films that finds a two order of magnitude increase in polymer viscosity and a shift from a no-slip to a slip in the polymer-substrate contact condition. Modeling and measurement of bit formation show that a cantilever temperature of 200 °C is required for bit formation at times greater than 1 ms, increasing to 500 °C for bit formation near 10 μs. A study of cantilever thermal reading sensitivity shows that the cantilever compares well with other AFM-based metrology approaches, with a sensitivity exceeding 10^5 μV/nm . This chapter reviews work on heat and mass transfer that governs the processes of thermomechanical data write, read, and erase.

1 Introduction

Atomic force microscopy (AFM), invented in 1986 [1], has emerged as the most common tool for sensing and actuating at the nanometer scale. The basic architecture of an AFM system consists of a micromachined tip at the end of a cantilevered beam. A motion control stage, attached to the cantilevered beam, brings the tip into contact with, and moves the tip laterally over, a surface. As the tip follows the shape of the surface, small changes in the vertical position of the cantilever beam are detected, shown in Figure 1. AFM can produce topographic maps of the surface with resolution limited by the shapes of the cantilever tip and the surface, and the quality of the scanning and sensing system, typically near 1 nm in all three dimensions. Schemes for vertical detection include piezoresistive activation of the beam [2], optical detection [1], and thermal detection [3]. A large number of local probe and cantilever-based instruments are based on this basic concept, including optical probes [4], force sensors [2], and biochemical sensors [5].

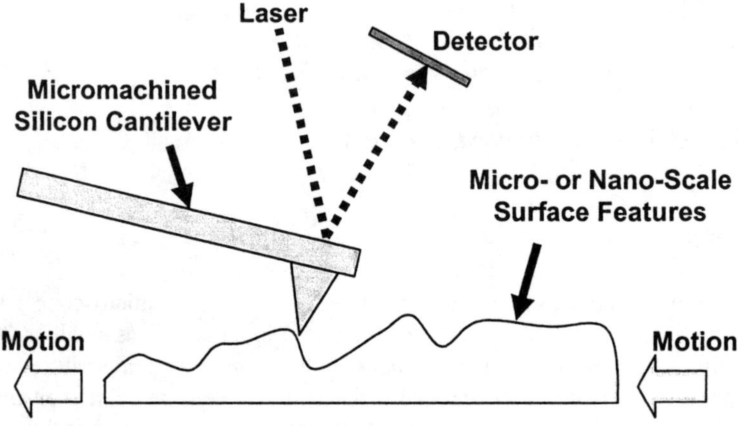

Figure 1: Schematic of AFM operation.

The cantilever tip can be used to intentionally modify the surface over which it scans. The various AFM surface modification research includes AFM cantilevers designed for guiding electromagnetic radiation into photoreactive polymer [6], cantilever tips which direct electrostatic discharge to locally oxidize a reactive surface [7], local chemical delivery with the AFM tip [8], direct indentation of soft materials [9], and the present approach of thermally-assisted indentation of soft materials [10].

1.1 Motivation for AFM data storage

The rapid advancement of magnetic data storage can be measured by continuous increases in the area density of data bits in commercial disk drives, currently growing at an annual rate of 100% [11]. The superparamagnetic effect, which governs the thermal stability of a magnetic data bit, will likely limit data density in current magnetic data storage technology near 100 Gbit/in^2 [12]. Several efforts show promise for expanding this limit even further, for example patterned magnetic media for perpendicular recording [13], which increase domain stability through domain separation, or thermally assisted magnetic recording [14], which increases the magnetic susceptibility of the recording media such that magnetic bits can be written only at high temperature. Despite the data storage densities possible with these technology advancements, it remains unclear which technology will permit data storage devices capable of 1 Tbit/in^2 and beyond.

Several AFM-based data storage technologies that employ surface modification can produce feature sizes that are significantly smaller than magnetic recording data bit sizes [3, 10]. A review of AFM-based data storage has been written by Mamin and co-authors [15]. Lithographic patterning of extremely small features also drives progress in technology surface modification, and in general many of the probe-based data storage approaches offer promise for lithography.

Despite the progress and promise of these many AFM data storage technologies, there remain pervasive challenges to transferring this technology to a commercial product platform. These challenges include:

- ***Speed***. Typical single-cantilever mechanical time constants are close to 100 kHz [16], which is small compared to a system-level data rate of 200 Mbit/sec or more for a commercially available disk drive [17]. Scanning probe data storage with competitive data rates are accomplished through single-cantilever optimization [18] and more recently the development of cantilever arrays [19].
- ***Ability to "close the loop"*** and detect the presence of or even the size and shape of the modified feature. The first design work for AFM cantilevers used in thermomechanical data storage wrote and read data bits with separate, individually optimized cantilevers [20]. Lithographic patterning that uses an AFM cantilever tip or STM tip as a waveguide requires wet chemistry to realize the features [6], and thus does not produce features that can be immediately read. In general, the best AFM cantilever for patterning features or writing data bits is not necessarily the best cantilever for sensing the features written.
- ***Reversibility***, as in erasing data or mending a lithographic mistake. While there exist successful write-once data storage approaches, such as CD-ROM, the magnetic-type recording scheme that offers many rewrites is often preferred.
- ***Robustness.*** Several approaches for probe-based surface modification and feature detection require high voltage [6], sophisticated feedback electronics, or careful alignment and preparation of a tunneling tip or a carbon nanotube [7]. The realization of a commercial product must overcome or avoid components that require laboratory equipment or high vacuum to operate reliably.

Very recent progress in thermomechanical data storage has provided promising solutions to each of these challenges. The present approach [21] aims to provide a large, functional array of AFM cantilevers, each capable of high data rates and high sensitivity. Elsewhere the same authors have reported data erasing [3, 22], thus satisfying the reversibility requirement. Continuous contact between the AFM cantilever and the polymer surface, the sharp onset of bit formation [10], and the reliable stability of the data bits at temperatures below the bit-writing temperature [23] make thermomechanical data storage very robust.

1.2 Review of thermomechanical data storage

In thermomechanical data writing, a resistively heated AFM cantilever is in contact with and scans over a substrate coated with a thin polymer film. Heat generated in the cantilever flows along the cantilever tip into the thin polymer film, locally raising the polymer film temperature and causing the polymer to soften. Force applied to the softened polymer from the heated cantilever tip causes the polymer to deform, thus forming an indentation, with a radius of curvature as small as 20 nm [3, 10]. Figure 2 illustrates thermomechanical data writing and Figure 3 shows data bits in a 35 nm thick film of polymethyl methacrylate (PMMA). At 20-50 nm in diameter, these small data bits correspond to a data density of up to 1 Tbit/in^2.

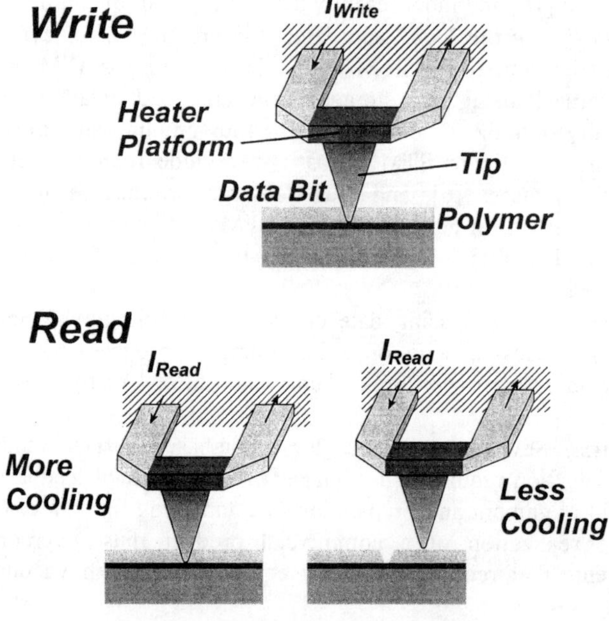

Figure 2: Schematic of thermomechanical data writing and thermal reading.

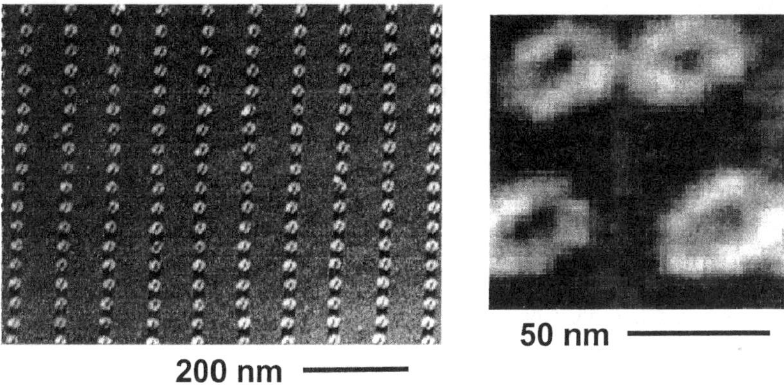

Figure 3: Indentations written and read with a heated AFM cantilever [3].

Several realizations of thermomechanical modification of polymer surfaces preceded the current approach. Mamin [24] first demonstrated thermomechanical data writing with the AFM by using an infrared laser to heat the cantilever. Further work used a commercial piezoresistive silicon cantilever that self-heated during brief electrical pulses [**25**]. While these resistively heated piezoresistive cantilevers could thermally write without the optical access of the first design, the entire cantilever was heated during an electrical heating pulse, resulting in cantilever thermal time constants as long as 0.45 ms. In order to reduce the thermal time constant for heating, Chui [20] designed and optimized AFM cantilevers with a heating time of less than 1 μs, and a cooling time of near 10 μs. The design improvement consisted of preparing the silicon cantilever with regions of differential impurity doping, such that only a small region of the cantilever near the tip was highly resistive, by creating a small heater with significantly reduced heat capacitance.

A resistively heated silicon cantilever can be used to thermally detect the presence of a previously written data bit, allowing data reading. In reading operation, the cantilever is heated to a temperature that is well below the threshold temperature for bit formation. As the warm cantilever follows the contour of a previously written data bit, the change in the thermal resistance between the cantilever and the substrate due to decreased separation produces a change in the temperature of the cantilever, measured through change in the electrical resistance of the cantilever heating element. By recording the dynamic temperature signal as the cantilever tip moves over the polymer surface, a surface contour map can be made. Figure 2 illustrates thermomechanical data reading. Binnig [3] measured the vertical sensitivity of thermal data reading to be between 10^{-4} and 10^{-5} per vertical nanometer for a cantilever that had not been optimized for thermal data reading.

This thermal data reading technique is distinct from techniques that use a thermocouple tip for topographic surface profiling [26] or make highly local temperature measurements [27]. Temperature mapping with a thermocouple-tipped AFM, known as scanning thermal microscopy (SThM), is significantly better than the present implementation for high-resolution thermometry, while the present thermal proximity sensing yields extremely high vertical sensitivity for topographic imaging.

A crucial stage in the development of thermomechanical data storage technology was the development of fabrication techniques to reliably make AFM cantilevers with uniform mechanical and electrical properties [20, 21, 28]. Figure 4 shows a recent cantilever, which is 50 μm in length and has a tip height of 1 μm. More recent versions of this cantilever have a smaller tip height and are thinner for improved reading sensitivity [29].

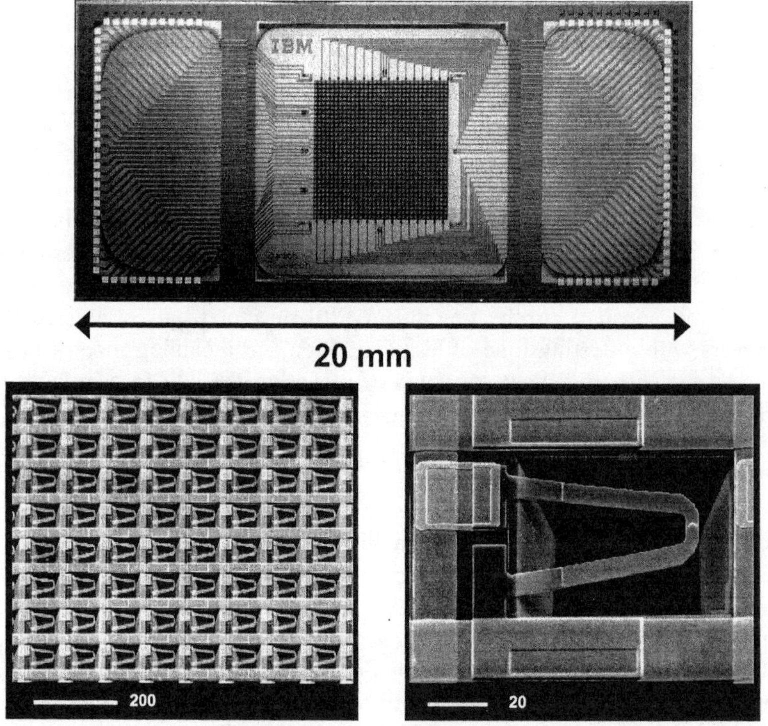

Figure 4: Picture of the Millipede cantilever array data storage chip and scanning electron microscope images of individual cantilevers in the array. Fabrication details can be found in refs. [21, 28] and successful operation of the array is reported in ref. [30].

A cantilever array can write and read data bits with higher speed than a single cantilever operating alone. While much work has been done to design and optimize individual AFM cantilevers, e.g., [18], only recently has there been significant progress in the development of cantilever arrays. Lutwyche [30] reports the fabrication and operation of a 5 × 5 cantilever array, which measured features of characteristic size 2 μm. A larger device with 1024 cantilevers in a 32 × 32 square array, known as the "Millipede" [21] and shown in Figure 4, demonstrated successful thermomechanical and thermal data reading at a data density of 100–200 Gbit/in^2 [29].

1.3 Chapter overview

This chapter reviews studies of heat and mass transport in the cantilever, cantilever tip, and polymer layer during data bit formation and detection. A major challenge in this study is understanding the flow of polymer films of thickness comparable to the polymer molecule radius of gyration. Section 2 reviews a fundamental study of intrinsic polymer viscosity at nanometer length scales that reveals remarkable increase in the polymer viscosity compared to bulk polymer. Performed on thin polymer films identical to those used as data storage media, this study informs bit stability at elevated temperatures. In section 3, a model of heat transfer in the cantilever tip accounts for aspects of subcontinuum heat conduction in the silicon cantilever tip and nearby air, and illustrates the basic mechanisms of data bit writing and reading. Measurement and modeling of thermomechanical data bit formation finds the time and temperature dependence of onset bit writing conditions. Predictions show the effect of changes in the tip shape and tip loading force on onset writing conditions. Section 4 describes modeling of cantilever operation and design of cantilevers for improved thermomechanical data bit writing and thermal data reading, and investigates the thermal data reading approach for topographic sensing. The conclusion describes potential future applications of this work, including AFM materials processing and metrology.

2 Relaxation kinetics in nanostructured polymer films

The physical properties of thin polymer films can differ from bulk properties, particularly at lengths comparable to or thinner than the radius of gyration of a molecule in bulk polymer [31]. Examples of polymer orientation, confinement, and frustration at small scales are observed as shifts in the measured glass transition temperature T_g, [32, 33, 34], changes in polymer thermophysical properties [35], or changes in mechanical viscoelastic properties [36, 37]. Of particular importance for understanding these observed phenomena are the chemical and mechanical influence of interfaces on polymer molecular motion. This section reviews the impact of confinement and bounding effects on thin polymer films.

2.1 Fundamentals of mass transport in confined soft materials

The length scale for a macromolecular material is the radius of gyration, R_g, which is the root mean squared distance between the mass centroid of the unperturbed molecule to each of the monomers in the macromolecular chain [31, 38]. Typical values for R_g range from 5 nm to 100 nm. For macromolecular films of thickness comparable to or smaller than R_g, [32] or for single molecules confined on a surface or in a microfluidic channel [39], the radius of gyration concept is less useful, as the molecule orients in the direction of confinement and will no longer be spherically symmetric.

Two relaxation mechanisms determine polymer molecular motion: a) monomer migration, which depends upon the free volume near each monomer and cooperative motion of nearby monomers, and b) reptation of the polymer chain, which requires migration of the polymer end. DeGennes [31] proposed a model of local macromolecular motion known as reptation, in which macromolecules in a condensed system move along the path of their own molecular backbone. More force or longer time is required for a macromolecule to locally move perpendicular to its path compared to motion through reptation. For free-standing or uncapped thin films, chain reptation is the dominant molecular transport mode [40, 38].

Changes of phase in macromolecular systems are second order transitions [41], rather than the first order phase transition of melting. Macromolecular materials do not have a discontinuity in the heat capacity-temperature relation, but rather have a slow, continuous change in heat capacity over a temperature range. The onset temperature for this heat capacity change is known as the glass transition temperature, T_g. Because the macromolecular materials are inhomogeneous at the molecular level, with many overlapping, intertwining molecules, there exists an energy barrier for each chain interaction with itself or another chain. The apparently continuous material change above T_g is in fact a large number of closely spaced energy barriers. The link between highly local molecular mobility and temperature allows glass transition phenomena to be observed through changes in mechanical, electrical, or optical properties.

Measurements of the glass transition temperature of thin polymer films show that molecular confinement and nearby surfaces can affect polymer molecular motion. Studies show a thickness-dependence of T_g in PMMA films, with a T_g increase for PMMA on silicon dioxide surfaces, and a T_g decrease for PMMA on gold surfaces [33]. Studies of free-standing polystyrene films show a large T_g depression below the bulk value of over 60 °C. This temperature depression is a function of film thickness and molecular weight, indicating the impact of confinement in the absence of a bounding surface [34]. The same films capped by a nearby bounding surface of silicon dioxide demonstrate a 10 °C depression of T_g, and a 4 °C depression for a sandwiched film.

Local measurements of glass transition in polymer films indicate the possibility of local glass transition effects even within the layer: the molecules closest to a bounding or free surface may have a different mobility or differing degrees of freedom than molecules farther from a surface [42]. For example, polymer tacticity has been shown to affect T_g, since variation in the density of polymer–surface interactions locally influence molecular mobility [43]. A positron lifetime spectroscopy study of thin polystyrene films on hydrogen-passivated silicon showed a glass transition temperature close to the surface that was lower than the glass transition in the film, which were both lower than the glass transition expected for bulk polystyrene [44].

Closely related to the glass transition temperature is the local mobility of polymer molecules in a thin film, which also depends upon molecular confinement and proximity to nearby surfaces. The diffusion of polymer layers in a supported thin film structure can be substantially reduced due to the presence of a nearby bounding surface. Indirect measurements of polymer diffusivity measured decreased activation energy required for phase separation in thin polymer films, due to surface energy provided by nearby surfaces [45]. Direct measurements of polymer interdiffusion made by fluorescent labeling of polymer sublayers showed that interpolymer friction coefficients in thin polystyrene films can be 10^2 to 10^3 times greater than the bulk coefficients, depending upon the support material [46]. The self-diffusivity of polymers is also thickness-dependent, demonstrating a factor of two reduction in PS films at thickness well above R_g [47]. Measurements made with neutron scattering show over an order of magnitude reduction in self-diffusivity of polystyrene films at distances up to 10 R_g away from a bounding interface [48], and a similar reduction in PMMA films, but only up to distances of 4 R_g from the bounding surfacc [49].

Simulations of macromolecuar motion allow for prediction of single molecular behavior near a surface and in confined geometries. Predictions show that even in the presence of a strongly attractive surface, molecular centroids are most closely bunched at a distance of R_g from the surface, as strong molecular alignment parallel to the surface prohibits molecules from moving any closer [50]. A simulation of polymer dynamics at a solid–melt interface showed that the interfacial region in the melt is extremely small at only a few tube diameters, and was independent of chain length. The direct impact of the interface was to anchor the molecular chains but this effect did not propagate far into the melt [51]. However, longer range surface forces and confinement effects did induce molecular anisotropy and affect diffusivity. A study of molecular chains of varying lengths [52] showed that while molecular anisotropy depends upon chain length and the energetics of the nearby surface, local chain–chain interactions are affected only weakly. Simulations of block copolymers showed that thermodynamic interactions between polymer blocks add a third complication which can locally affect molecular mobility [53].

Mechanical motion in thin polymer films is also restricted by confinement and bounding effects [54]. Films of polyphenylmethylsiloxane sandwiched between atomically flat mica surfaces undergoing small oscillatory motions demonstrate strong rubber elasticity at thickness below five to six R_g [36]. This elasticity could be due to 'bridging' of polymer molecules with portions of individual molecules near both sandwiching surfaces [55].

A two-fluid model of polymer motion near interfaces proposes that some molecules are trapped near a bounding surface while molecules away from the bounding surface or near a free surface have a different mobility [56] and experiments support this model [57]. More detailed models consider a local mobility, varying near interfaces but not discontinuously [34, 48].

2.2 Measuring flow characteristics in thin, nanostructured polymer films

In one recent measurement [58], a light scattering technique locally probes the temperature-dependent relaxation characteristics of a surface wave structure fabricated in a thin polymer film. Electron beam lithography forms spatially periodic three-dimensional polymer nanostructures in poly methyl methacrylate (PMMA). This polymer has an unperturbed radius of gyration of 35 nm, and experiments are conducted for thickness at and above 35 nm. Heating induces a surface-tension driven flow that results in decay of the surface wave amplitude, observed through measurement of the scattering signal from an incident laser. Atomic force microscope (AFM) scans of the polymer surface aid interpretation of the scattering intensity signal evolution. The measurement shows long-range polymer cooperative motion during flow onset, and larger than expected viscosity for the thinnest polymer structures.

The sample material is a standard electron beam poly methyl methacrylate (PMMA) photoresist, with a number average molecular weight of 3.5×10^5 g/mol, and a measured bulk sample reveals a glass transition temperature at 120 °C. The substrate for the PMMA optical grating structures is a silicon wafer that is first prepared with a spin-coating of 100 nm hard-baked SU8 epoxy. A grating pattern on PMMA is fabricated using an electron microscope exposure. The pattern consists of periodic rectilinear trenches spaced at 360 nm. Polymer samples are prepared at thicknesses 35 nm, 70 nm, 130 nm, and 300 nm, which correspond to approximately 1, 2, and 3, and 10 times the unperturbed radius of gyration.

Figure 5 summarizes the experimental setup. Hamdorf [59] used a similar technique to measure the surface properties of thick polystyrene films. The polymer grating structure is mechanically attached to a temperature-controlled heating stage, and is interrogated with a laser. The scattered light intensity is proportional to the spatial Fourier transform of the density autocorrelation function of the optically probed sample volume. The grating may be represented by a surface density modulation and the scattered intensity $S(\boldsymbol{k})$ is just proportional to the 2-dimensional Fourier spectrum of the local film topography. To first order, the intensity of the scattered signal is proportional to the square of the trench volume.

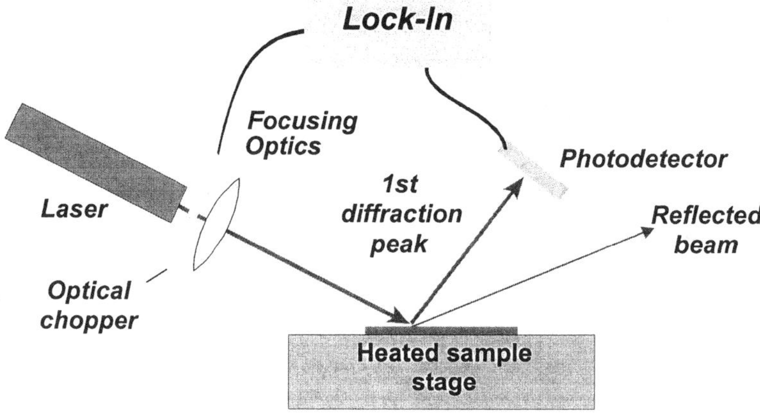

Figure 5: Schematic of experimental setup for polymer relaxation measurement.

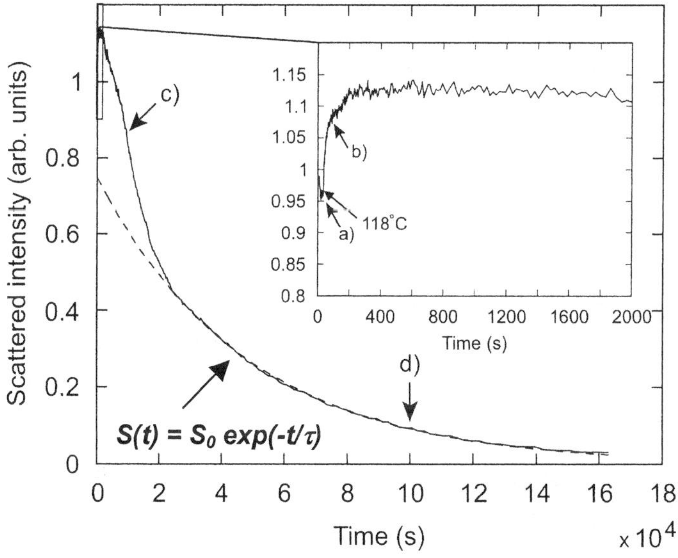

Figure 6: Typical optical signal time trace for the decay of a polymer corrugation structure [58]. This data is for 50 nm wide lines in 70 nm of PMMA.

Figure 6 shows a typical example of the intensity signal evolution for a polymer grating structure. The heater is turned on at t = 0 sec and the temperature is ramped to the final temperature, 125 °C within 50 seconds. The time-trace characteristics common to all of the samples measured are: (a) a lack of change until the polymer is heated to its glass transition temperature T_g, (b) an almost instantaneous increase in the scattering signal at T_g, (c) a period with an almost constant intensity, a transition regime with a rapid drop of the intensity by 30% to 50%, and then eventually (d) an exponential decay. For each of the samples measured, the relaxation (b) was extremely fast and always corresponded to a scattering signal increase of 10-15% over the initial intensity. An exponential decay $S(t) = S_0 \exp(-t/\tau)$ well represents the long time scale decay (d), also shown in Figure 6.

Assisting interpretation of the intensity data, AFM snap shot images of the grating topography are recorded, shown in Figure 7. For these, heat is applied to a fresh sample for a fixed time interval and then rapidly quenched to room temperature. Figure 7a shows the topography of an example grating (60 nm wide lines in 130 nm PMMA) before heating. The sharp edges of the etched trenches exhibit perfectly vertical side walls. Panel 7b shows the same type of grating that was quenched after 100 seconds of heating to 120 °C, visualized shortly after the sudden increase of the scattered intensity had been observed. A drastic initial relaxation can be observed as evidenced in the rounded edges. Also, the PMMA surface is smoother than in Figure 7a and one can discern a slight upward bow of the originally flat terraces. This overall rounding effect induces a redistribution of the Fourier components of the grating – high spatial frequency harmonics are convoluted onto the fundamental grating period as a result of total volume conservation – which corresponds to the initial rise in the scattered intensity. The polymer sample retains this overall shape for longer times in images 7c and 7d, with the exponential decay of the scattering signal corresponding to the polymer structure being most sinusoidal in shape.

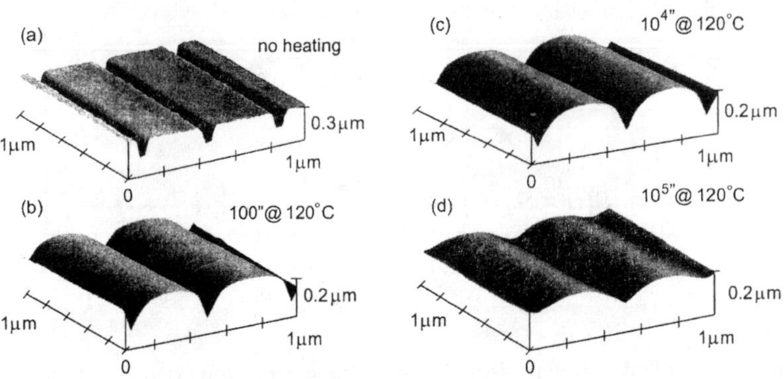

Figure 7: AFM 'snapshot' images of the melting polymer corrugation structures a) before heating, b) after 10^2 sec at 120 °C, c) after 10^4 sec at 120 °C, and d) after 10^5 sec at 120 °C [58].

The initial relaxation shown by the difference in Figures 7a and 7b can be interpreted in terms of a surface force-driven elastic response. The average initial stress acting on the PMMA film due to the sharp rectangular edges of the grating is of the order of $\sigma = \gamma l \approx 0.1$ MPa where $\gamma \approx 0.03$ N/m is the surface tension of PMMA at T_g [60] and l is a relaxation length of the order of the grating period. Below T_g the shear modulus is of the order of 1 GPa and therefore relative elastic strains are below 10^{-4}. At T_g, the PMMA softens dramatically as evidenced by a drop of the shear modulus by three orders of magnitude to values of typically 0.5 to 1 MPa in the rubbery regime [61]. Accordingly, substantial elastic strains of the order of 10^{-1} must develop in order to balance out surface tension forces. This relaxation effect can in fact be used to determine the glass transition temperature in thin film samples in general.

The AFM images of Figure 7 show that the surface of the polymer is much smoother after a short heating time. Scanning probe studies of the decay of the surface roughness of polystyrene and PMMA films at elevated temperatures found increased molecular diffusivity for roughness features of size less than the polymer radius of gyration [62, 63, 64]. The observed sharp temperature transition of surface wave morphology in Figure 7 is markedly different from the measurement of Kerle [64], who observed in thick polystyrene films a broad temperature transition in the decay of roughness features of similar size to our corrugation trenches. It is possible that confinement effects in the thin films contribute to a sharper transition.

Viscous flow, which is also surface tension-driven, provides a second much slower relaxation mechanism which lets the PMMA film recover a completely stress-free final state. The measured relaxation time τ can then be related to the decay rate Γ [65]

$$\Gamma = \frac{1}{2\tau} = \beta \Gamma_{\text{inf}} = \beta \frac{\gamma k}{2\eta} \qquad (1)$$

The parameter β accounts for the impact of the nearby substrate on the polymer flow. For an infinitely thick film $\beta = 1$ and $\Gamma_{\text{inf}} = \Gamma$. The influence of the nearby surface must be considered in order to describe the relaxation of the thin film samples. Imposing a no-slip boundary condition at the polymer-substrate yields $\beta = 2/3(kh)^3$. Imposing a slip boundary condition at the polymer-substrate interface yields $\beta = 1/2kh$.

A line-shape analysis of the polymer surface further aids in interpreting the optical measurement data. According to eqn. 1, the relaxation constant Γ_{inf} is proportional to the polymer sample corrugation wave number k. Thus, correct selection of the parameter β yields a linear relationship between Γ and k. Comparing a spatial Fourier transform of the snapshot images of Figure 7 with measured Γ yields the expected relationship between Γ and k for a slip polymer–substrate boundary condition for the 35 nm thick samples, and for a no-slip polymer–substrate boundary condition for all of the other samples. It is possible that the change from no-slip to slip boundary conditions for the thinnest samples is due to molecular orientation in the thinnest films. Molecular-level simulations [66, 67] show that the polymer chains are substantially more oriented near a surface compared to polymer chains in the bulk. Furthermore, the density of chain segments near the wall is substantially reduced in the presence of a weakly attractive surface compared to a strongly attractive surface. This can be explained by chain packing and confinement effects that prohibit chain segments from approaching the wall to within a distance of less than R_g unless there is an additional driving force.

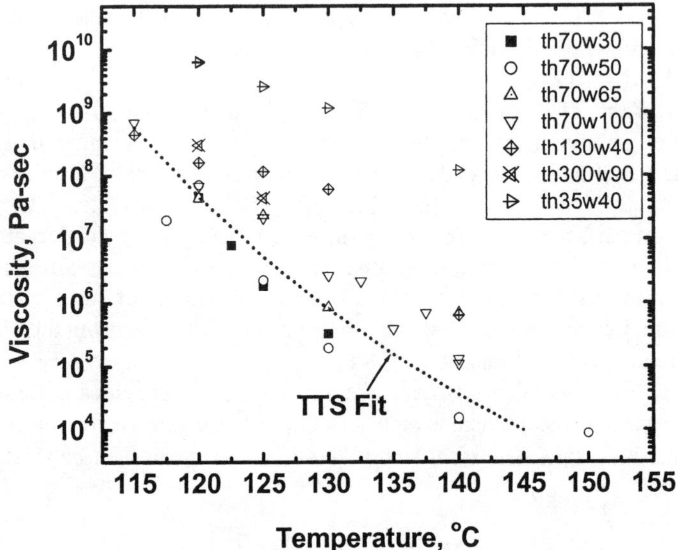

Figure 8: Polymer film viscosity as a function of temperature [58]. Data are rescaled with β chosen for the polymer–substrate interface condition measured with AFM. Solution of eqn. 2 yields the time-temperature superposition (TTS) fit. The nomenclature thXwY refers to a polymer film thickness of X nm and a corrugation line width of Y nm. A time temperature superposition fit is shown to the th70w100 data.

Figure 8 shows calculated polymer viscosity, based on relaxation time measurements and appropriate selection of the rescaling parameter β. For constant viscosity and surface tension at a given temperature, the data should collapse onto a single master curve. Time–temperature superposition [68] can be used to predict the temperature-dependence of polymer viscosity. The temperature dependence of the polymer viscosity is given by

$$\log\left(\frac{\eta}{\eta_g}\right) = -\frac{-17.44\left(T - T_g\right)}{51.6\left(T - T_g\right)} \tag{2}$$

where η_g is the viscosity at the glass transition temperature, T_g is the glass transition temperature and T is the measurement temperature. Within experimental scatter, the calculated fit compares well with the data for the thickest films. However, for the thinnest polymer films there is more than one order of magnitude increase in the relaxation time of the grating, suggesting that the relaxation dynamics are substantially different in these very thin films.

In general, the macromolecular kinetics of polymers at interfaces and in confined geometry depend upon chemical interaction with the surface, macromolecular confinement, and details of the chemistry of the polymer such as tacticity. The lack of a strong chemical interaction at the PMMA-SU8 interface for the above indicate that the results of Figure 8 can be largely attributed to the impact of macromolecular confinement.

The slip effect at the polymer–substrate interface is not necessarily a function of the mechanical confinement, as it has been observed in bulk polymer flows [69]. However, molecular orientation and slip are linked through polymer–substrate interfacial chemistry. For a confined polymer film on a weakly adsorbing substrate such as SU8, modeling predicts a slip boundary condition [70], compared to a no-slip boundary condition for a confined polymer film on a strongly adsorbing substrate [71]. This modeling also predicts a linear relation between viscosity η and polymer thickness h for values of h comparable to and less than the radius of gyration R_g [70]. While eqn. 1 reports a substantially larger viscosity thickness-dependence of h^3, the modeling of refs. [70] and [71] are for sandwiched polymer layers, as opposed to the uncapped films of the present study.

The possibility of increased polymer viscosity of Figure 8 can be viewed in light of highly local measurement of polymer diffusion. Measurement of interdiffusion of PMMA in layers of thickness near R_g shows a two order of magnitude increase in polymer interdiffusion, comparing well with the results of this chapter [72]. In general, for polymer films of thickness comparable to R_g, the overall polymer chain mobility is reduced [73, 74] while the polymer chain mobility can be increased within R_g of the surface [74], thus motivating a two-fluid model of thin-film polymer flow [74, 73]. The two-fluid model assumes a given molecular mobility within R_g of the substrate, and a second molecular mobility farther away from the substrate. One recent experiment [75] measured the shear modulus of the polystyrene films and found that the shear modulus was increased by nearly a factor of 30% for films of thickness R_g compared to the bulk polymer. However these authors made one measurement at 0.3 R_g and found the shear modulus *decreased* by approximately 15% compared with the bulk polymer.

Scanning probe measurements of temperature-dependent mechanical properties provide contradictory reports with some reporting thickness and substrate dependence [76, 77], and others reporting a lack of thickness or substrate dependence [78, 79, 80]. These studies in general leave open questions as to tip–polymer interaction, and it is possible that these studies probed the polymer properties only at the polymer surface. For example, one study reported a T_g depression measured with a controlled-temperature AFM stage on *bulk* polymer samples [81].

3 Modeling and simulation of nanometer-scale thermo-mechanical data bit formation

Two recent developments in the operation of AFM cantilevers for thermomechanical data storage have significantly advanced its viability as a commercial technology. The first is the ability of the cantilever to thermally detect the presence of previously written data bits via conduction across the cantilever–substrate air gap [3, 82]. The second recent development in the operation of AFM cantilevers for thermomechanical data storage is the use of ultra-thin polymer films as the data substrate to produce extremely small data bits [3, 10]. While these data bits were extremely small at 10-50 nm in diameter, a detailed engineering understanding of these developments had not been made, thus prohibiting optimization.

Thermomechanical data storage technology yields a spatial resolution and maximum writing rate that are governed by heat transfer from the cantilever. Past work [20] investigated thermal conduction within the cantilever with the goal of minimizing the time required for temperature changes near the tip region. Chui [20] used finite element simulations to predict transient temperature fields along the cantilever legs during Joule-heating pulses of duration near one microsecond. Electrothermal simulations and measurements were also made to characterize cantilever operation. Ashegi [83] found that the decreased thermal conductivity of the thin, heavily doped silicon cantilever legs, compared with bulk, pure silicon, decreased the thermal time constant of the cantilever cooling. Research on high-resolution thermal microscopy motivated several studies of thermal and thermoelectric transport at the contact of a probe tip and a surface [84, 85, 86]. A recent study [87] reports detailed simulations of heat transfer within the AFM cantilever tip and polymer data layer during thermomechanical data bit formation.

Published work on thermomechanical data storage noted the temperature, time, and loading force required to form a data bit [3, 30], but until recently [10] did not examine the process of data bit writing in detail and instead studied system-level figures of merit. Binnig [3] and Lutwyche [29] wrote data bits with loading forces in the range of 100 nN to 1 μN and heating pulses near 20 μs. These studies did not have optical access to the cantilever and could not make a direct measurement of the cantilever deflection required to determine loading force. It was possible, however, to estimate loading force from the motion of the cantilever mounting package after tip–polymer contact. The temperature of the cantilever heater region can be found by measuring the electrical resistance of the cantilever, which is a function of temperature. The two studies above [3, 29] found the minimum temperature in the cantilever heater region required to induce thermomechanical bit writing in thin layers of PMMA was 350 °C, and that below this temperature bit writing was not possible. This threshold writing temperature of 350 °C is significantly higher than the glass transition temperature of PMMA, which is 105 °C in the bulk. The large temperature difference between the glass transition temperature of bulk PMMA and the cantilever heater could be in part due to an elevated glass transition temperature in the thin polymer layer [32, 33, 34]. However, any shift from the bulk value is likely to be no more than 10 to 15 °C for PMMA, even in the presence of a strong polymer layer-substrate surface interaction [88]. Other possible explanations are that there exists a temperature drop across the length of the cantilever tip, or that the local temperature gradient in the polymer near the cantilever tip is sufficiently steep that not enough polymer is heated to melting temperatures in order to allow tip motion. One goal of thermomechanical data storage system development is to decrease device power consumption in data bit writing [92], and it is therefore important to identify the lowest temperature at which data bits could be written.

One study [90] used a modeling and simulation approach to predict the modes of heat transfer during data bit writing and reading, and finds that significantly more heat travels across the cantilever-substrate air gap than travels through the length of the cantilever tip [83, 89, 90]. Thus, it is the thermal impedance of the cantilever–substrate air gap, not the thermal impedance at the cantilever tip–polymer interface, which governs thermal data reading. This metrology approach, while thermal in nature, is distinct from scanning thermal microscopy [27], where the thermopower measured in small thermocouple scans over a heated, structured surface to measure both temperature and feature size. The differences between scanning thermal microscopy and the thermal impedance imaging technique of this paper emerged after the first demonstration of the present thermal data reading approach. Modeling of heat transfer in the cantilever [91] and cantilever tip and measurement validated the designs [82]. This modeling approach has since been used to design cantilevers for improved thermal data reading performance [92].

An important feature of an accurate model of bit formation is the ability to predict the spread of heat in the polymer data layer, and to predict the corresponding polymer temperature-dependent mechanical properties. Softening of the heated polymer in the neighborhood of the tip is a function of the cantilever tip temperature, the tip–polymer thermal contact resistance, and the thermophysical properties of the polymer. The polymer glass transition temperature [32, 33], thermal conductivity anisotropy [35], surface tension [59], and viscosity [93] are all expected to have values for thin polymer layers different than the values for bulk polymer. The relationships between mechanical relaxation time, temperature, and mechanical properties are likely different as well for thin layers. Theoretical treatments exist for the prediction of polymer properties at small scales [93], but progress yet to be made in the fundamental understanding of the physics of polymers at small scales does not justify their application here, particularly for a complicated geometry with time and temperature dependence. Because of the lack of accurate mechanical property values, the present work employs property values for bulk PMMA, and modifies them with the best information available for size effects. The lack of accurate property values will induce inaccuracies in bit formation simulation. Therefore, the modeling and simulation approach follows an approach that preserves the qualitative link between model and experiment, in order to understand the relative impact of the many parameters on bit writing.

This section presents the analysis of heat and mass transfer during data bit formation through a steady-state analysis of temperature in the cantilever tip, a prediction of the onset of bit writing, and prediction of bit formation time and bit size.

3.1 Thermal analysis of the cantilever tip and polymer layer

The total heat generated in the highly resistive heater region of the cantilever is transported along the cantilever, into the environment, and across the cantilever–substrate gap. The total power conducted down the tip is very small compared to the power transferred along other transport paths. The power density in the cantilever tip is substantially higher than the power density along any other heat transfer path from the cantilever heater, and provides several hundred degrees of heating at the tip–substrate interface. Thus, it is the presence of the cantilever heater region that performs the thermal data reading, and the presence of the heated cantilever tip that performs the data bit indentation. The temperature at the tip–polymer interface is a key parameter in engineering the bit-writing process.

Phonon-boundary scattering in the silicon cantilever tip reduces heat transfer along the length of the cantilever tip below the value predicted using bulk properties. Several approaches exist for estimating temperatures with constrictions of size comparable to or smaller than the phonon mean free path. Previous work on phonon-boundary scattering in single-crystal silicon layers solved the Boltzmann transport equation with a frequency-dependent phonon mean free path and specularly reflecting film boundaries [94], and found good agreement with experimental data. One study of heat transport in silicon films of thickness near 100 nm [95] found that thermal conductivity data fit very well to a model that assumes only longitudinal acoustic phonons (LAP) carry heat. This study predicted an average mean free path for the LAP was approximately 300 nm. A study of heat transport in a more complicated geometry [96] separated phonons into heat-carrying LAP and reservoir transverse acoustic and optical phonon modes. This study found substantial reduction in heat transport near a heated source in a crystalline film, both having critical dimensions small compared to the LAP mean free path.

The most rigorous study of heat conduction in the cantilever tip would provide a detailed account for phonon dispersion, and scattering on the tip walls and other phonons. Several unknown parameters reduce the usefulness of the most rigorous possible approach. As silicon is self-passivating, a thin layer of silicon dioxide certainly covers the tip. We speculate that the thickness of this oxide layer depends upon the overall time during which the tip is heated. There may also be contamination on the tip due to long scanning times on an organic substrate. The degree of thermal conductivity anisotropy in the polymer layer is not known, and the surface tension and viscoelastic properties at length scales comparable to the polymer radius of gyration have not been measured, nor do good physical models exist for their prediction. And finally, while there is good tip shape uniformity over an entire cantilever array [29], reports of tip shape between arrays or after long scanning times have not been published, and it is possible that tip wear could be important in thermomechanical data storage. Thus, the modeling of thermomechanical data bit formation employs a more approximate model of phonon transport in the cantilever tip that leverages the physical insights derived from measurements made on thin silicon layers [95] while accepting a larger overall modeling error closer to experimental uncertainties.

Predictions of the temperature distribution along the cantilever tip are made with an analysis that divides the tip into many finite volume elements. Models calculate thermal conduction between adjacent volume elements in the tip, conduction to the nearby air, and radiation to the nearby surfaces. When the diameter of the cantilever tip is locally comparable to or smaller than the mean free path of the heat carriers in the tip, thermal conductivity is reduced as phonons scatter against the tip walls. The silicon thermal conductivity is therefore calculated using Matthiessen's rule [97]

$$k = \frac{1}{3} Cv \left(\frac{1}{\Lambda^{-1} + d^{-1}} \right) \tag{3}$$

where d is the local tip diameter, k is the local thermal conductivity of the tip, C is volumetric heat capacity, v is the average phonon speed, and Λ is the phonon mean free path in the bulk material [97]. A value of 1.8×10^9 W/m^{2}K is used for the product Cv [98]. The model assumes constant Cv over the range of temperatures considered, which will produce an error of less than 15% [98]. Eqn. 3 is a modified exact solution for ballistic transport in radially symmetric media [99]. The temperature dependence of the bulk thermal conductivity of silicon is modeled using a fit to room temperature data for bulk samples [100].

Heat is transferred away from the side of the tip through radiative exchange with the heater and the substrate and conduction to the surrounding air. This steady-state model assumes that conduction across the cantilever-polymer air gap produces a constant, linear temperature gradient across the gap. Typical heating times are 1–100 μs [3, 10, 29, 82] which are much longer than the time of 100 ns for heat to diffuse across a typical cantilever–substrate air gap of thickness 0.3–1 μm. Thus the air between the cantilever and substrate is at thermal steady-state. Even at the smallest gap thickness of 300 nm, the gap is large compared to the mean free path of air molecules at room temperature, which is approximately 60 nm [101]. Thus the temperature gradient across the cantilever–substrate air gap is nearly constant. The resistance to thermal conduction between the tip and the air depends only on the Knudsen slip resistance, which accounts for the diameter of the tip being smaller than the mean free path of the air molecules [102].

For typical loading forces, the tip–substrate contact diameter is near 10 nm. For a perfectly crystalline silicon–silicon point contact, phonon transport across the tip–substrate would be fully ballistic. However, the silicon tip is in contact with a highly disordered polymer medium. Thus the model enforces a scattering site at the end of the tip. The total thermal impedance between the tip and the polymer is given as this phonon scattering site at the end of the tip and a spreading resistance into the polymer. The total thermal impedance, $R_{contact}$, into the polymer is given as

$$R_{contact} = \frac{12}{C v \pi d_{contact}^2} + R_{spread} \tag{4}$$

Where $d_{contact}$ is the diameter of the tip–polymer contact area, $k_{polymer}$ is the thermal conductivity of the polymer layer, and R_{spread} is the thermal resistance due to the spread of heat in the polymer layer. The first term in eqn. 4 enforces a phonon scattering site at the tip–polymer interface [103]. The second term is extracted from solution of the heat conduction equation in the polymer layer.

It is clear from eqn. 4 that the diameter of tip–polymer contact strongly affects the tip–polymer thermal impedance, and will therefore impact the temperature at the tip–polymer interface. Typically, contact between an AFM tip and a surface is approximated by a Hertzian model, sometimes including terms that account for adhesion forces or viscoelasticity [104]. This typical approach is most appropriate for tip load forces comparable to or less than ~ 5 nN, where adhesion is important; and for hard materials where there is a nearly flat tip–substrate interface [105]. While the PMMA polymer data layer is relatively hard with a Young's modulus of 3.0×10^9 Pa [106], it is much softer than the silicon tip, and complies with the shape of the end of the silicon cantilever tip, thus violating the criteria for a simple Hertzian contact. There exists an exact solution [107] for the indentation of a half sphere into a compliant surface, which relates the loading force, F_{load}, to the contact diameter, $d_{contact}$. The relation is given as

$$F_{load} = \frac{G}{1-\chi}\left\{\left(r_{contact}^{\ 2} + r_{tip}^{\ 2}\right)\cdot \log\left(\frac{r_{tip} + r_{contact}}{r_{tip} - r_{contact}}\right) - r_{tip} r_{contact}\right\} \tag{5}$$

where G is the shear modulus, $r_{contact}$ is the radius of contact and half the value of $d_{contact}$, and r_{tip} is the tip radius of curvature. Poisson's ratio, χ, is 0.5 for an incompressible material.

Typical tip loading forces are in the range of 5 nN – 300 nN [10] and the typical tip radius of curvature is near 20 nm [21]. However, the thickness of the cantilevers will vary with variations in the fabrication process and will also vary across an array or cantilever array, thus inducing a variation in the loading force. The cantilever spring constant varies as a function of the cube of the cantilever thickness [108]. Furthermore, while there is anecdotal evidence that tip wear is minimal for thermomechanical data writing [109], it is possible for tips to wear as they are in contact with and scan over a substrate for a long time [110]. Therefore, the cantilever tip radius of curvature varies from 10 nm to 50 nm in the present work to cover the range of possible practical values.

A range of tip loading forces has been reported to successfully write data bits. While some experimental bit writing data [10] has been reported for tip loading forces near 5 nN, the present contact model does not accurately predict tip-contact diameter below several times the lattice constant of silicon, or in the range of 1 nm, corresponding to loading forces near 30 nN. The maximum tip loading force will be approximately governed by the spring constant of the cantilever, which can be as stiff as 1.0 N/m [3], or as low as 0.02 N/m [10]. Experimentally, the tip loading force can be calculated as the product of the cantilever spring constant and the displacement of the base of the cantilever after the tip has first made contact with the polymer. As sharper tips have a larger dependence of contact diameter on loading force than rounder tips, so too will there be a stronger variation of tip–polymer interface temperature.

The tip is discretized into 1000 volume elements and the temperature distribution is found using Gauss–Seidel iteration. Temperature predictions are validated by an energy balance on each volume element and on the entire tip, and compare well with solutions made with twice the number of volume elements. The calculated value of R_{spread} compares well with a shape factor for a disk source on an infinite half space [111]. The transient analysis uses time steps equal to 1% of the total simulation time. The numerical solution of heat transfer in the polymer layer yields a solution that agrees to within 2% of the analytical solution for the transient temperature field induced by constant temperature boundary conditions in both the radial and normal directions.

3.2 Bit writing analysis

The conditions required to initiate bit formation include the time, temperature, and force with which the tip presses into the polymer data layer. Transport phenomena in both the cantilever and in the polymer govern the thermomechanical data bit writing process. Heat is generated in the cantilever, and cantilever motion is required for bit formation. The electrical and mechanical design of the cantilever impacts both the thermal and mechanical time constants of bit formation. The tip couples the cantilever beam to the polymer layer both mechanically and thermally. As mentioned above, for times longer than 20 ns, the tip is in thermal equilibrium with the cantilever heater region and acts as a simple thermal resistance. Analogously, the cantilever tip is rigid and therefore has no impact on the mechanical time constants. The polymer layer has a thermal time constant governed by the polymer layer thickness, and a mechanical time constant governed by the temperature field in the polymer near the tip. For a bit to be fully formed, the cantilever must be hot long enough for heat to spread from the tip into the polymer, the cantilever must have sufficient time to move into the bit, and the temperature of the tip must be sufficiently high for the polymer to be soft enough to move away from the tip in this time.

By coupling the mechanical and thermal analysis of bit formation, it is possible to predict the onset of data bit formation and the ultimate data bit size for a given heater temperature, heating time, and loading force. For cantilever heating times that are comparable to the cantilever mechanical time constant or the polymer mechanical relaxation time, equations of motion describe the forces acting on data bit formation, allowing an estimate for the threshold temperature for bit formation. For heating times that are very large compared to the cantilever heating time and the cantilever mechanical time constant, a quasi-steady model based on the contact model above allows estimate of the same.

In the measurements of Gotsmann [112] and Cross [113], the cantilever has a tip of height 300 nm, tip radius of curvature of 20 nm, spring constant of 0.055 N/m, and a mechanical resonance frequency of 100 kHz, or a mechanical time constant of 10 μs. The polymer data substrate is a bilayer consisting of 35 nm of PMMA on 80 nm of SU8 epoxy. The polymer bilayer resides on a silicon substrate. The purpose of the SU8 layer is to stop the tip from penetrating to the silicon, and for thermal insulation [3]. In the experimental analysis of data bit writing [10], a cold cantilever tip is brought into contact with a cold polymer surface. After tip–polymer contact, the base of the cantilever is brought closer to the substrate such that the cantilever tip is pressed into the substrate. The cantilever tip loading force, F_{load}, is the product of this displacement and the cantilever spring constant, k. The cantilever heats at a fixed voltage for a fixed time. Several measurements are made for varying heating voltages and heating times. The cantilever is then calibrated for the cantilever heating temperature produced for each heating time and voltage, as in [91, 92].

The temperature-dependent mechanical shear modulus of the polymer is calculated as a function of temperature, pressure, and time. The temperature dependence takes the form of the WLF shift parameter [114], which is used to extract polymer properties from tabulated bulk values [61]. The Reynolds number of tip motion through the polymer is always less than 0.01, and therefore inertial forces inside the polymer can be neglected.

For times comparable to and longer than 10 ms, the cantilever is in thermal and mechanical equilibrium with the polymer data layer. A simple calculation of the mechanical restoring force of the polymer as a function of temperature and time shows that for the longest heating pulses, a temperature of ~ 120 °C in the polymer layer is required to induce data bit formation. Figure 9 shows predictions for the tip–polymer interface temperature as a function of loading force for various heater temperatures. This model predicts that the lowest heater temperature at which a bit could be written is near 200 °C for a loading force above approximately 75 nN.

It is possible to predict the motion of the cantilever tip through the equation of motion of the cantilever, which is

$$m_{cant}\ddot{x} + kx = F_{load} - F_{pl} \tag{6}$$

where m_{cant} is the mass of the cantilever, x is the vertical position of the cantilever tip and F_{pl} is the force with which the polymer resists tip motion. The force with which the polymer resists tip motion is a function of the temperature field in the polymer near the tip.

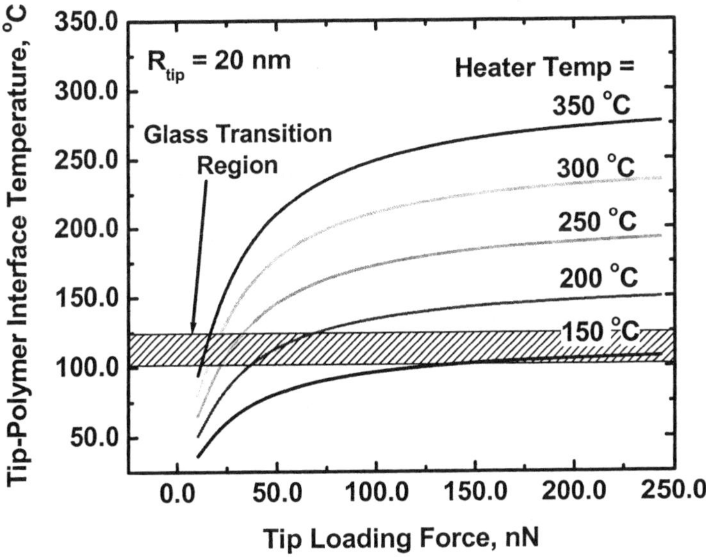

Figure 9: Predicted steady-state tip–polymer interface temperature as a function of loading force for a range of heater temperatures and a tip radius of curvature of 20 nm [87]. The shaded region represents the glass transition region, above which a bit will be written for long heating pulses.

Simulations of heat transfer from the cantilever tip for loading forces less than 500 nN and cantilever heater temperatures less than 700 °C show that the thickness of the softened polymer near the cantilever tip is always less than the cantilever tip radius of curvature. The depth of the softened polymer is very thin because of the superposition of two steep gradients: the temperature gradient near the cantilever tip is very large since the polymer thermal conductivity is very low at 0.2 W/mK, and the temperature-dependence of the shear modulus is logarithmic. Thus, the very thin region of melted polymer near the cantilever tip allows for modeling of the tip penetration into the polymer as a lubrication problem.

An analytical solution to the force required to displace the lubrication layer links the tip motion to polymer relaxation. At every time step the simulation tests whether the tip loading force is sufficient to move the tip one finite volume element thickness into the polymer layer, according to eqn. 6. The force with which the polymer resists tip penetration F_{pl} can be divided into the viscous resistance of the polymer, $F_{pl,viscous}$ [115] and the elastic resistance of the polymer, $F_{pl,elastic}$ [116]

$$F_{pl} = F_{pl,viscous} + F_{pl,elastic} = 3\pi \frac{\eta_{pl} r_{tip}^{4}}{t_{pl}^{3}} \dot{x} + 3\pi G' r_{tip}^{2} \varepsilon \quad (7)$$

where η_{pl} is the polymer viscosity, t_{pl} is the average thickness of the softened polymer near the tip, r_{tip} is the effective radius of the tip–polymer contact, or half of $d_{contact}$, x is the vertical position of the tip, G' is the elastic part of the shear modulus and ε is the strain in the melted polymer. If the criteria for tip motion are met, then the tip moves one element into the polymer. The new position of the tip serves as a position of the constant-temperature boundary in the solution of the heat equation. The polymer from the elements replaced by the tip are thrown away as they are assumed to have flowed away from the tip. The simulation stops when the tip has penetrated to a depth of 10 nm, which corresponds to the threshold for experimental bit writing.

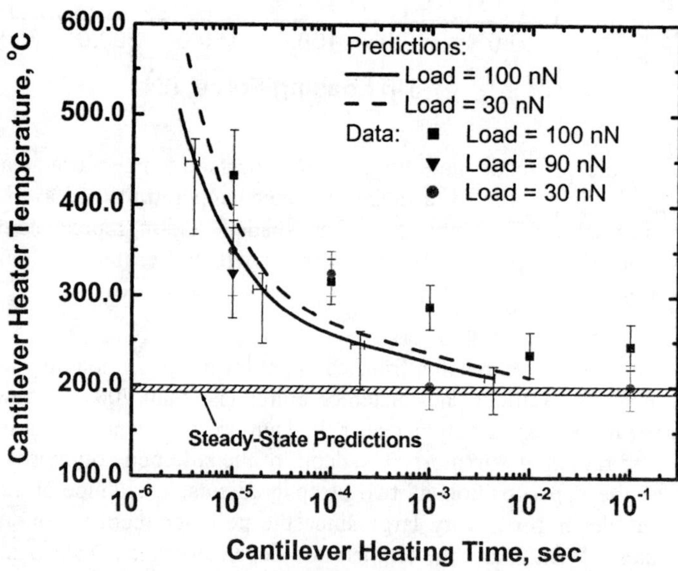

Figure 10: Measurement [112, 113] and prediction [87] of the threshold conditions of time, temperature, and tip loading force required to produce a data bit for a 20 nm tip radius of curvature. The gray area shows predictions for equilibrium conditions, bounded by loading forces of 30 nN and 100 nN.

Figure 10 shows measurement and prediction results. The measurement results are for loading forces of 30 nN, 90 nN, and 100 nN, and the prediction results for 30 nN and 100 nN. The measurements are from Gotsmann [117] and Cross [118]. Both measurements and predictions are for a cantilever tip radius of curvature of 20 nm. Predictions from the modeling of the present study are shown for both a moving tip and for steady-state thermal conditions. To within experimental error and measurement scatter, the modeling and simulation of the present work compares well with measurements.

The drop in threshold writing temperature for longer writing times is due to three factors: the inertia of the cantilever which must be overcome for tip motion, heat diffusion into the polymer, and the viscoelastic response of the polymer to tip penetration. First, the cantilever mechanical time constant of 100 kHz prohibits a data bit from being written much faster than approximately 1 μs, so due only to the cantilever mechanical time constant, we expect an asymptotic increase in threshold writing temperature as heating times decrease to near 1 μs. The steep increase in temperature required to write a data bit for short times also corresponds to the time required for heat to diffuse into the polymer layer away from the tip, and the time- and temperature-dependence of the polymer properties away from the tip. As shown in eqn. 7, the force with which the polymer resists tip penetration has a cubic dependence on the thickness of the soft polymer near the penetrating tip. The long time behavior of tip penetration, which asymptotically approaches a threshold bit-writing temperature of approximately 200 °C, is a function of only the polymer temperature-dependent shear modulus.

The two loading forces of 30 nN and 100 nN for which predictions are made in Figure 10 bound the reasonable practical loading window of the present thermomechanical data storage cantilever [3, 10, 29, 117, 118]. Measurements are shown for these two loading forces, plus a single measurement at a loading force of 90 nN. The 30 nN and 90 nN measurements were made as a group [118] and the 100 nN measurements were made at a separate time [117]. The large experimental scatter can be attributed to the difficulty of calibrating the cantilever heating temperature and loading force, as well as the difficulty of accurately measuring features of size near 10 nm. In practice, the data storage is extremely robust and the tip usually penetrates through the polymer layer. The predictions show a 20–50 °C difference in threshold writing temperature between the 30 nN and 100 nN loading forces at a given heating time. This is consistent with the difference between the measurements made at 30 nN and 90 nN, which were made at the same time with the same cantilever. However, the scatter between the two sets of data cannot further support this prediction. The predictions of Figure 10 also show that for a constant heating temperature, the difference in loading force between 100 nN and 30 nN will yield a 50–100 % difference in heating time required to produce a data bit.

Figures 10 and 11 provide insight into the previously published experimental results that a threshold temperature of 350 °C writes a data bit [3, 29]. One possible explanation is that, as shown in Figure 9, 350 °C is the temperature at which a bit is *always* written, regardless of loading force. Another important parameter is that these reports always wrote bits at 10–20 μs, for which the predictions of Figure 9 show that the threshold writing temperature is 300–350 °C.

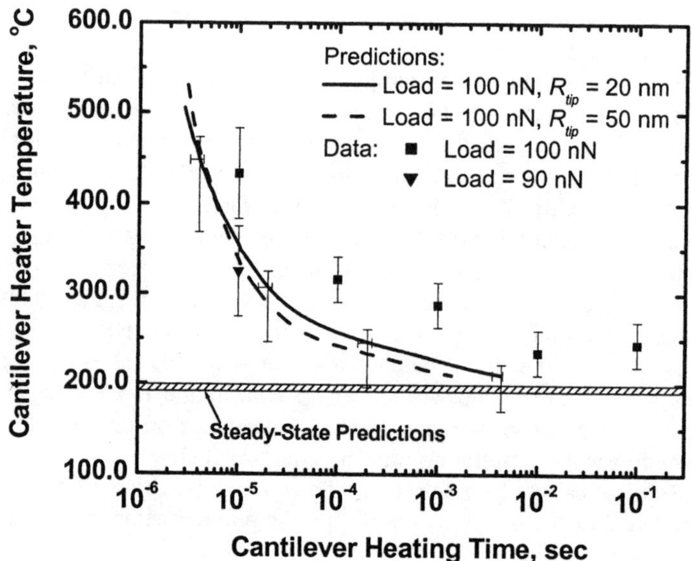

Figure 11: Measurement [112, 113] and prediction [87] of the threshold conditions of time, temperature, and tip loading force required to produce a data bit. The data is for a measured 20 nm tip radius of curvature. The predictions are for two values of tip radius of curvature, at 20 nm and 50 nm. The gray area shows predictions for equilibrium conditions, bounded by loading forces of 30 nN and 100 nN.

The shape of the cantilever tip can impact the onset writing conditions. Previous work has investigated the wear of cantilever tips during AFM data storage [18, 110, 119]. These studies have focused on AFM reading of features much larger than the present thermomechanical data storage bits at 100 nm or more. It was possible for these authors to fabricate AFM cantilever tips with a radius of curvature of 4 nm [18]. Long-term experiments with the tip scanning large distances over polymer media, in one case as high as 16 km of tip travel [110], found that the tip wears to a radius of curvature of 100 nm [110, 119]. The limits of data density in a thermomechanical data storage system depend upon the sharpness of the cantilever tip, and thus it is desirable to reduce the wear of the tip as much as possible. The tip shape affects not only the size of the ultimate bit, but also the time and temperature required for bit formation. Figure 11 displays the 100 nN and 90 nN measurements shown in Figure 10, as well as predictions for bit writing with cantilevers with tip curvature radii of 20 nm and 50 nm. The measurements were made with cantilever tips of 20 nm radius. The predictions show that a less sharp tip will have a higher threshold writing temperature at short times, but at long times the less sharp tip can penetrate the polymer with a lower heating temperature. These predictions are reasonable consequences of the model of heat transfer between the cantilever tip and the polymer substrate. The less sharp tip reduces the thermal impedance between the cantilever tip and the substrate, reducing the tip–polymer contact temperature for a given cantilever heater temperature. Thus, a higher heater temperature is required in order to form a data bit. For long heating times, the contact area between the tip and the sample increases for the less sharp tip over the contact area of the sharp tip, increasing the spread of heat into the polymer substrate. Because more polymer is heated underneath the less sharp cantilever tip at long times, the temperature required to write a data bit is reduced.

Errors in the analysis of bit writing originate from three sources: differences between tabulated properties for bulk PMMA and the actual properties of PMMA in an ultra-thin film, errors in modeling the value of F_{pl}, and numerical errors. The largest of these are modeling errors for F_{pl} and the errors originating from unknown property values. The sum of the property errors, numerical errors, and modeling errors induce a total error of +7%/–22% in time and +6%/–23% in temperature [87]. Figures 10 and 11 display the overall error of the present modeling approach through error bars on the smooth prediction curves in several places.

The modeling and simulation approach and results could be used to improve writing rate and data density of a thermomechanical data storage device. The two most easily modified cantilever design parameters are the tip shape and the cantilever resonance frequency/loading force. One possible improvement in the tip shape would be to flatten the tip–polymer interface, or even to create cylindrical rather than conical tips, in order to reduce the tip–polymer interface temperature. One possible improvement in the cantilever design would be to increase the cantilever resonance frequency while preserving the cantilever spring constant, which would allow high loading forces and improved cantilever mechanical response time.

4 Thermal data reading and topography mapping

A major consequence of the model of data bit writing described in Section 3 is its aid in understanding the mechanism of thermal data bit writing versus thermal data reading. The first report of thermal data reading [3] noted the change in thermal impedance between the cantilever and the polymer data substrate as the mechanism that allowed measurement of the nanostructured polymer surface. However, it was unclear whether the change in thermal impedance was due to thermal conduction through the cantilever tip or due to conduction across the air gap. It was further unclear as to whether the softening of the polymer required for data bit formation was due to heating from the cantilever tip alone or heat delivered from the cantilever heater across the air gap. A comparison of the modified continuum model of conduction through the cantilever tip represented by eqns. 3–5 to a one-dimensional model of heat transfer across the air gap shows that the temperature rise at the tip–polymer interface is much higher than the temperature rise in the polymer layer due to conduction and radiation across the cantilever–polymer gap. Predicted typical values for the temperature of the tip–polymer interface are several hundred degrees Celsius, and the predicted temperature rise in the polymer due to heating from the cantilever alone is less than 20 °C. Therefore, it is the heat conduction through the cantilever tip that induces polymer melting and the resulting data bit formation. However, comparison of the quantity of heat that is transferred through the cantilever tip with the quantity of heat that is transferred across the air gap shows that a factor of approximately 10^3 more heat goes across the air gap than through the cantilever tip. It is therefore the change in thermal impedance between the cantilever and substrate that allows thermal data reading. The realization that separate heat transfer mechanisms facilitate data bit writing and thermal data reading allowed simultaneous improvement of cantilevers for thermal data reading and writing [82] and also for cantilever array operation [92].

The thermal detection of data bits could be expanded as a general topography mapping technique. Typically, AFM cantilevers designed for topographic sensing employ optical detection of cantilever deflection [1], or a piezoresistive cantilever that acts as a strain gauge [2]. The advantage of the piezoresistive cantilever is that optical access is not required, thus allowing the operation of cantilever arrays. This section [120] compares the theoretical sensitivity and resolution limits of piezoresistive and thermal cantilevers for topographic mapping.

The thermal conductance from the cantilever governs thermal reading sensitivity. Through operation in a Wheatstone bridge where each leg of the bridge has an electrical resistance equal to the electrical resistance of the cantilever, the voltage displacement sensitivity of the thermal cantilever is [120]

$$\frac{\partial V}{\partial z} = -\alpha \frac{{V_b}^3}{8C(z)^2 R}(\partial C(z)/\partial z) \tag{8}$$

where V is the voltage read from the bridge, z is the distance between the cantilever and the substrate, α is the temperature coefficient of electrical resistivity, V_b is the bridge bias voltage, C is the thermal conductance from the cantilever, and R is the temperature-dependent cantilever electrical resistance.

The ability to detect small displacements is limited by several factors, including amplifier noise, Johnson noise, thermomechanical noise, and 1/f noise. It is possible to provide analytical expressions for all of these noise sources, and carry out a quantitative analysis of their relative importance. While there are several proposed mechanisms behind 1/f noise, it has been shown that a model for this noise offered by Hooge provides an excellent empirical fit to data [2], and that it competes with Johnson and amplifier noise only for very lightly doped piezoresistors in very thin cantilevers, or at low frequencies. For the purposes of this analysis, it is neglected. Amplifier noise is also neglected, although it will be an important practical consideration for both the piezoresistive and thermal cantilevers. The total thermal and Johnson noise in the cantilevers will have a magnitude

$$V_{noise} = \sqrt{(\alpha V_b T)^2 \frac{k_B}{C(z)} + 4k_B TR} \tag{9}$$

The first part of eqn. 9 refers to the thermal noise, and the second part accounts for Johnson noise. The minimum detectable feature size is then

$$z_{\min} = \frac{Noise}{Sensitivity} = \frac{V_{noise}}{\frac{\partial V}{\partial z}} \tag{10}$$

The thermal conductivity of the thin, doped silicon cantilever is less than the thermal conductivity of pure, bulk silicon due to phonon-impurity and phonon-boundary scattering. While the thermal conductivity of doped, thin silicon layers is not known for a wide range of temperature, thickness and doping concentration, a nominal value of k_{cant} = 50 W/mK can be taken from the available literature [121] for a cantilever doped to its saturation point near room temperature and of thickness 300 nm. The thermal conductivity of the air is k_{air} = 0.03 W/mK at room temperature. At z = 300 nm, $k_{air} = k_{air,bulk}$. However, the mean free path of air molecules at standard temperature and pressure is approximately Λ_{air} = 60 nm [101] and thus for values of z comparable to and less than Λ_{air} the thermal conductivity will be reduced from its bulk value.

In practice, the design space for a thermal cantilever is large and extends considerably beyond the sizes and shapes of thermal cantilevers suggested previously. For convenience, the dimensions for both the thermal cantilever and the piezoresistive cantilever are selected such that the cantilever length, width, and thickness are the same as the Millipede cantilever used in thermomechanical data storage.

Figure 12 shows the predicted sensitivity for the thermal cantilever over a range of operating power, which corresponds to a range of operating voltage and temperate. The sensitivity increases with decreasing cantilever thickness. The sensitivity shown is the absolute value: for temperatures near room temperature, the temperature coefficient of resistivity is negative. The sharp decrease and subsequent rise of sensitivity near 1 W is the point where the temperature coefficient of resistance changes numerical sign. The measurements in Figure 12 were not explicitly made for differential sensitivity, but rather a step change in cantilever height [92]. However, the trend and order of magnitude in the model compares very well with the data, thus building confidence in the analysis.

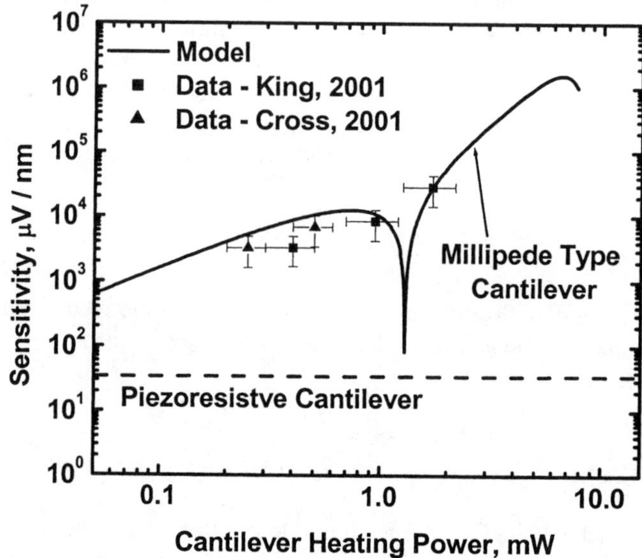

Figure 12: Voltage displacement sensitivity for the Millipede cantilever in thermal operation and for an identically sized piezoresistive cantilever [120]. Data is from refs. [10, 82]. The piezoresistive cantilever operates at a constant bias voltage.

Figure 12 also shows the predicted operation of the piezoresistive cantilever at a constant bridge voltage of 2 V. The piezoresistive coefficient of doped silicon decreases with increasing temperature, and it is therefore desirable to operate the piezoresistive cantilever as close to room temperature as possible. The models for the behavior of piezoresistive cantilevers are well established and while no data is provided for the piezoresistive cantilever sensitivity, matching measurements and models has been demonstrated elsewhere [2]. Overall, the thermal cantilever is between two and four orders of magnitude more sensitive than the piezoresistive cantilever.

Figure 13 shows predictions for the minimum detectable resolution of the thermal cantilever and piezoresistive cantilever. The model for the behavior of the piezoresistive sensor is taken from established measurements and models [2]. While sensitivity is important, it is the overall cantilever resolution that is most important for measuring small features. The lowest possible resolution is desired, which corresponds to thinner cantilevers and higher heating powers. While more experimental work remains to more fully understand the thermal topography sensing, the remarkable advantage of thermal sensing over piezoresistive sensing offers great potential for metrology with heater-cantilevers.

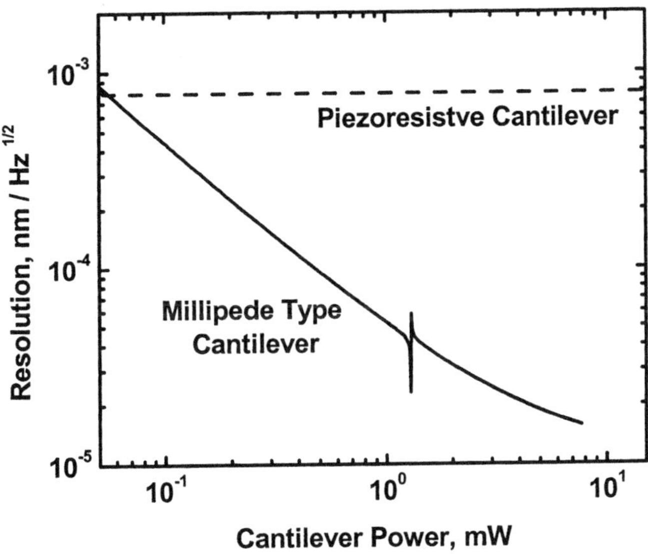

Figure 13: Predicted minimum detectable feature size for the Millipede cantilever in thermal operation and for an identically sized piezoresistive cantilever [120]. The piezoresistive cantilever operates at a constant bias voltage.

5 Summary and conclusions

This chapter reviews fundamental and applied work that aids understating the basic physical processes in the thermomechanical formation of polymer nanostructures, and improves the designs of devices for their formation and detection. Significant open scientific questions address the physics of soft materials at length scales comparable to molecular size, described in a review of fundamentals of subcontinuum mass transport in soft materials. A fundamental investigation of mechanical relaxation in ultra-thin nanostructured polymer films shows the thickness-dependence of polymer mechanical properties at dimensions comparable to the polymer radius of gyration. The nanostructures show remarkably fast local relaxation compared to the relaxation of the whole polymer. Analysis of thermomechanical data bit formation shows the temporal and spatial limits of thermomechanical features formed in polymer, and the dependence of the processing on cantilever tip geometry. The large temperature drop across the cantilever tip accounts for the high temperatures required in the cantilever heater region in order to induce onset bit writing. Predictions show that future tips could be optimized such that the tip cone is very sharp but the tip–polymer interface would be broadened. Thermal analysis of the cantilever and cantilever tip improves understanding of the heat transfer paths in thermomechanical writing and reading. Cantilevers are optimized simultaneously for thermomechanical data bit writing and thermal data reading, allowing design improvements for cantilevers, cantilever arrays, and data storage systems. Finally, an analysis of the thermal cantilever for topographic sensing shows that the thermal cantilever has one to four orders of magnitude more sensitivity and one to two orders of magnitude higher resolution than piezoresistive sensors of comparable size and shape.

Technical challenges remain in the development of a thermomechanical data storage system. While the basic mechanisms of thermomechanical data bit write, read, and erase are understood, a detailed relationship between tip shape, bit shape, and operating parameters remains to be established, either theoretically or experimentally. The subcontinuum models of heat and mass transport in the cantilever tip and in the polymer layer should be improved with detailed numerical studies supported by measurements. Mechanical wear of the tip and of written bits has not been investigated in detail, nor has possible chemical wear in the polymer. The overall accuracy of the manufacturing process will also affect the window of cantilever operation.

The thermomechanical formation of polymer nanostructures offers an inexpensive technique for the manufacture of mechanical structures with dimensions near 1 nm. The thermomechanical manufacturing platform consists of a heated AFM cantilever, which can form polymer nanostructures as well as thermally map the modified polymer surface. Additionally, micromachined imprinting masters mold large areas with nanometer-sized features [122]. This manufacturing technique allows the formation of small mechanical features for MEMS and micromechanical devices. Engineering a polymer nanostructure will not only address the thermomechanical process by which it is formed, but also the chemical nature of the polymer and the imprinting device, to produce mechanical structures which interact with biological and chemical systems. The growth of the organic electronics market offers opportunities for direct mechanical formation of nanometer-scaled quantities of bandgap-engineered polymers for organic light emitting diodes and other organic electronics [123]. MEMS structures could be produced where the polymer mechanical elements are also optically [123] or biochemically [124] active.

Acknowledgements

The authors would like to thank the Micro/NanoMechanics Group at the IBM Zurich Research Laboratory, for the collaboration and technical exchanges that have made much of this work possible. In particular, we thank Urs Duerig, Gerd Binnig and Peter Vettiger. WPK was supported in his graduate studies by IBM Zurich Research and the IBM Graduate Research Fellowship Program. WPK thanks Brent Nelson for helpful comments on the manuscript.

References

[1] Binnig, G., Quate, C.F. & Gerber, C., Atomic Force Microscope. *Physical Review Letters*, **56**, pp. 930–933, 1986.

[2] Harley, J.A. & Kenny, T.W., High-Sensitivity Piezoresistive Cantilevers Under 1000 Angstrom Thick. *Applied Physics Letters*, **75**, pp. 289–291, 1999.

[3] Binnig, G., Despont, M., Drechsler, U., Häberle, W., Lutwyche, M., Vettiger, P., Mamin, H.J., Chui, B.W. & Kenny, T.W., Ultrahigh-Density Atomic Force Microscopy Data Storage with Erase Capability. *Applied Physics Letters*, **76**, pp. 1329–1331, 1999.

[4] Duering, U., Pohl, D. & Rohner, F., Near-Field Scanning Optical Microscopy. *Journal of Applied Physics*, **59**, pp. 3318–3327, 1986.

[5] Fritz, J., Baller, M.K., Lang, H.P., Rothuizen, H., Vettiger, P., Meyer, E., Guentherodt, H.-J., Gerber, C. & Gimzewshi, J.K., Translating Biomolecular Recognition into Nanomechanics. *Science*, **288**, pp. 316–318, 2000.

[6] Wilder, K., Quate, C., Adderton, D., Bernstein, R. & Elings, V., Noncontact Nanolithography using the Atomic Force Microscope. *Applied Physics Letters*, **73**, pp. 2527–2529, 1998.

[7] Cooper, E.B., Manalis, S.R., Fang, H., Dai, H., Matsumoto, K., Minne, S.C., Hunt, T. & Quate, C.F., Terabit-per-square-inch Data Storage with the Atomic Force Microscope. *Applied Physics Letters*, **75**, pp. 3566–3568, 1999.

[8] Piner, R.D., Zhu, J., Xu, F., Hong, S. & Mirkin, C.A., 'Dip-Pen' Nanolithography. *Science*, **283**, pp. 661–663, 1999.

[9] Heyde, M., Rademan, K., Cappella, B., Greuss, M., Sturm, H., Spangenberg, T. & Niehus, H., Dynamic Plowing Nanolithography on Polymethylmethacrylate using an Atomic Force Microscope. *Review of Scientific Instruments*, **72**, pp. 136–141, 2001.

[10] Cross, G.L.W., Despont, M., Drechsler, U., Dürig, U., Vettiger, P., King, W.P. & Goodson, K.E., Thermomechanical Formation and Thermal Sensing of Nanometer-Scale Indentations in PMMA Thin Films for Parallel and Dense AFM Data Storage. In: *Fundamentals of Nanoindentation and Nanotribology II, Materials Research Society Symposium Proceedings*, **649**, 2001.

[11] Avoiding a Data Crunch: Special Industry Report on Data Storage. In: *Scientific American*, 2000.

[12] Grochowski, E. & Hoyt, R.F., Future Trends in Hard Disk Drives. *IEEE Transactions Magnetics*, **32**, pp. 1850–1854, 1996.

[13] Lohau, J., Moser, A., Rettner, C.T., Best, M.E. & Terris, B.D., Writing and Reading Perpendicular Magnetic Recording Media Patterned by a Focused Ion Beam. *Applied Physics Letters*, **78**, pp. 990–992, 2001.

[14] Ruigrok, J.J.M., Coehoorn, R., Cumpson, S.R. & Kesteren, H.W., Disk Recording Beyond 100 Gb/in.2: Hybrid Recording? *Journal of Applied Physics*, **87**, pp. 5389–5403, 2000.

[15] Mamin, H.J., Ried, R.P., Terris, B.D. & Rugar, D., High-Density Data Storage Based on the Atomic Force Microscope. *Proceedings of the IEEE*, **87**, pp. 1014–1027, 1999.

[16] Park Scientific 'Piezolever' data sheet.

[17] IBM Travelstar Disk Drive series data sheet.

[18] Ried, R.P., Mamin, H.J., Terris, B.D., Fan, L.S. & Rugar, D., 6-MHz 2-N/m Piezoresistive Atomic Force Microscope Cantilevers with INCISIVE tips. *Journal of Microelectromechanical Systems*, **6**, pp. 294-302, 1997.

[19] Chow, E.M., Soh, H.T., Lee, H.C., Adams, J.D., Minne, S.C., Yaralioglu, G., Atalar, A., Quate, C.F. & Kenny, T.W., Integration of Through-Wafer Interconnects with a Two-Dimensional Cantilever Array. *Sensors and Actuators*, **83**, pp. 118–123, 2000.

[20] Chui, B.W., Stowe, T.D., Ju, Y.S., Goodson, K.E., Kenny, T.W., Mamin, H.J., Terris, B.D. & Ried, R.P., Low-Stiffness Silicon Cantilever with Integrated Heaters and Piezoresistive Sensors for High-Density Data Storage. *Journal of Microelectromechanical Systems*, **7**, pp. 69–78, 1998.

[21] Vettiger, P., Despont, M., Drechsler, U., Dürig, U., Häberle, W., Lutwyche, M., Rothuizen, H.E., Stutz, R., Widmer, R. & Binnig, G., The "Millipede" – More than One Thousand Tips for Future AFM data storage. *IBM Journal of Research and Development*, **44**, pp. 323–340, 2000.

[22] King, W.P. & Goodson, K.E., Thermal Writing and Thermal Nanoimaging with a Heated Atomic Force Microscope Cantilever. *ASME Journal of Heat Transfer*, **124**, p. 597, 2002.

[23] Dürig, U., Cross, G., Despont, M., Drechsler, U., Haeberle, W., Lutwyche, M.I., Rothuizen, H., Stutz, R. Widmer, R., Vettiger, P., Binnig, G.K., King, W.P. & Goodson, K.E., Millipede – An AFM Data Storage System at the Frontier of Nanotribology. *Tribology Letters*, **9**, pp 25–32, 2000.

[24] Mamin, H.J. & Rugar, D., Thermomechanical Writing with an Atomic Force Microscope Tip. *Applied Physics Letters*, **61**, pp. 1003–1005, 1992.

[25] Mamin, H.J., Thermal Writing Using a Heated Atomic Force Microscope Tip. *Applied Physics Letters*, **69**, pp. 433–435, 1996.

[26] Williams, C.C. & Wickramasinghe, H.K., Scanning Thermal Profiler. *Applied Physics Letters*, **49**, pp. 1587–1589 1986.

[27] Majumdar. A., Scanning Thermal Microscopy. *Annual Review of Materials Science*, **29**, pp. 505–585, 1999.

[28] Despont, M., Brugger, J., Drechsler, U., Dürig, U., Häberle, W., Lutwyche, M., Rothuizen, H.E., Stutz, R., Widmer, R., Binnig, G. &

Vettiger, P., VLSI-NEMS Chip for Parallel AFM Data Storage. *Sensors and Actuators*, **A80**, pp. 100–107, 2000.

[29] Lutwyche, M.I., Despont, M., Drechsler, U., Durig, U., Hablerle, W., Rothuizen, H., Stutz, R., Widmer, R., Binnig, G.K. & Vettiger, P., Highly Parallel Data Storage System Based on Scanning Probe Arrays. *Applied Physics Letters*, **77**, pp. 3299–3301, 2000.

[30] Lutwyche, M., Andreoli, C., Binnig, G., Brugger, J., Drechsler, U., Haberle, W., Rohrer, H., Rothuizen, H., Vettiger, P., Yaralioglu, G. & Quate, C., 5×5 AFM Cantilever Arrays a First Step Towards a Terabit Storage Device. *Sensors and Actuators*, **A73**, pp. 89–94, 1999.

[31] deGennes, P.-G., *Scaling Concepts in Polymer Physics*. Cornell University Press: Ithaca, New York, 1979.

[32] Keddie, J.L., Jones, R.A.L. & Corey, R.A., Size-Dependent Depression of the Glass Transition Temperature in Polymer Films. *Europhysics Letters*, **27**, pp. 59–64, 1994.

[33] Keddie, J.L., Jones, R.A.L. & Corey, R.A., Interface and Surface Effects on the Glass-Transition Temperature in Thin Polymer Films. *Faraday Disscussions*, **98**, pp. 219–230, 1994.

[34] Forrest, J.A., Dalnoki-Veress, K., Stevens, J.R. & Dutcher, J.R., Effect of Free Surfaces on the Glass Transition Temperature of Thin Polymer Films. *Physical Review Letters*, **77**, pp. 2002–2005, 1996.

[35] Kurabayashi, K., Asheghi, M., Touzelbaev, M. & Goodson, K.E., Measurement of the Thermal Conductivity Anisotropy in Polyimide Films. *Journal of Microelectromechanical Systems*, **8**, pp. 180–191, 1999.

[36] Hu, H.-W. & Granick, S., Viscoelastic Dynamics of Confined Polymer Melts. *Science*, **258**, pp. 1339–1342, 1992.

[37] Wyart, F.B. & deGennes, P.-G., Viscosity at Small Scales in Polymer Melts. *European Physical Journal E.*, **1**, pp. 93–97, 2000.

[38] Doi, M. & Edwards, S.F., *The Theory of Polymer Dynamics*. Clarendon Press: Oxford, 1986.

[39] Smith, D.E., Babcock, H.P. & Chu, S., Single-Polymer Dynamics in Steady Shear Flow. *Science*, **288**, pp. 1724–1727, 1999.

[40] deGennes, P.-G., Glass transitions in Thin Polymer Films. *European Physical Journal E.*, **2**, pp. 201–205, 2000.

[41] Jackle, J., Models of the Glass Transition. *Reports of Progress in Physics*, **49**, pp. 171–231, 1986.

[42] Jones, R.A.L. & Richards, R.W., *Polymers at Surfaces and Interfaces*. University Press: Cambridge, 1999.

[43] Grohens, Y., Brogly, M., Labbe, C., Davis, M.-O. & Schultz, J., Glass Transition of Stereoregular Poly(methyl methacrylate) at Interfaces *Langmuir*, **14**, pp. 2929–2932, 1998.

[44] DeMaggio, G.B., Frieze, W.E., Gidley, D.W., Zhu, M., Hristov, H.A. & Yee, A.F., Interface and Size Effects on the Glass Transition in Thin Polystyrene Films. *Physical Review Letters*, **78**, pp. 1524–1527, 1997.
[45] Dalnkoi-Veress, K., Forrest, J.A. & Dutcher, J.R., Mechanical Confinement Effects on the Phase Separation Morphology of Polymer Blend Thin Films. *Physical Review E*, **57**, pp. 5811–5817, 1998.
[46] Zheng, X., Sauer, B.B., Van Alsten, J.G., Schwarz, S.A., Rafailovich, M.H., Sokolov, J. & Rubinstein, M., Reptation Dynamics of a Polymer Melt Near an Attractive Solid Surface. *Physical Review Letters*, **74**, pp. 407–410, 1995.
[47] Frank, B., Gast, A.P., Russel, T.P., Brown, H.R. & Hawker, C., Polymer Mobility in Thin Films. *Macromolecules*, **29**, pp. 6531–6534, 1996.
[48] Zheng, X., Rafailovich, M.H., Sokolov, J., Ystrzhemechny, Y., Schwarz, S.A., Sauer, B.B. & Rubinstein, M., Long-Range Effects on Polymer Diffusion Induced by a Bounding Interface. *Physical Review Letters*, **79**, pp. 241–244, 1997.
[49] Lin, E.K., Wu, W. & Satija, S.K., Polymer Interdiffusion Near an Attractive Surface. *Macromolecules*, **39**, pp. 7224–7231, 1997.
[50] Mansfield, K.F. & Theodorou, D.N., Interfacial Structure and Dynamics of Macromolecular Liquid: A Monte Carlo Simulation Approach. *Macromolecules*, **22**, pp. 2143–3152, 1989.
[51] Bitsanis, I. & Hadziioannou, G., Molecular Dynamics Simulations of the Structure and Dynamics of Confined Polymer Melts. *Journal of Chemical Physics*, **92**, pp. 3827–3847, 1990.
[52] Winkler, R.G., Matsuda, T. & Yoon, D.Y., Stochastic Dyanamics Simulations of Polymethylese Melts Confined Between Solid Surfaces. *Journal of Chemical Physics*, **98**, pp. 729–736, 1993.
[53] Van Giessen, A.E. & Szleifer, I., Monte Carlo Simulation of Chain Molecules in Confined Environments. *Journal of Chemical Physics*, **102**, pp. 9069–9076, 1995.
[54] Granick, S., Motions and Relaxations of Confined Liquids. *Science*, **253**, pp. 1374–1379, 1991.
[55] Subbotin, A., Semenov, A. & Doi, M., Friction in Strongly Confined Polymer Melts: Effect of Polymer Bridges. *Physical Review E*, **56**, pp. 623–630, 1997.
[56] Bruinsma, R., Slow Spreading of Polymer Melts. *Macromolecules*, **23**, pp. 276–283, 1990.
[57] Silberzan, P. & Leger, L., Spreading of High Molecular Weight Polymer Melts on High-Energy Surfaces. *Macromolecules*, **25**, pp. 1267–1271, 1992.
[58] King, W.P., Ph.D. Thesis, Stanford University, 2002.
[59] Hamdorf, M. & Johannsmann, D., Surface-Rheological Measurements of Glass Forming Polymers Based on the Surface Tension Driven Decay of

Imprinted Corrugation Gratings. *Journal of Chemical Physics*, **112**, pp. 4262–4270, 2000.
[60] Mark, J.E., *Polymer Data Handbook*. Oxford University Press: New York, 1999.
[61] Fuchs, K., Friedrich, C. & Weese, J., Viscoelastic Properties of Narrow-Distribution Poly(methyl methacrylates). *Macromolecules*, **29**, pp. 5893–5901, 1996.
[62] Kajiyama, T., Tanaka, K. & Takahara, A., Surface Molecular Motion of the Monodisperse Polystyrene Films. *Macromolecules*, **30**, pp. 280–285, 1997.
[63] Li, Z., Tolan, M., Hohr, T., Kharas, D., Qu, S., Sokolov, J., Rafailovich, M.H., Lorenz, H., Kotthaus, J.P., Wang, J., Sinha, S.K. & Gibaud, A., Polymer Thin Films on Patterned Si Surfaces. *Macromolecules*, **31**, pp. 1915–1920, 1998.
[64] Kerle, T., Lin, Z., Kim, H.-C. & Russell, T.P., Mobility of Polymers at the Air/Polymer Interface. *Macromolecules*, **34**, pp. 3484–3492, 2001.
[65] Landau, L.D., *Fluid Mechanics*, Oxford University Press: Boston, 1958.
[66] Mansfield, K.F. & Theodorou, D.N., Interfacial Structure and Dynamics of Macromolecular Liquid: A Monte Carlo Simulation Approach. *Macromolecules*, **22**, pp. 2143–3152, 1989.
[67] Bitsanis, I. & Hadziioannou, G., Molecular Dynamics Simulations of the Structure and Dynamics of Confined Polymer Melts. *Journal of Chemical Physics*, **92**, pp. 3827–3847, 1990.
[68] Ferry, J.D., *Viscoelastic Properties of Polymers*. Wiley: New York, 1980.
[69] Mehtar, V. & Archer, L.A., Slip in Entangled Polymer Flows. *Macromolecules*, **31**, pp. 6639–6649, 1998.
[70] Subbotin, A., Semenov, A., Hadziioannou, G. & Brinke, G., Rheology of Confined Polymer Melts under Shear Flow: Weak Adsorbtion Limit. *Macromolecules*, **28**, 1995.
[71] Subbotin, A., Semenov, A., Manias, E., Hadziioannou, G. & Brinke, G., Rheology of Confined Polymer Melts under Shear Flow: Strong Adsorbtion Limit. *Macromolecules*, **28**, pp. 1511–1515, 1995.
[72] Lin, E.K., Wu, W. & Satija, S.K., Polymer Interdiffusion near an Attractive Surface. *Macromolecules*, **39**, pp. 7224–7231, 1997.
[73] Frank, B., Gast, A.P., Russel, T.P., Brown, H.R. & Hawker, C., Polymer Mobility in Thin Films. *Macromolecules*, **29**, pp. 6531–6534, 1996.
[74] Zheng, X., Rafailovich, M.H., Sokolov, J., Ystrzhemechny, Y., Schwarz, S.A., Sauer, B.B. & Rubinstein, M., Long-range Effects on Polymer Diffusion Induced by a Bounding Interface. *Physical Review Letters*, **79**, pp. 241–244, 1997.
[75] Wolff, O. & Johannsmann, D., Shear Modulii of Polystyrene Thin Films Determined with Quartz Crystal Resonators in the Sandwich Configuration. *Journal of Applied Physics*, **87**, pp. 4182–4188, 2000.

[76] Fryer, D.S., Nealey, P.F. & dePablo, J.J., Thermal Probe Measurements of the Glass Transition Temperature for Ultrathin Polymer Films. *Macromolecules*, **33**, pp. 6439–6447, 2000.

[77] Ge, S., Pu, Y., Zhang, W., Rafailovich, M.H., Sokolov, J., Buenviaje, C., Buckmaster, R. & Overney, R.M., Shear Modulation Force Microscopy Study of Near Surface Glass Transition Temperature. *Physical Review Letters*, **85**, pp. 2340–2342, 2000.

[78] Nakajima, K., Yamaguchi, H., Lee, J.-C., Kageshima, M., Ikehara, T. & Nishi, T., Nanorheology of Polymer Blends Investigated by Atomic Force Microscopy. *Japan Journal of Applied Physics*, **36**, pp. 3850–3854, 1997.

[79] Ge, S., Pu, Y., Zhang, W., Rafailovich, M.H., Sokolov, J., Buenviaje, C., Buckmaster, R. & Overney, R.M., Shear Modulation Force Microscopy Study of Near Surface Glass Transition Temperature. *Physical Review Letters*, **85**, pp. 2340–2342, 2000.

[80] Oulevey, F., Burnham, N.A., Gremaud, G., Kulik, A.J., Pollock, H.M., Hammiche, A., Reading, M., Song, M. & Hourston, D.J., Dynamic Mechanical Analysis at the Submicron Scale. *Polymer*, **41**, pp. 3087–3092, 2000.

[81] Hammerschmidt, J.A., Gladfelter, W.L. & Haugstad, G., Probing Polymer Viscoelastic Relaxations with Temperature-Controlled Friction-Force Microscopy. *Macromolecules*, **32**, pp. 3360–3367, 1999.

[82] King, W.P., Kenny, T.W., Goodson, K.E., Cross, G.L.W., Despont, M., Durig, U., Rothuizen, H., Binnig, G. & Vettiger, P., Atomic Force Microscope Cantilevers for Combined Thermomechanical Data Writing and Reading. *Applied Physics Letters*, **78**, pp. 1300–1302, 2001.

[83] Asheghi, M., King, W.P., Chui, B.W., Kenny, T.W. & Goodson, K.E., Thermal Engineering of Doped Single-Crystal Silicon Microcantilevers for High Denstiy Data Storage. Presented at Transducers 1999, the 10th International Conference on Solid-State Sendors and Actuators, Sendai, Japan, 1999.

[84] Weber, L., Gmelin, E. & Queisser, H.J., Thermal Resistance of Silicon Point Contacts. *Physical Review B*, **40**, pp. 1244–1249, 1989.

[85] Phelan, P.E., Nakabeppu, O., Ito, K., Hijikata, K., Ohmori, T. & Torikoshi, K., Heat Transfer and Thermoelectric Voltage at Metallic Point Contacts. *Journal of Heat Transfer*, **155**, pp. 757–762, 1993.

[86] Nakabeppu, O., Igeta, M. & Hijikata, K, Experimental-Study on Point-Contact Transport Phenomena Using the Atomic Force Microscope. *Microscale Thermophysical Engineering*, **1**, pp. 201–213, 1997.

[87] King, W.P. & Goodson, K.E., Modeling and Simulation of Thermomechanical Nano-Indentation in Polymer. *Proceedings to the ASME/JSME Joint Thermal Engineering Conference*, Hawaii, March 2003.

[88] Forrest, J.A., Dalnoki-Veress, K. & Dutcher, J.R., Interface and Chain Confinement Effects on the Glass Transition Temperature of Thin Polymer Films. *Physical Review E*, **56**, pp. 5705–5715, 1997.

[89] King, W.P., Santiago, J.G., Kenny, T.W. & Goodson, K.E., Modeling and Simulation of Sub-Micrometer Heat Transfer in AFM Thermomechanical Data Storage. *ASME MEMS*, **1**, pp. 583–588, 1999.

[90] King, W.P. & Goodson, K.E., Modeling and Simulation of Nanometer-Scale Thermomechanical Data Bit Formation. Presented at ASME National Heat Transfer Conference, Anaheim, CA, 2001.

[91] King, W.P., Kenny, T.W., Goodson, K.E., Cross, G., Despont, M., Duerig, U., Lutwyche, M., Rothuizen, H., Binnig, G. & Vettiger, P., Design of AFM Cantilevers for Combined Thermomechanical Data Writing and Reading. Presented at Solid State Sensors and Actuators Workshop, Hilton Head, SC, 2000.

[92] King, W.P., Kenny, T.W., Goodson, K.E., Cross, G.L.W., Despont, M., Durig, U., Rothuizen, H., Binnig, G. & Vettiger, P., Design of Atomic Force Microscope Cantilevers for Combined Thermomechanical Writing and Thermal Reading in Array Operation. *Journal of Microelectromechanical Systems*, **11**, pp. 765–774, 2002.

[93] Wyart, F.B. & deGennes, P.-G., Viscosity at Small Scales in Polymer Melts. *European Physical Journal E.*, **1**, pp. 93–97, 2000.

[94] Asheghi, M., Touzelbaev, M.N., Goodson, K.E., Leung, Y.K. & Wong, S.S., Temperature-Dependent Thermal Conductivity of Single-Crystal Silicon Layers in SOI Substrates. *Jounal of Heat Transfer*, **120**, pp. 30–36, 1998.

[95] Ju, Y.S. & Goodson, K.E., Phonon Scattering in Silicon Films of Thickness below 100 nm. *Applied Physics Letters*, **74**, pp. 3005–3007, 1999.

[96] Sverdrup, P.G., Ju, Y.S. & Goodson, K.E., Sub-Continuum Simulations of Heat Conduction in Silicon-on-Insulator Transistors. *Journal of Heat Transfer*, **123**, pp. 130–137, 2001.

[97] Ziman, J.M., *Electrons and Phonons*, Oxford University Press: Oxford, 1960.

[98] Kittel, C., *Introduction to Solid State Physics*, Wiley: New York, 1996.

[99] Casimir, H.B.G., Note on the Conduction of Heat in Crystals. *Physica*, **5**, pp. 495–500, 1938.

[100] Touloukian, Y.S., *Thermal Conductivity: Nonmetallic Solids*. IFI/Plenum: New York, 1970.

[101] Vincenti, W.G. & Kruger, C.H., *Introduction to Physical Gas Dynamics*. Krieger: Malabar, Florida, 1965.

[102] Rohsenow, W.M. & Choi, H.Y., *Heat, Mass, and Momentum Transfer*, Prentice-Hall: Englewood Cliffs, NJ, 1961.

[103] Schwartz, E.T. & Pohl, R.O., Thermal Boundary Resistance. *Review of Modern Physics*, **61**, pp. 605–668, 1989.

[104] Cappella, B. & Dietler, G., Force-Distance Curves by Atomic Force Microscopy. *Surface Science Reports*, **43**, pp. 1–104, 1999.

[105] Johnson, K.L., *Contact Mechanics*. Cambridge University Press: Cambridge, 1987.

[106] Mark, J.E., *Polymer Data Handbook*. Oxford University Press: New York, 1999.

[107] Sneddon, I.N., The Relation Between Load and Penetration in the Axisymmetric Boussinesq Problem for a Punch of Arbitrary Profile. *International Journal of Engineering Science*, **3**, pp. 47–57, 1965.

[108] Roark, R.J. & Young, W.C., *Roark's Formulas for Stress and Strain*, 6th edn. McGraw-Hill: New York, 1989.

[109] Personal conversation with U. Duerig, IBM Zurich Research Laborarory.

[110] Terris, B.D., Rishton, S.A., Mamin, H.J., Ried, R.P. & Rugar, D., Atomic Force Microscope-Based Data Storage: Track Servo and Wear Study. *Applied Physics A*, **66**, pp. S809–S813, 1998.

[111] Incropera, F.P. & DeWitt, D.P., *Introduction to Heat Transfer*, 4th edn. Wiley: New York, 2002.

[112] Gotsmann, B. & Duerig, U., unpublished.

[113] Cross, G. & Duerig, U., unpublished.

[114] Ferry, J.D., *Viscoelastic Properties of Polymers*. Wiley: New York, 1980.

[115] Wineman, A.S. & Rajagopal, K.R., *Mechanical Response of Polymers*. Cambridge University Press: Cambridge, 2000.

[116] Roark, R.J. & Young, W.C., *Roark's Formulas for Stress and Strain*, 6th edn. McGraw-Hill: New York, 1989.

[117] Gotsmann, B. & Duerig, U., unpublished.

[118] Cross, G. & Duerig, U., unpublished.

[119] Terris, B.D., Shishton, S.A., Mamin, H.J., Best, J.D., Logan, J.A. & Rugar, D., Atomic Force Microscope-Based Data Storage Using Replicated Media. *Journal of Vacuum Science and Technology B*, **14**, pp. 1584–1587, 1997.

[120] King, W.P., Kenny, T.W. & Goodson, K.E., Comparison of Piezoresistive and Thermal Detection Approaches to Atomic Force Microscopy Topography Measurement. *ASME-IMECE MEMS and Nanotechnology Symposium*, 2002.

[121] Asheghi, M., Ph.D. Thesis, Stanford University, 1999.

[122] Chou, S.Y., Krauss, P.R. & Ronstrom, P.J., Imprint Lithography with 25-Nanometer Resolution. *Science*, **272**, pp. 85–87, 1996.

[123] Yang, P., Wirnsberger, G., Huang, H.C., Cordero, S.R., McGehee, M.D., Scott, B., Deng, T., Whitesides, G.M., Chmelka, B.F., Buratto, S.K. & Stucky, G.D., Mirrorless Lasing from Mesostructured Waveguides Patterned by Soft Lithography. *Science*, **287**, pp. 465–467, 2000.

[124] Heeger, P.S. & Heeger, A.J., Making Sense of Polymer-Based Biosensors. *Proceedings of the National Acadamy of Sciences*, **96**, pp. 12219–12221, 1999.

CHAPTER 5

Two-phase flow microstructures in thin geometries: multi-field modeling

R. Kumar
University of Central Florida, Orlando, FL, USA.

Abstract

The high heat flux density in miniaturized passages causes two-phase flow regime transitions similar to conventional passages. However, in these microchannels, surface tension plays an important role and changes the flow regime transitions, pressure drop and heat transfer, which are not predicted well by the available correlations. Therefore, a mechanistic approach to modeling is needed, which relies on a fundamental understanding of two-phase flow in microchannels. A fundamental modeling approach uses multiple fields, where the flow regimes are modeled as combinations of several idealized fields (e.g., continuous liquid, continuous vapor, dispersed liquid and dispersed vapor). By developing mechanistic closure models in the interfaces between these fields, researchers have shown that the accuracy of the flow topology, pressure drop and heat transfer can be improved. Predictions from the three-field and four-field approaches have been compared with available void fraction and velocity measurements in the gap. In adiabatic annular flow, the three fields (continuous liquid, continuous vapor and dispersed liquid) are shown to adequately predict the thin film and the droplet field. The four-field model is also shown to be able to predict heated, two-phase flow from the inception of bubbly flow to annular flow through a slug-churn-turbulent transition without the use of flow regime maps in all geometries including circular tubes, rectangular ducts and annular gaps. The closure models in different flow regimes and the experiments that support model development are discussed.

1 Introduction

There has been a recent need for a fundamental understanding of two-phase flow topology, pressure drop and critical heat flux in tubes and channels of the size of

10 μm to about 5 mm. This interest is primarily industry-driven as very small diameter tubes are now employed in residential air-conditioners, and thin compact evaporators are used in automotive and aerospace industries. Triangular, square and rectangular geometries in the millimeter range find their applications in the electronic cooling industry. The high heat flux density used in miniaturized electronic circuits requires efficient cooling techniques. In another application of two-phase flow in millimeter and micron size channels, Proton Exchange Membrane (PEM) fuel cells are currently being developed as nonpolluting energy sources which combine hydrogen and oxygen to produce electrical current and only water as a by-product. The polymer membrane in the fuel cell must be well hydrated to facilitate proton transfer, but if too much water is present, flooding will occur which restricts reactant oxygen from reaching catalytic sites at the membrane surface. Therefore, it is desirable to maintain a very high superficial gas velocity in the flow channels that distribute oxygen (usually from air) for the cathode reaction. As outlined by Wheeler *et al.* [1], the stability of fuel cell operation is dependent on the annular two-phase flow regime existing in the cathode flow channels.

Reviewing the needs of the various industries, Kandlikar [2, 3] classifies thin channels based on their hydraulic diameters. According to this classification, microchannels (10 μm to 200 μm) are employed in inkjet printers and many MEMS (Micro-Electro-Mechanical Systems), minichannels (200 μm to 3 mm) in compact plate-fin evaporators, and conventional channels (>3 mm) in steam tube boilers, evaporator tubes, etc. This type of classification is loosely defined and will certainly undergo further changes.

In microchannels, nucleating bubbles of size ~5 μm have been photographed [2]. These bubbles grow in size and quickly span the gap, and therefore the slug and churn-turbulent regimes are dominant in microchannels. This increases the pressure gradient and the heat transfer coefficient of these thin channels compared to the large channels. The existing correlations that have been benchmarked for larger channels (≥ 8 mm) do not predict the thin channel data. The reason is that these correlations took into account only the gravity and vapor shear forces that are important for large channels. For thin channels, the surface tension effect becomes significant. It is not clear whether a critical channel size exists at which the surface tension effect begins to dominate. However, there is enough evidence, as we shall see later, that the microstructures of the fluid flow and heat transfer for a micro-sized channel behave in a similar manner as a millimeter-size channel. Since there is enough data in minichannels (as categorized by Kandlikar [2]), it may be possible to extrapolate that information to microchannels to support model development. Indeed, it will be shown that all the flow regimes visualized in a 2 mm tube from bubbly to annular flow have also been photographed in a 50 μm tube.

One of the issues with microchannels is the geometric shape. Although conventional belief is that heat transfer should be more efficient in rectangular geometries (e.g., compact evaporators) compared to tubes of the same hydraulic diameter, there is no such evidence in minichannels (Tran *et al.* [4]). Whether this fact could be extended to microchannels is not clear at this time. We do know that the flow in rectangular channels is more complex and less well understood than in circular geometries. Circular geometry measurements tend to limit our ability to adequately account for three-dimensional effects. The behavior of bubbles and slugs in a confined space between two walls has been observed to be different from that in a circular geometry. The flow regime transitions, pressure drop and heat transfer in thin geometries will be discussed in section 2, Global characteristics.

Thus, performance improvement in system components depends on the size and shape of the flow passage, system pressure, mass flux and fluid properties. Since the number of flow and geometric variables has increased for microchannels, it is essential to focus more on the fundamental understanding of the two-phase flow behavior and to rely less on empirical correlations. Although the two-fluid technology has progressed significantly, there is little precedence describing the interfacial forces, heat and mass transfer in difficult transition flow regimes. In the slug and churn-turbulent regimes, where all types of interfaces exist, not only do transport equations and closure models need to be developed for each phase, but also each phase should accommodate the existence of that phase in different forms and shapes. For example, a bubble can be spherical, distorted, confined and/or elongated. In this regard, several fields can exist within the same phase. Section 3, Local flow characteristics, will deal with the multi-field approach, their transport equations, closure conditions and sample results. This type of representation would encompass the various flow regimes, which typically include single-phase subcooled flow at the inlet to completely separated annular flow at the exit. Multi-field formulation of two-fluid equations is state-of-the-art and a major step towards improving design, reducing expensive tests and providing numerical data in complex flow situations where tests are not possible.

2 Global characteristics

2.1 Flow patterns in thin channels

In his discussion on the importance of the flow regimes in two-phase flows, Hosler [5] writes that *the understanding of the various flow configurations is analogous to knowing whether the flow is laminar or turbulent in single-phase flow*. The knowledge of flow patterns is very important in the design of various components such as plate-fin heat exchangers, cooling channels and microelectronics, where boiling occurs.

In heated flows in thin channels, single-phase subcooled liquid enters a channel and heats up until sufficient wall superheat is reached to initiate nucleation. Following the onset of nucleation, where the bubble sizes of 5 μm have been visualized in microchannels (Kandlikar [2]), the vapor fraction increases as more vapor is formed. As the vapor fraction increases, the bubbles grow by coalescence and interfacial evaporation until they become large enough to be constrained by the walls of thin circular and noncircular tubes. This results in the transformation of dispersed vapor (dv) to continuous vapor (cv), as the flow enters the slug flow regime, and thereafter the churn-turbulent regime. As more continuous vapor is formed, the coalescence rate increases dramatically, resulting in rapid transition to an annular flow regime. At this point, a continuous high velocity vapor core bounded by a thin liquid film on the wall is formed. Thus the wall plays a major role in the modeling of thin separated films.

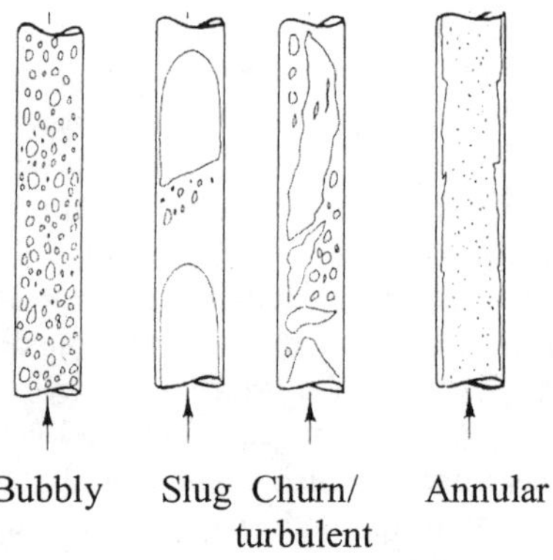

Figure 1: Flow patterns in thin tubes.

The annular flow regime is an important regime to model since it is a precursor to critical heat flux (CHF). This flow regime typically occurs for void fractions in excess of 70% depending on the system pressure and mass flux. In annular flow, the liquid phase flows in part as a film along the walls and in part as droplets entrained in the central gas core. Among the different types of droplet generation associated with entrainment, liquid bridge disintegration in the transition regime and roll wave shearing of the separated films are considered to be the most important mechanisms. The ratio of the liquid phase in the film to that in droplet form varies according to the fluid flow rates and the density ratio. The interface between the liquid film and the vapor core is characterized by disturbance waves. This interfacial wave motion is an important mechanism affecting the interfacial shear stress and the primary contribution to the increase in shear in annular flow is from the disturbance waves. Along the length of the channel, the liquid film may drop from a few hundred microns in thickness to a few microns. Consequently, the droplet size in the vapor core decreases from a few hundred microns to a few microns along the length of the channel. These droplets are deposited back on to the film to keep the interface convoluted and in nonequilibrium. Thus, the deposition process maintains a wetted wall and postpones any possible catastrophe of the system component through rapid temperature excursion.

Flow visualization studies in large and small tubes are abundant. The flow regime maps generally differ based on the orientation of the tube and the fluid pressure. Flow patterns are different in channels that are less than 3 mm in hydraulic diameter because surface tension is dominant. Most of these studies have been performed in air–water flows and some in refrigerant fluid.

Damianides and Westwater [6] found that for air–water flows in small horizontal tubes (1–5 mm), the stratified regime decreases in extent in the regime map until it completely vanishes at 1 mm. This has been generally proven to be true by other investigators [7, 8] in horizontal and vertical tubes. Serizawa and Feng [9] visualized air–water two-phase flows in a 25 μm horizontal tube and steam–water flows in a 50 μm tube. They noted that in addition to a slug flow, liquid rings appeared between long slugs followed by droplet flow, when the heat input was increased. Some of the studies mentioned above also compared the flow patterns with Taitel and Dukler's [10] analysis for large tubes and found the transition lines to vary significantly. Of note is the study of Yang and Shieh [11] in air–water and R-134a in 1–3 mm horizontal tubes. They found that the R-134a transition from the slug regime is narrower and shifted toward lower superficial gas velocity compared to air-water. They attributed this behavior to the surface tension which acts to maintain the liquid hold up between the two walls. They also reported that the R-134a transition to annular flow was so sharp that a slight increase in vapor quality or vapor velocity changed the flow to annular.

While some transition lines for small tubes vary slightly compared to large diameter tubes, significant deviations in the transition lines are found for rectangular channels [12, 13]. In one of the earlier, beautiful collections of two-phase flow photographs in a thin vertical rectangular channel, Jones and Zuber [14] observed that the slug regime seldom contains Taylor-type bubbles. Instead, the bubbles in the slug flow oscillated wildly from side to side and snakewise along the upward flow direction. They observed the Karman vortex street behavior by the entrained bubbles behind major voids using the high-speed motion pictures in a 3.0 mm gap air–water channel. Under the influence of the trailing vortex street, the nose of the elongated bubbles moved from side to side. They visualized confined bubbles of different size. In rectangular channels, as the bubbles coalesce and are confined, they generally start growing in the width direction and become slugs. This is the early transition regime. The lift force exerted from the edges at different points on a slug is nonuniform because of the bubble shape that tends to keep the slug oscillating. The vortex shedding seen in their photographs is associated with the side-to-side fluctuation of the planar cap. They noted that slug-like flows occur for the void fraction, α, between ~0.2 and ~0.8. Later from geometric considerations, Mishima and Hibiki [15] showed that the possibility of collisions and coalescence would increase if the maximum distance between two bubbles were less than the projected diameter of a flat bubble. This transition was shown to occur at $\alpha = 0.2$ in thin rectangular channels, similar to the observations of Jones and Zuber [14].

The other papers of importance in air–water flows in thin rectangular channels are by Troniewski and Ulbrich [16] (10 channels with aspect ratios of 0.1 to 12), Ali and Kawaji [17] (150 μm × 80 mm), Mishima *et al.* [18] (1 mm × 40 mm, 2.4 mm × 40 mm), and Wilmarth and Ishii [19] (1 mm × 20 mm, 2 mm

× 15 mm). All these studies used air-water and could not agree on a consensus transition line. As Wilmarth and Ishii [19] point out, the difference in the slug–churn turbulent transition may be caused by the different widths used for the same gap as well as the distribution parameter, C_0, in the drift flux model. The distribution parameter tends to increase with decreasing channel gap, causing the transition boundaries to shift toward lower superficial gas velocity. In addition, the drift velocity is a function of the aspect ratio as given by Griffith [20]:

$$V_{gj} = 0.23 + 0.13(t/w)\sqrt{\Delta\rho g w/\rho_l} \qquad (1)$$

where t and w are the thickness and width of the channel, and ρ_l and $\Delta\rho$ are the liquid density and density difference respectively.

In diabatic flows in thin channels, with increasing pressure, the boundaries between bubbly slug and slug annular flow patterns shift to higher local bulk enthalpy [5]. This is due to the decrease of the liquid-to-vapor density ratio, ρ_l/ρ_g, at high pressures. Increasing the flow rate shifts the boundaries to lower bulk enthalpy. At very high pressures beyond a certain mass flux, the slug may disappear completely. Kumar *et al.* [21] measured a significant increase in bubble frequency when the system pressure in a refrigerant fluid is increased at high mass fluxes. Their measurements show that a bubbly-type regime can exist at high void fractions well up to 60% in certain flow conditions. At these levels, the bubbles bridge the narrow space and are pancake-shaped.

In general, there are considerable differences in the transition lines developed by researchers. The reasons could be due to the subjective nature of flow visualization and the method of injection of the two fluids. Considering the vast difference in flow regime maps between their study and those of others, Wambsganss *et al.* [13] noted that *the flow pattern transitions in small rectangular channels are best determined independently*. This continues to be so even after a decade, and hence only the general trends of transition in thin channels are dependable.

2.2 Pressure drop and heat transfer

In thin channels, flows are characterized by large masses of continuous vapor that fill the gap. Under these conditions, the thin liquid layer separating these large bubbles from the heated wall evaporates rapidly, increasing pressure drop and the heat transfer coefficient.

2.2.1 Pressure drop

Tran *et al.* [22] noted that the pressure gradient in high quality R-134a flow through a 2.4 mm tube is more than 3 times higher than that of Eckels *et al.*'s [23] 8 mm tube data. However, the trends in the data agreed with those of a large tube. For example, the two-phase flow pressure drop increased with the increase in exit quality and mass flux, and increased with a decrease in system pressure. For thin channels, they developed a new correlation for frictional pressure drop taking into account the effects of surface tension and channel size. They accounted for these effects through a dimensionless number, earlier defined by Cornwell and Kew [24], called the confinement number,

$$N_{conf} = \left(\sigma / g\Delta\rho\right)^{1/2} / D. \tag{2}$$

where D, σ and $\Delta\rho$ are the tube diameter, surface tension and the density difference respectively. Later we will see that the Eotvos number, Eo, is an important parameter in distorted and confined bubbly flow that uses the same variables as given in eqn. (2), except an equivalent bubble diameter is used instead of the hydraulic diameter.

When the bubbles coalesce and a separated annular regime is formed, the confinement number would not be an appropriate parameter. This was seen to be the case by Lee and Lee [25] who found Tran *et al.*'s [22] correlation to significantly overpredict their R-113 data in channels with gap sizes of 400 μm, 1 mm and 3 mm and a width of 20 mm. Their film thicknesses were estimated to be between 20 μm and 120 μm. Thus, although Tran *et al.* [22] built their correlations using small circular tube and rectangular channel data, it is not clear whether their correlations are applicable in all flow regimes. Other separated flow models have been found to correlate narrow annular gaps of 35–110 μm well in R-113 flow using the Martinelli parameter [26]. Air-water pressure drop data in horizontal tubes of diameter ~1-1.5 mm have been predicted well by the homogenous model in bubbly and slug flow and not in the annular flow regime [27]. It is important to note that the transition from laminar flow can occur at lower Reynolds numbers than in conventional channels. There is enough evidence that in microchannels of hydraulic diameters varying from ~300 μm to ~750 μm, the Reynolds number changes from 700–1000 at the inlet to about 1800–2000 at the exit [28, 29]. It was conjectured that the variations in the heating rate or wall temperature were the cause of the large changes in the thermal properties and hence the Reynolds number increased along the streamwise direction to the exit. Warrier *et al.* [29] showed that the laminar friction factor predicted the measurements well for Re ~ 500 and 800, and underpredicted the data for Re ~ 1100–1400.

Kim and Bang [30] did an interesting comparison study between a horizontal circular and square tube with the same hydraulic diameter of 1.66 mm in refrigerant fluid R-22. This work brings out the effect of the liquid wetting area on pressure drop. As the vapor quality increases and the flow is separated and annular, the liquid tends to collect in the corners of the square tube, while the wetted area is more uniform for the circular tube. Less wetted area decreases the pressure drop for the square channels.

The cornering effect seen in sharp-edged cross-sections becomes more pronounced for higher aspect ratios in annular flows. Vassallo *et al.* [31] compared the frictional pressure gradient in adiabatic and wall-heated flows at two different pressures in R-134a in a thin rectangular duct. As shown in Figure 2, the frictional pressure gradient at high pressures shows a dip at a certain void fraction before it rises again. They attribute this dip to the high interfacial friction in the churn-turbulent regime compared to the annular flow regime. As the void fraction increases beyond this point, the superficial gas velocity increases in the vapor core, the liquid film becomes thinner, and the frictional pressure gradient increases again. They showed that the point at which the frictional pressure gradient has a local minimum is also the point where the void distribution changes from center-peaked to a wall-peaked profile in thin film annular flows. These local characteristics of void fraction change due to the presence of the droplets will be discussed later in section 3.6, Droplet models.

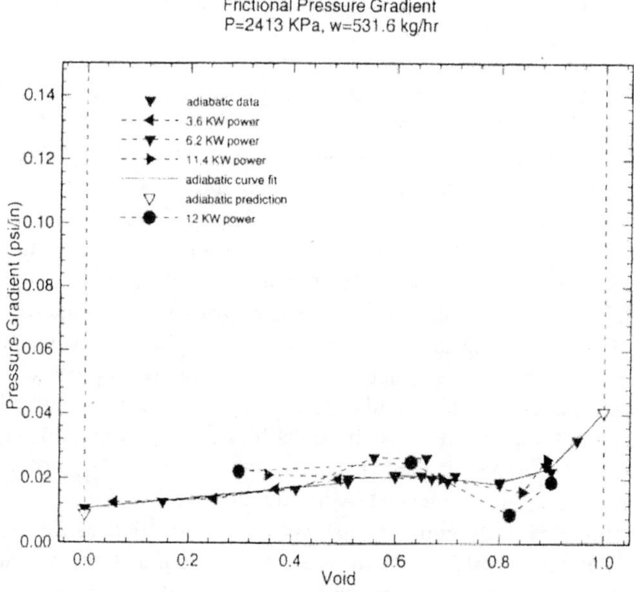

Figure 2: Frictional pressure gradient in adiabatic and wall heated channels [31].

2.2.2 Heat transfer

As in the case of pressure drop discussed in section 2.2.1, in the presence of slugs in thin channels, the heat transfer coefficient is higher than in conventional channels. Compared to the larger diameter channels, heat transfer coefficients in thin rectangular and circular channels of $D \sim 2.5$ mm [4] and annular gap of 1-1.5 mm [32] were enhanced. Although there seemed to be a potential for enhancing heat transfer by optimizing the cross-sectional geometry, the heat transfer was about the same for circular and rectangular channels.

Depending on the channel size and the flow regimes, the functional dependence of heat transfer on quality, mass flux and wall heat may vary. Tran *et al.* [4] measured heat transfer with R-12 and R-134a in a small circular channel, and found that nucleate boiling dominated when the wall superheat was high. Their heat transfer coefficient was effectively independent of quality and mass flux over a wide quality range, but was dependent on wall heat flux. For small channels, they found nucleate boiling to dominate, and replaced the Reynolds number in Lazarek and Black's [33] correlation by the Weber number. This correlation predicted the data in small circular and rectangular channel data very well for three different refrigerant fluids. However, in the same quality range in annular flow, Lee and Lee [25] saw the opposite effect in rectangular channels of gap 400 μm, 1 mm and 2 mm. Their two-phase heat transfer coefficient, h_{tp}, increased with quality and mass flux. The effect of mass flux on h_{tp} was found to be small, as the gap size became small. Based on the estimated film thickness, they separated the data for the three channels for $Re_{lf} \leq 200$ and $Re_{lf} > 200$ as shown in Figure 3. The data for $Re_{lf} \leq 200$ clearly shows that the film is laminar, and the heat is transferred in the conduction mode, and $h_{tp}=k_l/\delta$, where k_l and δ are the liquid conductivity and film thickness respectively. The film thickness for the 400 μm channel ranges between 20 μm and 70 μm. Even in this thin channel, turbulent heat transfer can take place for $Re_{lf} > 200$.

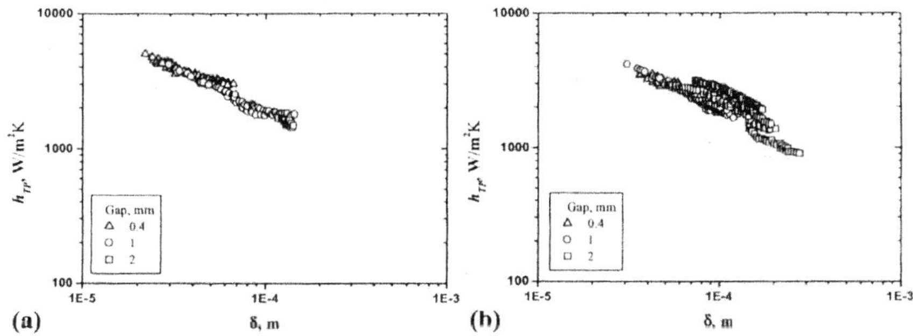

Figure 3: Boiling heat transfer coefficient in rectangular channels [25] for a) $Re_{lf} \leq 200$; b) $Re_{lf} > 200$. *lf* refers to the liquid film.

It is difficult to provide any depth in the review of heat transfer because of the inconclusive work in the area of flow boiling in microchannels. Although some survey articles [2, 3] are available in the literature, there is a definite need for measurements to understand the boiling mechanisms as well as to enhance the CHF database.

2.3 Summary

Significant progress has been made in describing the global parameters of two-phase flows in thin channels. However, more data are needed in microchannels, and more physical parameters would be needed to correlate the microchannel data. It is clear that pressure drop, heat transfer coefficient and critical heat flux depend on the local characteristics of the flow. Therefore, local data would be needed to analyze the flow and to develop models for three-dimensional two-fluid codes. It is important to understand the dynamic processes and microstructures involved in thin channels to accurately predict the local and global quantities, without relying on empiricism.

3 Local flow characteristics

The two-phase flow analyses have evolved over the years from a fluid mixture approach where a drift-flux is used [34] to a two-fluid approach, which is readily adaptable to multi-dimensional analysis. The fluid mixture analysis is one-dimensional and is based upon either homogeneous or separated flow models. By contrast, in the two-fluid approach, each phase has its own set of conservation equations. This approach has been successful so long as the flow regime is bubbly. In thin channels, it is difficult to model the interfacial forces and heat transfer when each phase exists both as dispersed as well as continuous flows. For example, vapor can exist as nucleating bubbles at the wall, distorted slugs, bubbles in the thin liquid film or as a continuous entity in the vapor core. Similarly, liquid can exist as a film on the wall or as droplets in the vapor core. They may all coexist at a certain streamwise location in a channel. This situation cannot be modeled with just two fields. Therefore, a multi-field approach is needed.

3.1 Ensemble averaging approach

Due to the complexities of the individual interfacial interactions between phases, it is impractical to solve the Navier–Stokes equations in the whole field containing multiple flow regimes using Direct Numerical Simulation (DNS). Instead an averaged set of flow equations is solved. During the averaging process, microscale flow phenomena (individual interface interactions) are filtered out and the resulting averaged equations are used to predict the desired macroscale flow behavior (such as local void distributions) once the appropriate closure conditions are included. As with all averaging processes, some physical information is lost during averaging. Therefore, closure conditions or constitutive equations must be introduced, the form of which must be determined through analysis of the interfacial physics. These relations or 'models' will be described in detail later.

Time, space and ensemble averaging are the three main types of averaging commonly used [35]. Ensemble averaging is more appropriate for general two-phase flows since it preserves both spatial and temporal flow variations. Here, the observed results at a given fixed location are summed for all possible flow realizations, and divided by the total number of observations. Hence closure conditions used in the current analysis were developed within the cell model averaging framework developed by Drew [36]. The cell model is an approximation to the ensemble average. The ensemble is the set of flows that can occur at $\underline{x}$ with the sphere (bubble) center occupying different positions $\underline{z}$ in the cell. If f is some microscopic process, the dependence on the ensemble for this case is denoted by $f(\underline{x},t;\underline{z})$. In this model, flow realizations of a single bubble contained within the averaging cell are calculated as

$$\bar{f} = \iiint_{V_{cell}} f(\underline{x}, t; \underline{z}) \wp(\underline{x}, t; \underline{z}) dV \tag{3}$$

where $\underline{x}$ is the fixed flow observation point located within the volume dV, $\underline{z}$ is the location of the bubble center located within the cell, and $\wp$ is the probability of a particular flow realization. A phase indicator function is defined as given in eqn. (4) below. The ensemble-averaged equations are obtained by first multiplying the single-phase equations by the phase-indicator function, and using the definitions given in eqn. (4). The cell model is valid for dispersed flows only, and for nondilute flows, the closure is done on a more phenomenological basis.

$$X_k(x,t) = \begin{cases} 1 & \text{if phase k at x,t} \\ 0 & \text{otherwise} \end{cases}$$
$$\bar{\rho}_k = \frac{\overline{\rho_k X_k}}{\overline{X_k}} = \frac{\overline{\rho_k X_k}}{\alpha_k} \tag{4}$$
$$\bar{f}_k = \frac{\overline{f_k X_k \rho_k}}{\alpha_k \bar{\rho}_k}$$

3.2 Multi-field modeling approach

The multi-field formulation of the two-fluid model has been chosen for its ability to predict the flow regimes and flow transitions that occur in two-phase flows [37–41]. In this formulation, various flow regimes are modeled as mixtures of the four idealized flow fields: continuous liquid (*cl*), dispersed liquid (*dl*), continuous vapor (*cv*) and dispersed vapor (*dv*). The three fields, *cl*, *cv* and *dl* can represent annular flow. This three-field formulation would be adequate to model a) the wavy interface between continuous liquid and a continuous vapor, and b) the droplet interface between a dispersed liquid and a continuous vapor field. In a boiling flow, a fourth field, *dv*, would be required, and therefore, a third interface, *cl-dv*, needs to be modeled. If more fields were chosen, modeling the interfaces would be simpler. However, it would also take more effort to develop an understanding of the flow distributions in various flow regimes, and model various interfaces. Based on the works of [37, 38, 41], in diabatic flows, two dispersed vapor fields and two dispersed liquid fields are the minimum number of fields required to capture all the dynamic interactions. For each field, conservation equations for mass, momentum, and energy are solved. Within the conservation equations are terms which represent the interactions among the four fields and between the fields and the walls. Within the continuous fields, dispersed fields of the opposite phase can exist. The large slug/churn-turbulent regime, also called the transition regime, exists between the pure bubbly and annular flow regimes. The transfer of mass between the fields occurs through hydrodynamic (coalescence, liquid bridge breakup, entrainment, deposition, etc.) and heat transfer (evaporation and condensation) mechanisms. This redistribution of mass among the different fields provides the predictive capability to handle flow regime transitions as continuous events rather than as discrete events.

3.3 Governing equations

Conservation of phasic mass:

$$\frac{\partial}{\partial t}(\alpha_i\rho_i)+\frac{\partial}{\partial x^k}\left(\alpha_i\rho_i u_i^k\right)=\sum_{j\neq i}\left(\Gamma_{ji}-\Gamma_{ij}\right) \tag{5}$$

Conservation of phasic momentum:

$$\begin{aligned}\frac{\partial}{\partial t}\left(\alpha_i\rho_i u_i^l\right)+\frac{\partial}{\partial x^k}\left(\alpha_i\rho_i u_i^k u_i^l\right)=-\alpha_i\frac{\partial p}{\partial x^l}+\frac{\partial}{\partial x^k}\{\alpha_i\mu_i(\frac{\partial u_i^l}{\partial x^k}+\frac{\partial u_i^k}{\partial x^l})\}+\\ M_i^l+\sum_{j\neq i}D_{ij}(u_j^l-u_i^l)+\Gamma_{ji}u_j^l-\Gamma_{ji}u_i^l\end{aligned} \tag{6}$$

Conservation of phasic energy:

$$\frac{\partial}{\partial t}(\alpha_i\rho_i h_i)+\frac{\partial}{\partial x^k}\left(\alpha_i\rho_i u_i^k h_i\right)=\frac{\partial}{\partial x^k}(\alpha_i k_i\frac{\partial T_i}{\partial x^k})+Q_i+\sum_{j\neq i}\left(\Gamma_{ji}h_j^{\text{int}}-\Gamma_{ij}h_i^{\text{int}}\right) \tag{7}$$

The subscript i denotes the field. The subscripts in the mass transfer term, Γ_{ij}, denote a quantity transferred from field i to field j. For example, an evaporative mass generation term in bubbly flow is indicated by *cl-dv* and coalescence from small to large bubbles by *dv-cv*, etc. The three interfaces are *dv-cl* (bubbly regime), *cv-cl* (transition and thin film interfaces) and *cv-dl* (droplet flow). The superscript k is the repeated index. $D_{ij}(u_j^l-u_i^l)$ represents the interfacial drag force; M_i^l, the nondrag forces including Reynolds stresses; Q_i, the net heat transfer to phase i due to interfacial, wall and volumetric heating sources. The thermodynamic variables, temperature and enthalpy are related by $C_{P_i}=\partial h_i/\partial T_i$. From the definition of volume fraction, $\Sigma\alpha_i=1$. If solutions to four fields in three dimensions are sought, 20 coupled transport equations need to be solved. If the two continuous fields (*cl* and *cv*) are turbulent, transport equations for k and ε would have to be solved in each of the *cl* and *cv* fields as explained in section 3.9.

The fields are coupled via interfacial transfer terms underlined in eqns. (5)–(7). These terms are related by the so-called jump conditions and are described in [38]. The continuity jump condition arises from the fact that mass cannot accumulate on the interface. This condition is met since the mass transfer terms are written as source and sink terms in the opposite fields (e.g., $\Gamma_{cl\text{-}dv} + \Gamma_{dv\text{-}cl} = 0$). The momentum jump condition dictates that the net force acting on the interface (summation of the underlined forces in all fields) is balanced by surface tension effects. In general, neglecting surface tension effects, no net momentum is stored at the interface. The mass is assumed to depart from the donor phase at the donor phase velocity, and arrive in the receptor phase with that same velocity. The energy jump condition dictates that the net energy flux to the interface is balanced by the latent heat and work done against surface tension. If the latter term is neglected, the interface is assumed to be saturated, and mass transfer will take place at saturation. The interfacial heat transfer converts mass at the phasic enthalpy of the donor phase to the saturation enthalpy of the receptor phase.

3.4 Forces acting on a bubble in a narrow space

3.4.1 Drag force

Most of the available correlations for drag have been developed for a single bubble in an infinite medium. These ideas can be extended to small tubes and channels for very small bubbles. However, for larger bubbles that are confined by the walls, new models need to be developed. When the bubbles are confined, they start elongating with a spherical cap and a cylindrical tail in the case of a tube, and a planar cap in the case of a thin rectangular channel. Beyond a certain size slug, the vapor will be considered to be a continuous vapor field as will be discussed later. Therefore, the bubbles may look like Taylor bubbles in the narrow dimension, whereas they may seem circular in the other two dimensions. This remains a concern since the cell model does not approximate ensemble averaging. In order to develop a drag model in the dispersed vapor (*dv*) field, the spherical equivalent size of the capped bubble will be considered. Where the bubble ceases to be dispersed and becomes a continuous field will of course depend on the dimensions of the channel, i.e., the hydraulic diameter. Even within the *dv* field, the bubbles go through distinct regimes and have different rise velocities. The rise velocity given as a function of the equivalent bubble diameter is an important component in the development of bubble drag.

The bubble diameter at which the bubble ceases to be spherical is proportional to $(\sigma/\Delta\rho g)^{0.5}$, where σ and $\Delta\rho$ are the surface tension and density difference respectively. This quantity simply comes from the force balance between the buoyancy and surface tension acting on the bubble. This can also be recast in terms of the Eotvos number, *Eo*, defined as $Eo = g\Delta\rho d^2/\sigma$, where d is the maximum horizontal dimension of the bubble. Tomiyama [42] classified bubbles in several groups based on their *Eo*, and solved multi-field equations similar to eqns. (5)–(7) in vertical tubes. For each field or group, he used a drag coefficient based on *Eo* and the Morton number. Based on the proposed drag coefficient, he calculated the terminal rising velocities of single bubbles in stagnant liquids, and plotted Re vs. Eo for various Morton numbers, Mo [defined in eqn. (9)] to compare with available data. He classified the bubble shapes for Mo $< 10^{-6}$ into spherical and ellipsoidal (oblate spheroid, $Eo < 40$) and spherical-cap ($Eo > 40$) with a flat rear surface. This is an effective approach in thin tubes, if the flow regime is predominantly bubbly and slug flow. It is not clear whether the drag relations given by [42] for air–water systems are applicable for other fluids and at high pressures. A similar approach to the drag model was taken by Kumar *et al.* [21] who measured bubble rise velocity and bubble sizes in different groups at high pressures (Figure 4).

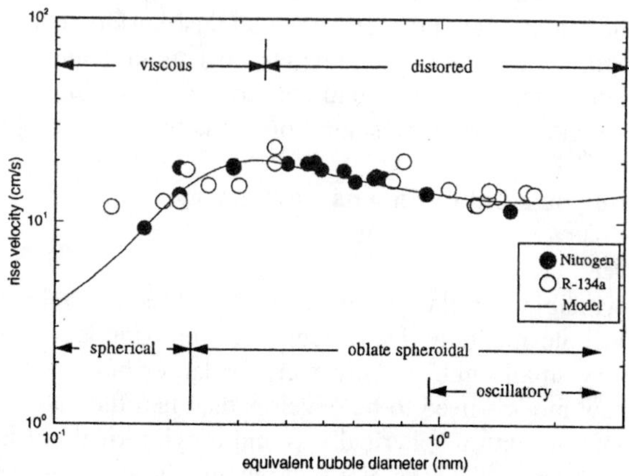

Figure 4: Relationship between bubble diameter and rise velocity using R-134a and nitrogen bubbles in a high aspect ratio duct [21].

The bubbly regimes are identified based on the bubble shape and rise behavior. The small bubbles obey the Stokes' law in the viscous regime so that the rise velocity can be expressed as

$$U_V = \frac{g\Delta\rho d_{dv}^2}{K_b \mu_{cl}}; \quad K_b = K_M \left(\frac{d_{dv}}{d_\infty}\right)^{0.425}; \quad d_\infty = 2\left\{\frac{\sigma}{\Delta\rho g}\right\}^{1/2} \tag{8}$$

$$K_M = 60(1 - e^{-5.3\times10^{10} Mo_{mb}}); \quad Mo_{mb} = \frac{Mo}{(1-\alpha_{dv})^4}; \quad Mo = \frac{g\mu_{cl}^4 \Delta\rho}{\rho_{cl}\sigma^3}. \tag{9}$$

where d_∞ is the diameter corresponding to the minimum rise velocity, and d_{dv} is the bubble diameter in the *cl-dv* (continuous liquid-dispersed vapor) regime. At larger bubble diameters, the bubbles become oblate and the rise velocity slightly decreases because the wake behind the bubble becomes turbulent. In this distorted-inertial regime, the rise velocity, U_{DI}, is given by [43]

$$U_{DI} = \left[g \frac{d_{dv}}{2} \frac{\Delta\rho}{\rho_{cl}} + \frac{2\sigma}{\rho_{cl} d_{dv}} \right]^{1/2} \tag{10}$$

using the analogy that exists between the propagation velocity of surface waves and rise velocity of bubbles in infinite media. In a multiple bubble system, the single bubble rise velocity becomes the local drift velocity. Using Fan and Tsuchiya's [44] model and data in Figure 4, the local drift velocity can be written as [21]

$$V_{gj}^{B} = \left\{ \left(V_{gj}^{V}\right)^{-8} + \left(V_{gj}^{DI}\right)^{-8} \right\}^{-1/8}; \quad V_{gj}^{V} = U_V (1 - \alpha_{dv}); \quad V_{gj}^{DI} = U_{DI} f_{DI}. \tag{11}$$

Here, f_{DI} accounts for the multiple bubble effects [45] and is given by

$$f_{DI} = \frac{(1-\alpha_{dv})^{5/2} \left\{ 1 + \dfrac{0.46}{Mo^{1/8}} \right\}}{1 + \dfrac{0.46}{Mo^{1/8}} (1-\alpha_{dv})^{9/7}} \tag{12}$$

The drag coefficient for a single bubble in a multiple bubble environment can be written by balancing the drag force to its buoyancy as

$$C_D = \frac{4}{3} \frac{d_{dv}}{V_{gj}^{B2}} \frac{\Delta\rho}{\rho_{cl}} g \tag{13}$$

Now, the drag force per unit volume in a two-phase bubbly flow can be expressed in terms of the drag coefficient as

$$F_D = C_D N A_P \rho_{cl} \frac{V_{gj}^2}{2} \tag{14}$$

where V_{gj} is the drift velocity in the multiple bubble system, and is different from V_{gj}^{B} given earlier. V_{gj} is related to the relative velocity between the phases in the following manner:

$$V_{gj} = (1-\alpha_{dv}) u_R . \tag{15}$$

u_R is the relative velocity between the *cl* and *dv* fields, N and A_P in eqn. (14) are the bubble number density and the projected area of the single bubble and are given by

$$N = \alpha_{dv} / V_B; \quad A_P / V_B = 3/2d_{dv} \tag{16}$$

where V_B is the bubble volume. Note that NA_P is proportional to the interfacial area density. The drag given in eqn. (14) appears to be explicitly independent of the bubble diameter; however, V_{gj}^B in the drag coefficient is dependent on the spherical equivalent diameter. This expression for bubbly drag can now be used in the local application of drag force in a multi-dimensional code.

3.4.2 Lift force

The spiraling motion of the bubble depending on its size is well established for large size tubes. In thin channels, the micron-size bubbles originating at the heated surface are at first spherical in shape. As they grow in size, they become slightly distorted, but continue to be three-dimensional. The motion of these bubbles tends to become oscillatory since the bubbles are distorted until they grow in size to be completely confined between the heated walls in the narrow space. The lift force captures the influence of the liquid velocity gradient on the bubbles, thereby impacting the transverse distribution of the void fraction profile. This force drives the small, spherical bubbles toward the walls. In wall-heated two-phase flow, the superheated liquid layer exists near the wall, while the bulk fluid may be subcooled. Bubbles segregate near the wall where they originate, and this situation is usually called wall peaking of void fraction. Lahey [45] predicted Serizawa *et al.*'s [46] wall-peaked bubbly flow data well in a vertical tube. Downstream of the onset of nucleate boiling, the bubbles become large, distorted and oscillatory, the lift force changes direction and the bubbles move away from the wall. This allows the vapor to concentrate at the center, giving rise to a center-peaked profile.

Zun [47] describes the intrinsic fluctuating bubble motion in terms of the Eotvos number, *Eo*, and reports that the magnitude of the fluctuation increases with *Eo*. Using an interface tracking method, Tomiyama [42] predicted single bubble motions to determine the effect of *Eo* (Figure 5). Their predicted velocity field indicates that the bubble fluctuation is closely related to periodic vortex shedding, and *Eo* is the most appropriate parameter to define the lateral movement of the bubbles. At low *Eo*, the tendency of the bubble is to move toward the wall, and at high *Eo*, the bubble migrates to the center. The lift force was developed using an inviscid flow field around a sphere [48, 49]. By integrating the interfacial pressure over the sphere's surface:

$$F_L = -C_L \rho_{cl} \alpha_{dv} u_R \times (\nabla \times u_R) \tag{17}$$

The functional form of the lift is the same for shear-induced lift in small *Eo* and lift induced by the vortex motion behind the deformed bubble in large *Eo* [42]. Kumar *et al.* [21] found a similar functionality for the lift coefficient in R-134a in a thin high aspect ratio duct, as given opposite.

$$C_L = \begin{cases} 0.1 & Eo < 3 \\ 0.33\left[1 - 0.41\sqrt{Eo}\right] & 3 \le Eo \le 9 \\ -0.075 & Eo > 9 \end{cases} \tag{18}$$

The Eotvos number, Eo in eqn. (18) is defined in terms of the cross-section averaged bubble diameter. A different functionality was developed for air–water in a circular tube by Tomiyama [42], who defined Eo in terms of the maximum horizontal dimension of the bubble. Note that the lift coefficient used here in R-134a for small bubbles is 0.1, much smaller than the value of 0.3 generally used in air–water systems [42]. The Eotvos number based lift coefficient allows good predictions of wall-peaked and center-peaked void profiles in high pressure systems in thin ducts of different orientations as will be shown in the integrated model predictions later.

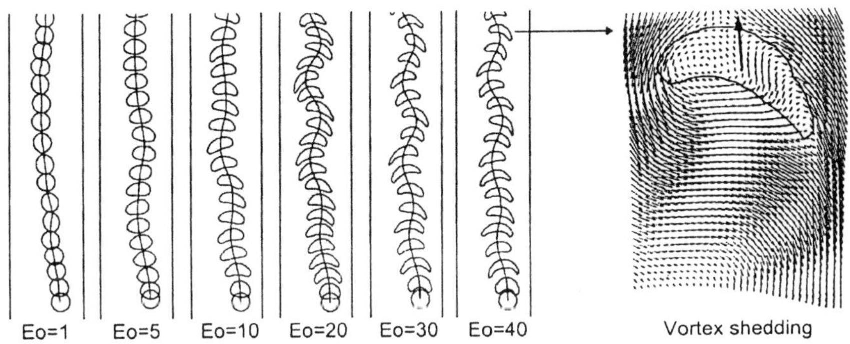

Figure 5: Effect of Eotvos number on fluctuating motion of a single bubble [42] in a multi-group bubble formulation.

3.4.3 Lubrication wall force

For bubbles flowing near a wall, a lateral force acts to move the bubbles toward the center of the channel. This wall force is derived from potential flow theory, and is developed to approximate the flow field interaction between a single two-dimensional circular bubble with a planar wall [50, 51].

$$-M_{cl}^{w} = M_{dv}^{w} = -[C_{w1} + C_{w2}(\frac{d_{dv}}{y})]\frac{\rho_{cl}\alpha_{dv}}{d_{dv}}\left|(u_{dv} - u_{cl}).n_x\right|^2 n_y \tag{19}$$

where x and y are the streamwise and the narrow space directions respectively, n_x, n_y are the unit normals in their respective directions, and C_{w1}, C_{w2} = -0.1, 0.12. A form of this equation was used in [21]. The model had a shortcoming in that a bubble located far from the wall would also be attracted to the wall. To correct this defect, Tomiyama *et al.* [52] modified this model based on measured bubble trajectories for single bubbles released from a near wall region in a stagnant liquid. For a circular tube, this force was corrected to, using the current notation,

$$M_{dv}^{w} = -C_{w}\frac{d_{dv}}{2}\left[\frac{1}{y^{2}} - \frac{1}{(D-y)^{2}}\right]\rho_{cl} \mid (u_{dv} - u_{cl}).n_{z} \mid^{2} n_{r} \tag{20}$$

where D is the diameter of the tube, n_z and n_r are the unit vectors in the axial and radial directions, respectively. They used an Eotvos number dependent coefficient as they had done in [42].

3.4.4 Wall shear

Since the continuous liquid stays in contact with the wall until the film dries out, the wall shear can be applied only to the liquid phase. The wall shear formulation uses a law-of-the-wall approach in two-phase flow, which has been validated in [53, 54]. The wall shear is given as

$$\tau_{w} = T_{m}u_{P}; \qquad T_{m} = \rho_{cl}C_{\mu}^{1/4}k_{P}^{1/2}\kappa / \ln(Ey_{P}^{+}) \tag{21}$$

where $E = 9.0$, the subscript P denotes the first node from the wall, y_P is the wall node distance, and k_P is the kinetic energy at the wall node calculated from the kinetic energy transport equation, κ and C_μ are the single phase coefficients of 0.4 and 0.09 respectively.

3.4.5 Effective viscosity

Sato *et al.* [55] proposed an effective viscosity model that calculates the two-phase turbulence viscosity in the continuous liquid field. This model superimposes shear-induced viscosity with bubble-induced turbulent viscosity as

$$\mu_{T} = \rho_{cl}C_{\mu}k_{cl}^{2} / \varepsilon_{cl} + \rho_{cl}C_{\mu b}d_{dv}\alpha_{dv}u_{R} \tag{22}$$

The notation used in eqn. (22) is consistent with those in single-phase flows. $C_{\mu b}$ is the bubble-induced coefficient, which is taken to be 1.2. In principle, the fluctuation of the liquid velocity from the mean is split into two parts: one due to the inherent liquid turbulence independent of relative motion of the bubbles and the other due to the additional turbulence caused by bubble agitation. The second term is similar in form to Prandtl's hypothesis and corresponds to the probability of vortex generation. This model is valid only for dilute bubbly flows, but has been applied with relative success at high void fractions in thin channels. The linear superposition of turbulent viscosities is questionable for high void fraction bubbly flows.

3.4.6 Turbulent dispersion

A phenomenological turbulent dispersion model was derived by de Bertadano [56] and is given by

$$M_{td} = -0.1\rho_{cl}k_{cl}\nabla\alpha_{dv}. \tag{23}$$

The impact of this model in dispersing the bubbles and affecting the inception of transition is significant, and the modeling coefficient was empirically determined and could vary depending on the channel shape, size and pressure.

3.4.7 Bubble size

Much has been discussed about the bubble size and shape in relation to lift and drag forces. In fact most of the models discussed so far are connected directly or indirectly with the bubble size. Therefore, it is crucial to determine the development of the bubble size and shape. This is usually done by solving a transport equation for bubble number density [57–59] or interfacial area density [60] using closure conditions for bubble coalescence and breakup, or bubble grouping as done in [42]. This will be discussed in a later section. To assess the predictability of the lateral distributions in void fraction in bubbly flows, expressions for both departure size and bulk size are needed. Although these two sizes depend on different parameters (i.e., the departure size depends on wall superheat and the bulk size depends on the local void fraction), Levy's [61] model can be used as the building block for model development.

Levy's model was built on the premise that bubble departure is controlled in part by competition between surface tension, which tends to hold the bubble to the nucleation site, and wall shear stress, which tends to pull the bubble off the nucleation site. When the pressure is increased, the saturation temperature increases, decreasing the surface tension. The decreased surface tension results in a smaller bubble departure diameter. Increasing the flow rate increases the wall shear stress, which similarly reduces the bubble departure diameter. Levy's model is given by

$$d_0 = 0.0315(\sigma d_h / \tau_w)^{1/2} \tag{24}$$

where τ_w is the wall shear stress. The bubbles away from the wall can be considerably larger than the departure diameter. Using d_o as the kernel, Kumar *et al.* [21] correlated the two bubble sizes as follows:

Departure bubble size:

$$d_{dv} = \min[d_0, d_0 \exp\{-(\Delta T_{sub} / 45\Delta T_{sonb})^{0.5}\}] \tag{25}$$

where $\Delta T_{sub} = T_s - T_{cl}$ and ΔT_{sonb} is the nucleation inception wall superheat given in section 3.8.

Bulk bubble size:

$$d_{dv} = d_0 \left\{ 1 + \frac{21.64\alpha_{dv}}{(\alpha_0 - \alpha_{dv})^{0.29}} \right\} \tag{26}$$

where α_0 is a critical void fraction, which is set arbitrarily at 0.8. At high void fractions, the bubble size grows to an incipient slug size, which is the minimum cylindrical planar cap bubble size in channels for which the slug terminal rise velocity limit is reached. At this point the dispersed vapor transitions into continuous vapor in a multi-field modeling approach, and different characteristic length scales would have to be used.

3.4.8 Interfacial area density (*cl-dv*)

The interfacial area density for the bubble is determined from the assumption of an equivalent spherical particle, and is given by

$$A'''_{dv-cl} = 6\alpha_{dv} / d_{dv} \tag{27}$$

3.5 Annular flow forces (*cl-cv*)

3.5.1 Wave characteristics

In annular flow, the interface between the liquid and the vapor core is characterized by interfacial waves, which play a dominant role in governing the system parameters. As described by Hall-Taylor *et al.* [62], there are basically two types of waves: large amplitude disturbance or roll waves and low amplitude high frequency surface ripples. The topology of the interface in annular flows was recently mapped using several conductivity probes by Gottmann [63] as shown in Figure 6. Recently, three-dimensional circumferential ring waves in air–water in tubes [64], and λ-shaped waves in R-134a in a high aspect ratio rectangular duct [65] were observed. All these waves travel at much higher velocities than the liquid films, increase the pressure drop and hence the interfacial shear stress, and are a source of entrained droplets in the vapor core.

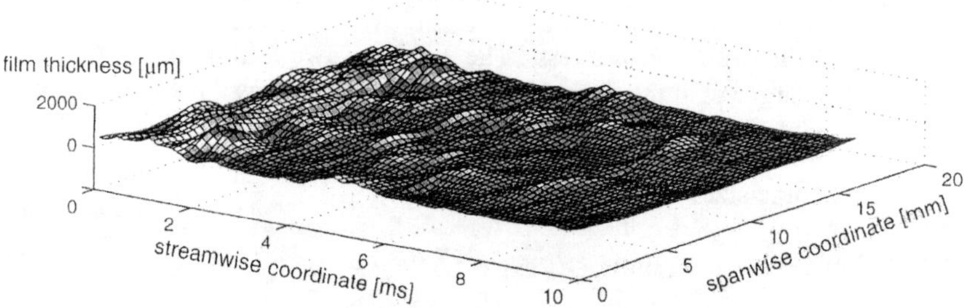

Figure 6: Interface reconstruction of film thickness measurements using nine conductivity probes in air–water [63]. The base film (substrate) thickness varied from 50 μm to ~350 μm.

One class of models is based on the premise that the waves on the interface act as roughness elements. This has led to the development of a two-phase friction factor analogous to the pipe friction factor for single-phase flow. The liquid film is treated as a rough surface to the gas core and friction at the interface can be expressed in terms of an equivalent sand roughness.

The most widely used model for interfacial shear using a friction factor was proposed by Wallis [66]. Using the notation *cl* and *cv* for continuous liquid and continuous vapor, Wallis' expression for the interfacial shear, τ_i, can be written in terms of the average velocity in the vapor phase, $\overline{u}_{cv}$, interfacial friction factor, f_i, as

$$\tau_i = \frac{1}{2} f_i \rho_{cv} \overline{u}_{cv}^2; \quad f_i = f_{sp}(1 + 300\frac{\delta}{D}) \tag{28}$$

where f_{sp} is the single-phase friction factor usually taken to be 0.005, and δ and D are the average film thickness and the tube diameter. This correlation was improved to include the pressure effects [67] and in the ripple-wave dominated air-water flow [68]. The surface with ripples cannot simply be replaced with a solid rough surface, and the interfacial shear stress should include the wave dynamics, i.e., wave frequency, wave velocity and wave shape [69].

3.5.2 Film thickness

3.5.2.1 Cornering effect near sharp edges Henstock and Hanratty [70] have correlated the liquid film thickness for low-pressure air-water annular flows in circular tubes. This correlation takes the form

$$\begin{aligned} h^+ &= \frac{h}{\nu_{cl}}\sqrt{\frac{\tau_c}{\rho_{cl}}} = \left[(0.707\,\mathrm{Re}_{lf}^{0.5})^{2.5} + (0.037\,\mathrm{Re}_{lf}^{0.9})^{2.5}\right]^{0.4} \\ \tau_c &= 2/3\tau_w + 1/3\tau_{cl-cv} \end{aligned} \tag{29}$$

The nondimensional film thickness, h^+, is given in terms of the wall shear, τ_w and interfacial shear, τ_{cl-cv}, using a weighting factor. This correlation overpredicts the data in thin rectangular ducts [71]. This is due to the "cornering" of the liquid. Due to the maldistribution of liquid film thickness and flow rate, the actual liquid film Reynolds number at the measurement location was smaller than the average value obtained from the liquid flow rate and measured entrainment fraction, which resulted in the overprediction.

Kumar's group [31, 65, 72, 73] measured the void profiles in the thin dimension and the width in adiabatic and wall-heated annular flows. In wall-heated flows, Kumar and Trabold [73] reported that as the pressure in R-134a is increased for the same mass flux and void fraction, the film thickness becomes narrower. At the lowest and highest pressures of 0.9 MPa and 2.4 MPa, their hot-film anemometer measurements showed that the void fraction profile in the narrow dimension (*Z*/*t*) changes from a parabolic type to a wall-peaked profile (Figure 7). The strong inversion in the narrow space at high pressure can be attributed to the increased Weber number (due to reduced surface tension), which increases the entrainment fraction. The film thickness in the edge is nearly the aspect ratio times the film thickness on the flat side. This redistribution of the liquid film to the edges results in larger drops being entrained.

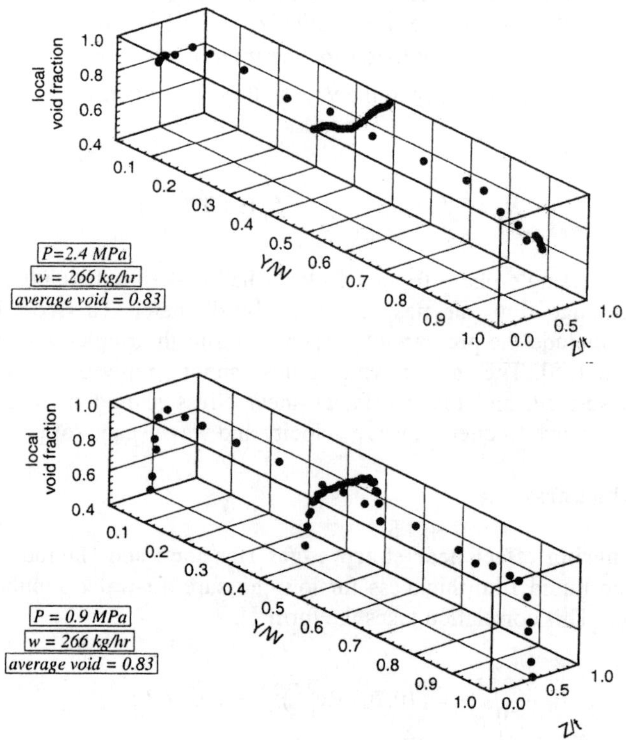

Figure 7: Void fraction profiles in the narrow and the width dimensions at two different pressures in R-134a [73].

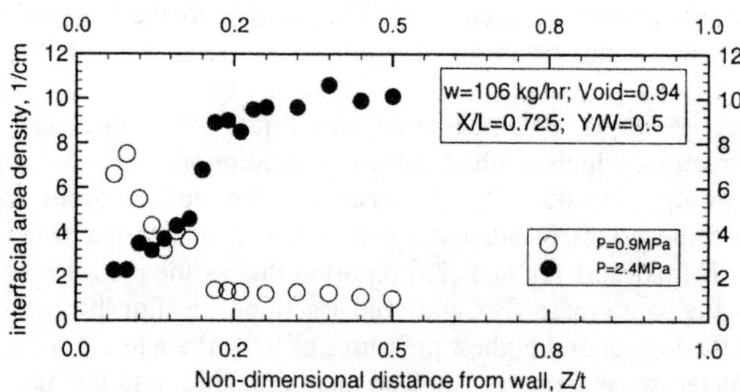

Figure 8: Interfacial area density profile at two different pressures in a vertical rectangular duct [73].

The interfacial area density of the droplets (*cv-dl*) for the flow conditions in [73] is seen to decrease from the wall to the core at low pressures, with the opposite effect at high pressures (Figure 8). The droplet diameter is an inferred measurement and carries a higher uncertainty than interfacial area density. However, both measurements have nearly the same trend. At high pressures, the droplet diameters range from 10 μm to 300 μm. The smaller drops are probably sheared off from the very thin film on the flat side, staying closer to the wall. The larger ones in the middle of the duct emanate either from the edges or from the liquid bridges in the transition regime. Careful measurements of drop size need to be made near the edges and in the transition regime in rectangular ducts.

3.5.2.2 Base film and wave amplitude The close-up view of the interface near the wall given in Figure 9 shows that the interface between the liquid film and the vapor core is separated but convoluted [74]. The film contains large size bubbles that are misshapen and confined through entrainment or evaporation. The vapor core contains droplets that are entrained from the film due to wave shearing or broken up from the liquid ligaments in the churn-turbulent regime below. The liquid film can be divided into two layers: *cl* that extends from the wall to the wave troughs and the disturbed wavy layer from the crests to the troughs [75]. This flow situation is depicted in Figure 9.

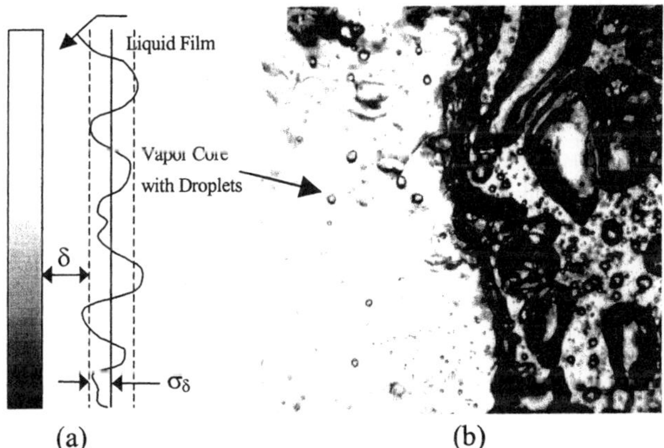

Figure 9: a) Schematic of the film with base film thickness, δ, and amplitude σ_δ; b) photograph of the film (3.8mm × 3.8mm) in annular flow in R-134a [74].

The ratio of the root mean square fluctuation to the base film thickness, σ_δ/δ, may be taken to be a good measure of the relative roughness, *K*. The friction factor correlations in the literature are based on the premise that the interface structure for a given ratio of film thickness to the diameter is independent of the phasic flow rates. However, the conductivity measurements made in a rectangular duct [76] show that the film thickness and the relative roughness are functions of Re_{cv} and Re_{cl}, which are written in traditional notation as Re_g and

Re_l respectively, based on average phasic velocities. The interface is treated as a rough surface to both the liquid film as well as the gas core. In addition, if the shear and velocity in each phase across the interface were equated, the friction factor would depend on the phasic Reynolds number raised to a different exponent. Accordingly, using the data in [76], the film thickness and roughness are fitted in terms of $Re_l^{0.15}/Re_g^{0.3}$ as shown in Figure 10. Note that the relative roughness peaks to nearly 4.0 and drops to a value of 1.0 for thicker base films. The ratio of the disturbance wave amplitude to mean film thickness is close to 5.0 as reported by Hewitt [77].

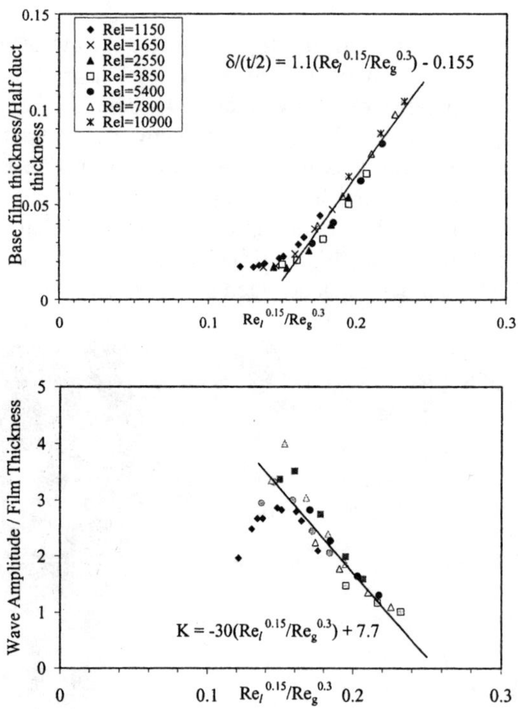

Figure 10: Base film thickness and wave amplitude in air–water flows [76].

Figure 10 shows that the wave amplitude to mean film thickness ratio is a strong function of the film thickness. When the Reynolds number ratio is less than 0.15, the film thickness is found to be a constant at 50 μm. However, a slight increase in the liquid flow rate results in a region dominated by ripples with pulses of disturbance waves. These spikes increase the film roughness, and have also been identified by Hall-Taylor *et al.* [62]. The void inversion and the subsequent thinning of the film are not expected in air–water flows. Consequently, the film thickness and roughness would have to be modeled for these special situations. It is possible that other parameters such as density ratio and Weber number play a role in correlating the data in all fluids.

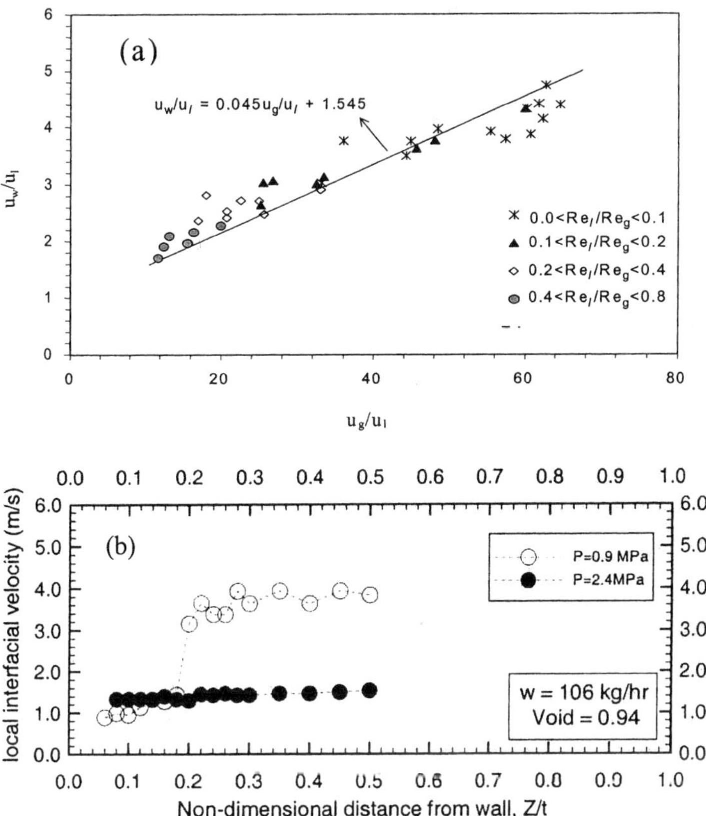

Figure 11. a) Conductivity measurements of wave velocity as a function of vapor velocity in air–water [76]; b) hot film measurements of droplet velocity in the narrow dimension in wall-heated R-134a flow at two different pressures [73].

3.5.3 Wave velocity

By equating the shear at the interface for both liquid and vapor, a simple relationship can be obtained for the average wave (or interfacial velocity) in terms of the average liquid and vapor, and assuming that each phase across the interface acts as a single-phase fluid [76]. The traditional subscripts, *g* and *l* are used to indicate vapor and liquid to denote average quantities.

$$u_i = u_w = \frac{Cu_g + u_l}{1+C}; \quad C \propto \left(\frac{\rho_g}{\rho_l}\right)^{1/2}\left(\frac{\mathrm{Re}_l}{\mathrm{Re}_g}\right)^{1/4} \tag{30}$$

The wave velocity measurements made with a conductivity probe [76] normalized by the liquid velocity, u_w/u_l, are plotted against u_g/u_l in Figure 11a,

and compared with the model in eqn. (30). Since only a limited scatter is seen for various Re_l / Re_g ranges, it is reasonable to assume that the wave velocity is a weak function of the Reynolds number ratio for the given density ratio.
C for air-water flows is $1.7(\rho_g/\rho_l)^{0.5}$. This constant is likely to change for different fluids and in boiling systems. The droplet velocity measurements in wall-heated annular flow made in the narrow dimension in R-134a by Kumar and Trabold [73] are shown in Figure 11b. The jump in the velocity profile at a nondimensional distance of 0.2 at 0.9 MPa offers evidence of the presence of disturbance waves. At 2.4 MPa, the profile is flatter, suggesting that the film is much thinner due to the cornering effect of square and rectangular channels. They used a value of $1.0(\rho_g/\rho_l)^{0.5}$ for C to calculate the average wave velocity in R-134a. The wave velocity obtained from the average liquid and vapor velocities qualitatively agrees with the measurements shown. Direct measurements of film thickness and wave velocity are needed to improve the model for high-pressure systems.

3.5.4 Interfacial shear

Much of the continuous vapor/continuous liquid modeling has been nonlocal and one-dimensional so far. We will continue along these lines at this time, and develop the interfacial shear for noncircular geometry incorporating the wave velocity measurements. For local application of interfacial shear, we will use the local interfacial area density. This type of modeling is only a slight improvement to what exists in the literature, and much work needs to be done in annular flow model development.

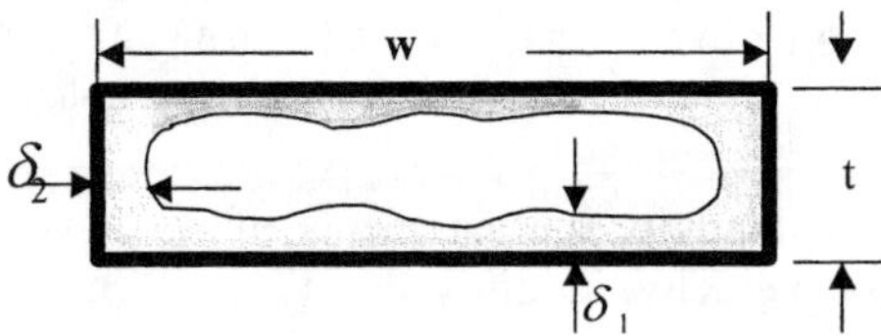

Figure 12: Schematic of the liquid distribution in a thin rectangular duct.

From Gottmann's [63] rectangular duct study, it is clear that (a) Wallis' [66] model for interfacial shear is valid, and (b) the average film thickness near the edges is approximately equal to the aspect ratio times that on the flat side of the duct. To implement Wallis' model locally in a rectangular duct, consider the liquid distribution in Figure 12. The average volume fraction of the continuous vapor can be written using $\delta_1 = \delta$, and $\delta_2 = (w/t)\delta$,

$$\bar{\alpha}_{cv} = \frac{(t-2\delta_1)(w-2\delta_2)}{tw} = 1 - 4\frac{\delta}{t} + 4\left(\frac{\delta}{t}\right)^2. \qquad (31)$$

For a very thin film, eqn. (31) can be approximated as

$$\frac{\delta}{d_h} = \frac{1-\bar{\alpha}_{cv}}{8}. \tag{32}$$

where d_h is the hydraulic diameter, and $\bar{\alpha}_{cv}$ is the cross-section average volume fraction in the continuous vapor field. For a circular tube, δ/d_h is twice the right hand side of eqn. (32). This equation should be viewed with caution. Although the average thickness appears to be thinner for a thin rectangular duct than for a circular tube, much of the liquid film collects near the edge. Now, the average interfacial area density for a thin duct is

$$\bar{A}''' = \left\{ \frac{2(w-2\delta_2)+2(t-2\delta_1)}{wt} \right\} = 2(\frac{1}{t}+\frac{1}{w})(1-2\frac{\delta}{t}) \tag{33}$$

From the definition of d_h and eqn. (31),

$$\bar{A}''' = \frac{4}{d_h}\sqrt{\bar{\alpha}_{cv}}. \tag{34}$$

The interfacial (*cl-cv*) drag in an annular flow film can now be written as

$$M_{D_f}^{cl-cv} = \frac{1}{2}\rho_{cv}(u_{cv}-u_i)^2(\frac{4}{d_h}\sqrt{\bar{\alpha}_{cv}})f_{sp}\left\{1+\frac{75}{2}(1-\bar{\alpha}_{cv})\right\} \tag{35}$$

where u_i is the interfacial velocity. Note that the expression in eqn. (35) includes an interface velocity and a cross-section averaged void fraction, which replaces the film thickness. Taking $C = (\rho_{cv}/\rho_{cl})^{0.5}$, $(u_{cv} - u_i)$ can be expressed in terms of the relative velocity, u_R, as

$$(u_{cv}-u_i) = \left(1+\sqrt{\rho_{cv}/\rho_{cl}}\right)u_R. \tag{36}$$

Using eqn. (36) in eqn. (35), the final expression for the interfacial shear is obtained in terms of the relative velocity, density ratio and average void fraction:

$$M_{D_f}^{cl-cv} = C_i\frac{1}{2}\frac{\rho_{cv}}{\left(1+\sqrt{\rho_{cv}/\rho_{cl}}\right)^2}\left[f_{sp}\left\{1+\frac{75}{2}(1-\bar{\alpha}_{cv})\right\}4\frac{\sqrt{\bar{\alpha}_{cv}}}{d_h}\right]u_R^2, \tag{37}$$

where C_i is a coefficient that needs to absorb the changes both in wave velocity as well as in the average film thickness for different fluids. Eqn. (37) expresses the interfacial drag in terms of the average *cv*-volume fraction. Due to the distortion of the interface by the waves, it may be more appropriate to use the cv-volume fraction rather than a local film thickness. Using a local interfacial area density, a localized form of eqn. (37) can be developed in a multi-dimensional formulation. This would ensure that when the *cl-cv* interface is not present locally, the interfacial shear is zero. Thus, the interfacial area density, in effect, could eliminate the need for a film thickness model to solve the equations.

3.5.5 Annular flow lateral forces

Near the *cl-cv* wavy interface in annular flow, there is a competition between restoring and disturbance forces such as the inertial force, surface tension, viscous, and pressure effects. Woodmansee and Hanratty [78] assumed that atomization is governed by a Kelvin–Helmholtz mechanism. The formation, destruction and maintenance of the wavy liquid film on the wall in annular flow results in a suction force caused by *cv* at the wave crest balanced by a retaining force [78]. The functional form of the suction force is obtained from a heuristic argument that the force is proportional to the product of the density and the square of the relative velocity between the phases. This suction force was used in its simple form in a three-field formulation by [40, 79]:

$$M_{cl-cv} = C_{coll}\rho_{cv}\left|V_R\right|^2 A'''_{cl-cv}\hat{n}_{wall} \tag{38}$$

It should be recognized that in eqn. (38), the use of $\hat{n}_{wall}$ makes the model nonlocal, and restricts the description to be valid only for thin films. The retaining force that balances the suction force in eqn. (38) is given in [78] in terms of surface tension, wave height, h_w, and the characteristic length of the wave, l_w. This was later approximated as a continuum surface force [80]. These expressions are given as follows:

$$M^{\sigma}_{cl-cv} = C_{\sigma}\sigma h_w / l_w^2 = \sigma\nabla\cdot\hat{n}. \tag{39}$$

It is assumed that the curvature term in eqn. (39) would model the liquid distribution near the corners well. However, results from this type of modeling have not been published so far. Kumar and Trabold [40] used a form of the suction or collective force in eqn. (38) in a three-field formulation to maintain the liquid on the wall, and predicted both local and global annular flow characteristics well. Antal *et al.* [79] showed that collective and dispersive type of lateral force (without the surface tension effects) were needed in maintaining the thin liquid film on the upper wall when the duct was tilted. It is recognized that a lateral force is necessary to maintain the liquid film on the wall, but what is available currently in the literature is at best a phenomenological model. Significant work needs to be done in this area considering the surface tension effects and wave speed.

3.5.6 Wall shear

It can be assumed that the continuous liquid is in contact with the wall in annular flow, and flows as single-phase Couette flow along the wall. Therefore, the wall shear is applied only to the *cl*-field and not to the *dl*, *dv* or *cv* fields. Wall shear stress is related to the turbulent kinetic energy, and is expressed similar to eqn. (21) in bubbly flow:

$$\tau_w = T_m u_{P,cl}; \quad T_m = \frac{\rho_{cl} C_\mu^{1/4} k_{P,cl}^{1/2} \kappa}{\ln(E y_c^+)}; \quad y_c^+ = \frac{\rho_{cl} y_c C_\mu^{1/4} k_{P,cl}^{1/2}}{\mu_{cl}} \tag{40}$$

where the subscript P denotes the first node away from the wall, y_c is half the thickness of the film, given by the product of the *cl*-volume fraction and the nodal distance, y_p. The constants, C_μ, E and κ are 0.09, 9.0 and 0.4 respectively, as in single-phase flows.

3.5.7 Interfacial area density

The wall is assumed to be completely wetted ($\alpha_{cl} = 1$ at the wall), and the *cl*-volume fraction decreases rapidly in the first few nodes away from the wall. The film thickness and the wave characteristics of the interface determine the steepness of the transition from the wall. Using a phase indicator function as given in eqn. (4), the interfacial area density can be calculated from the gradient normal to the wall [36] as

$$A'''_{cl-cv} = \nabla \alpha . \hat{n} \tag{41}$$

where $\hat{n}$ is a vector normal to the wall.

3.6 Droplet models

3.6.1 Droplet entrainment and deposition

One of the most important parameters that govern the transport process in transition and the annular flow is droplet entrainment. Together with the deposition of droplets, it controls the film thickness. Among the different types of generation of droplets, roll wave shearing in the annular flow and liquid bridge disintegration mechanisms in the transition regime are primarily associated with entrainment. The film entrainment and deposition of droplets are linked in that the entrained flux and the deposited flux should be balanced in equilibrium. The entrainment fraction under equilibrium conditions, E_∞, is the droplet fraction that includes both the net droplets entrained from the film and the droplet flow that is convected:

$$\dot{m}_{dl} = E_\infty (\dot{m}_{cl} + \dot{m}_{dl}) \tag{42}$$

3.6.2 Entrainment rate

The entrainment fraction was correlated based on the entrainment inception criterion and a force balance at the wavy interface by Ishii and Mishima [81]:

$$E = \tanh(7.25\text{x}10^{-7} We^{5/4}\, \text{Re}_{cl}^{1/4}). \tag{43}$$

The entrainment fraction obtained in low-pressure air–water, nitrogen–water and Freon-113 by [71] in thin rectangular ducts compared to the air–water tube data well. For $We^{5/4}\,\mathrm{Re}_{cl}^{1/4} > 10^6$, where most of the Freon and steam–water data exist in the range of $0.3 < E < 0.9$, the correlation overpredicted the data. By including the effect of Weber number in a correlation developed by Dallman *et al.* [82], they predicted the data better at all pressures. A simpler form of this correlation was provided in [83] as

$$E = f(We_c) = (1+\frac{3845}{We_c})^{-1};\quad We_c = \frac{\rho_m j_g^2 D}{\sigma};\quad \rho_m = \alpha_{cv}\rho_{cv} + \alpha_{dl}\rho_{dl}. \tag{44}$$

where We_c is the core Weber number and j_g is the average superficial vapor velocity. This modified Weber number dependence in the entrainment model seems to be appropriate. An increase in superficial vapor velocity shears off the liquid film more, thus increasing the entrainment rate. In addition, as the pressure decreases in the system, the surface tension increases, thereby decreasing the Weber number. This, in turn, reduces the entrainment fraction.

Going back to eqn. (42), it is convenient to generalize the entrainment rate in terms of the Reynolds number of the film and the droplets,

$$\mathrm{Re}_{cl} = \frac{\rho_{cl}u_{cl}\alpha_{cl}\delta}{\mu_{cl}};\quad \mathrm{Re}_{dl} = \frac{\rho_{dl}u_{dl}\alpha_{dl}d_h}{\mu_{dl}}. \tag{45}$$

Rewriting eqn. (45) in terms of the nondimensional entrainment rate,

$$\frac{\dot{\varepsilon} d_h}{\mu_{cl}} = \begin{cases} f(We_c)(\mathrm{Re}_{cl}+\mathrm{Re}_{dl}) & \mathrm{Re}_{cl} \geq \mathrm{Re}_{cl\,crit} \\ 0 & \text{otherwise} \end{cases} \tag{46}$$

$\mathrm{Re}_{cl\,crit}$ is generally set to 80 since below this critical Reynolds number, entrainment does not occur [71]. In eqn. (46), $f(We_c)$ may be replaced by eqn. (44). The entrainment rate, $\Gamma_{cl\text{-}dl}$, is written in terms of the interfacial area density between the continuous liquid and continuous vapor as

$$\Gamma_{cl-dl} = \dot{\varepsilon} A'''_{cl-cv} \tag{47}$$

3.6.3 Deposition rate

The deposition rate per unit area of the interface is given as [84]:

$$\dot{m}_{dl} = k\rho_{dl}\alpha_{dl},\ \text{where}$$

$$k = \begin{cases} 0.18\sqrt{\sigma/\rho_{cv}D} & \rho_{dl}\alpha_{dl}/\rho_{cv} < 0.3 \\ 0.083\sqrt{\sigma/\rho_{cv}D}\left(\rho_{dl}\alpha_{dl}/\rho_{cv}\right)^{-0.65} & \rho_{dl}\alpha_{dl}/\rho_{cv} > 0.3 \end{cases} \tag{48}$$

Alternatively, the nondimensional deposited flux of droplets can also be written in terms of the droplet Reynolds number as

$$\dot{d}d_h / \mu_{cl} = C_1 \, \mathrm{Re}_{dl} \, . \tag{49}$$

Re_{dl} is usually defined in terms of the droplet velocity component normal to the plane of deposition. However, the droplets could deposit on the film due to the vapor core turbulence even in the absence of a normal component. Therefore, accounting for this perturbed velocity, the deposition rate can be written as

$$\Gamma_{dl-cl} = \rho_{cl}\tilde{u}_{dl}\alpha_{dl}A'''_{cl-cv}, \tag{50}$$

where $\tilde{u}$ is the sum of the mean normal component and the turbulence component. The entrained and droplet flux should be equal in adiabatic equilibrium flows. Through numerical experiments, the turbulence component of the droplet velocity can be evaluated in terms of u_{cv}. A typical value for adiabatic equilibrium flows is 3%. The deposition model given in eqn. (50) does not explicitly account for the motion of the interface in nonequilibrium flows. It is assumed that the interface recedes from the droplets very slowly.

3.6.4 Droplet size

It is a consensus view that the drop size depends on the superficial gas velocity in two-phase annular flow, and takes general form, $d_{dl} \sim j_g^{-n}$, where j_g is the superficial vapor velocity. However, the value of n has been a subject of debate. Data in different fluids show that n can be 1.0 [85] or 0.3 [86]. The standard Weber number criterion suggests that the mean drop diameter is proportional to j_g^{-2}. A more sophisticated model that accounts for the interface between the film and vapor core proposed by Kataoka *et al.* [87] shows that the mean drop diameter is proportional to $j_g^{-4/3}$:

$$d_{dl} = K \frac{\sigma}{\rho_{cv}(\alpha u_{cv})^2} \mathrm{Re}_{cv}^{2/3} \left(\frac{\rho_{cl}}{\rho_{cv}}\right)^{1/3} \left(\frac{\mu_{cv}}{\mu_{cl}}\right)^{2/3} \tag{51}$$

This model approximates the relative velocity between the mean gas flow and film flow by the superficial gas velocity, j_g. This assumption is generally valid for thin liquid films. More recently, Kocamustafaogullari *et al.* [88] developed an expression for the Sauter mean droplet diameter, by considering the maximum stable droplet size in the turbulent vapor core:

$$\frac{d_{dl}}{d_h} = 0.65[C_w^{4/15} We_m^{-3/5} (\frac{\mathrm{Re}_{cv}^4}{\mathrm{Re}_{dl}})^{1/15} (\frac{\rho_{cv}}{\rho_{dl}} \frac{\mu_{cv}}{\mu_{dl}})^{4/15} (\frac{\Delta\rho}{\rho_{dl}})^{-3/5}]$$

$$C_w = \begin{matrix} 0.028 N\mu^{-0.8} & N\mu \le 1/15 \\ 0.25 & N\mu > 1/15 \end{matrix} ; \qquad N\mu = \mu_{cl} / \rho_{cl}\sigma(\sigma / g\Delta\rho)^{1/2} \tag{52}$$

The quantity in brackets is K in Figure 13. Trabold and Kumar [65] and Beus *et al.* [71] found eqn. (52) to predict the average drop size data well in low and high-pressure fluids in rectangular ducts. Beus *et al.* [71] also recommended that the leading constant be changed to 1.3. Neither paper suggested improvements to the model to account for the three-dimensional cornering effect.

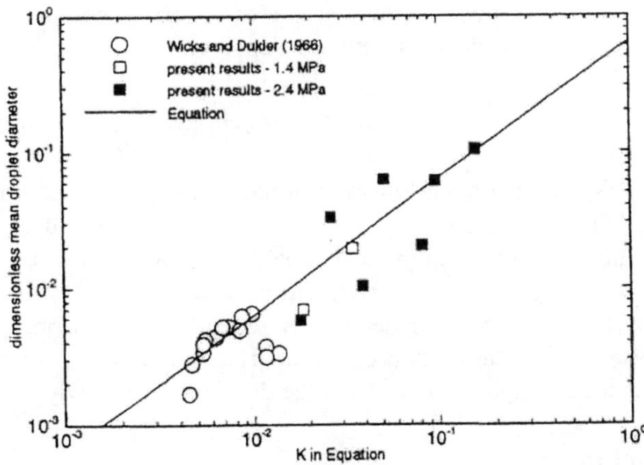

Figure 13: Average drop size measurements in R-134a flow in a thin rectangular duct [65] compared to the correlation given in eqn. (52).

3.6.5 Droplet drag

This model was originally proposed by Ishii and Zuber [89] and updated by Ishii and Mishima [90], who developed it for the viscous and the distorted particle regime from similarity criteria and incorporated the multiple particle behavior:

$$C_{D_{dl}} = \begin{cases} \dfrac{24}{\mathrm{Re}_{dl}}\left(1+0.1\mathrm{Re}_{dl}^{3/4}\right) & \mathrm{Re} < 1000 \\[2ex] \dfrac{2}{3} d_{dl}\left(\dfrac{g\Delta\rho}{\sigma}\right)^{1/2}\left\{\dfrac{1+17.67(1-\alpha_{dl})^{18/7}}{18.67(1-\alpha_{dl})^{3}}\right\}^{2} & \mathrm{Re} > 1000 \end{cases} \qquad (53)$$

$$\text{where} \quad \mathrm{Re}_{dl} = \frac{\rho_{cv}(u_{cv}-u_{dl})d_{dl}(1-\alpha_{dl})^{2.5}}{\mu_{cv}}.$$

In the viscous regime for $\mathrm{Re} < 1000$, the distortions of fluid particles are negligible. The factor $(1 - \alpha_{dl})^{2.5}$ modifies the viscosity and accounts for the increased drag in the nonspherical droplet. It also accounts for the effect of the interaction of multiple droplets on each other.

3.6.6 Droplet lift

This force was originally derived for bubbles using the potential theory as given in eqn. (17). In the droplet field, the fields are switched, and the droplet lift impacts the transverse distribution of the droplets. This model is given by Kumar and Trabold [40] who used a value of 0.2 for C_L:

$$\begin{aligned} M_{cv}^{L} &= C_L \rho_{cv} \alpha_{dl} u_R \mathrm{x} (\nabla \mathrm{x} \mathrm{u}_{cv}) \\ u_R &= u_{cv} - u_{dl} \end{aligned} \quad (54)$$

3.6.7 Droplet turbulent dispersion

The model for dispersion of droplets in the annular core is adapted from the bubbly model given in eqn. (56), and is given by

$$M_{dl}^{d} = -0.1 \rho_{cv} k_{cv} \nabla \alpha_{dl} \quad (55)$$

3.6.8 Droplet interfacial area density

The droplet interface is assumed to arise around a spherically shaped dispersed liquid (*dl*) field whose diameter is locally defined. Based on the spherical equivalent drop size, the interfacial area density is given by

$$A''' = 6\alpha_{dl} / d_{dl}. \quad (56)$$

3.7 Flow regime transition modeling

Coalescence and breakup have a significant effect on void distributions in the thin spacing and in determining flow regime transitions. In adiabatic, turbulent flows, equilibrium dictates that coalescence and breakup rates balance one another. The equilibrium population contains a range of different size bubbles which continuously break up and coalesce. As the vapor fraction increases, a nucleating bubble grows in size and fills the narrow space, the fraction of vapor that exists as continuous vapor increases, leading to a smooth transition from bubbly flow to slug flow. The transition regime consists of a mixture of dispersed vapor (*dv*) and continuous vapor (*cv*) with the corresponding bubbly and annular flow forces. This presents a problem. The continuous vapor is in the form of large vapor bubbles that fill the space rather than a truly continuous vapor core in the transition regime as seen in large tubes and channels. The models applied to this regime should reflect this difference. Therefore, the models that are applicable to the large vapor bubbles are used in the current approach.

The transition to annular flow is characterized by the breakdown of the large slugs via liquid bridge breakup into a vapor core containing droplets and an annular film. Interfield mass transfer characterizes both the transitions from bubbly flow to churn-turbulent, and from churn-turbulent to annular flow. The calculation of interfield mass transfer, interfacial forces, interfacial area density and heat transfer must capture this transition.

3.7.1 Coalescence (*dv-cv*) and breakup (*cv-dv*)

Edwards [59] extended the coalescence and breakup models of [57, 58] to the four-field formulation. The breakup and coalescence rates, *B* and *C* are evaluated by considering interactions with the bubble population. The transition regime is constructed as a combination of *dv* consisting primarily of bubbles of critical size and *cv* comprised of bubbles that are equal to or greater than the critical size. In this regime, the formulation of *cv* occurs both through the interaction of critical size bubbles (*dv*) with each other and the interaction of critical size bubbles with large bubbles (*cv*). In this method, the bubble number probability density function (pdf) can be split into

$$f(v) = f_{dv}(v) + f_{cv}(v) \tag{57}$$

Considering only the coalescence, *C*, the number density $f(v)$ of *v* sized *dv*-bubbles will be increased by the fraction (η) of coalescence of *v'* and *v'-v* smaller sized bubbles and decreased by the coalescence of *v* and *v'* sized bubbles. This is mathematically represented by

$$\begin{aligned} C(v) = {} & \frac{(\eta-1)}{2}\int_0^v c(v', v-v') f_{dv}(v', r, t) f_{dv}(v-v', r, t) dv' \\ & - \int_0^\infty c(v', v) f_{dv}(v', r, t)\left(f_{dv}(v, r, t) + f_{cv}(v, r, t)\right) dv' \end{aligned} \tag{58}$$

In the four-field framework, coalescence events can occur not only between two *dv* fields but also between the *dv* and *cv* fields. It is assumed that the probability of coalescence is proportional to the product of each bubble distribution function. The multiple-bubble effect was incorporated [58] by multiplying by $Y(\alpha) = 1/(1-\alpha)$. A similar expression to eqn. (57) can be developed for the breakup. By taking the different order moments, Edwards [59] showed that the net mass transfer rate is given by

$$\Gamma_{dv-cv} + \Gamma_{cv-dv} = \frac{(c_{0_{dv-dv}} 2\eta\alpha_{dv}^2 + c_{0_{dv-cv}} \alpha_{dv}\alpha_{cv})\rho_{cv}}{(1-\alpha_{dv}-\alpha_{cv})} - b_0\zeta\rho_{cv}\alpha_{cv}. \tag{59}$$

Note that α in the $Y(\alpha)$ expression is split into $\alpha_{dv}+\alpha_{cv}$. The coalescence and breakup rates, c_0 and b_0, remain to be specified and will depend upon the specific mechanisms involved. η and ζ are fractions of all coalescence events that result in *cv*, and the breakup events that result in *dv*, respectively.

The coalescence rate can be predicted if the bubble collision frequency and the coalescence probability are known. In a simple scenario, the product of this collision frequency and the bubble mass will then give the coalescence rate.

Using the analogy to molecular collisions in the kinetic theory of gases, the collision frequency can be calculated as the product of the number density, collision cross-section and collision velocity [91]. For two bubble populations, i and j, the coalescence rate is

$$c_0(v_i + v_j) = P_{c_j} A_{c_{ij}} V_c; \qquad A_{c_{ij}} = \frac{\pi}{4}(d_{b_i} + d_{b_j})^2 \tag{60}$$

i represents dv bubbles, while j can represent both dv and cv bubbles. The collision velocity, V_c, depends on the specific coalescence mechanisms such as due to turbulence, wake capture, velocity gradient and rise velocity. The first two are seen to be the most important to model, and have been given [91] as

$$V_c = 1.4\varepsilon^{1/3} d_{\max}^{1/3} + u_{rise} F_w. \tag{61}$$

The first term is due to the fluctuating velocity at the same length scale as the bubble. The second term results due to wake capture when a small bubble is entrained in the wake behind a large bubble. In this wake, the small bubble accelerates toward the trailing edge of the larger bubble where it collides and subsequently coalesces. F_w is used as an empirical geometry factor.

For a rectangular duct analysis, Maneri *et al.* [92] used the incipient slug size as $d_{max} = 0.527(w^2 t)^{1/3}$, where t and w are the thickness and width of the rectangular geometry respectively. The rise velocity was taken to be the slug velocity with an empirical function from the data provided for a rectangular geometry [20]:

$$\begin{aligned} u_{rise} &= f(t/w)(gw\Delta\rho/\rho_{cl})^{1/2} \\ &\text{where } f(t/w) = 0.245\{1 + 0.82t/w - 0.368(t/w)^2\} \end{aligned}. \tag{62}$$

The two bubble volumes in dv and cv are given by

$$v_{dv} = \pi d_{\max}^3 / 6; \qquad v_{cv} = \pi l_{cv}^2 t / 4; \qquad l_{cv} = (2d_{\max}^3 / 3t)^{1/2} \tag{63}$$

where l_{cv} is the diameter of the idealized cylindrical planar incipient slug. The projected areas are

$$A_{c_{dv-dv}} = 2l_{cv} t; \qquad A_{c_{dv-cv}} = (l_{cv} + w)t. \tag{64}$$

The probability, P_c, in each population is empirically determined.

3.7.2 Interfacial area density and characteristic length

From the geometry given in Figure 14, the interfacial area density for the planar slug is given as

$$A'''_{cl-cv_t} = S_b N = S_b\left(\frac{\alpha_{cv}}{V_b}\right) = \pi l_{cv} t\left(\frac{\alpha_{cv}}{\pi l_{cv}^2 t / 4}\right) = 4\alpha_{cv} / l_{cv} \tag{65}$$

where S_b and N are the bubble surface area and the number density. Since the *cl-cv* interface characterizes both transition and the annular flow regimes, it would be prudent to divide the large planar bubbles into two parts; one for the thin wavy film along the sides and one for the interface bridging the narrow space which forms the liquid bridges. Combining the interfacial area density eqn. (65) for transition and eqn. (41) for annular flow,

$$A'''_{cl-cv} = A'''_{cl-cv_f} + A'''_{cl-cv_t} = \nabla \alpha . \hat{n} + 4\alpha_{cv} / l_{cv} \tag{66}$$

As l_{cv} increases with α_{cv}, the first term in eqn. (66) becomes negligible as the annular flow is approached. For the spherical equivalent slug, the diameter, d_{cv} can be calculated. However, d_{cv} is a function of α_{cv}, and needs to be empirically obtained as in bubbly flow. d_{cv} was correlated by Maneri *et al.* [92]:

$$d_{cv} = \alpha_c^2 d_{\max} / (\alpha_c - \alpha_{cv})^2 \,. \tag{67}$$

α_c is the critical void fraction where the liquid bridges rupture forming a continuous vapor core or an infinitely long slug. d_{max} is the maximum bubble size at which *dv* becomes *cv*.

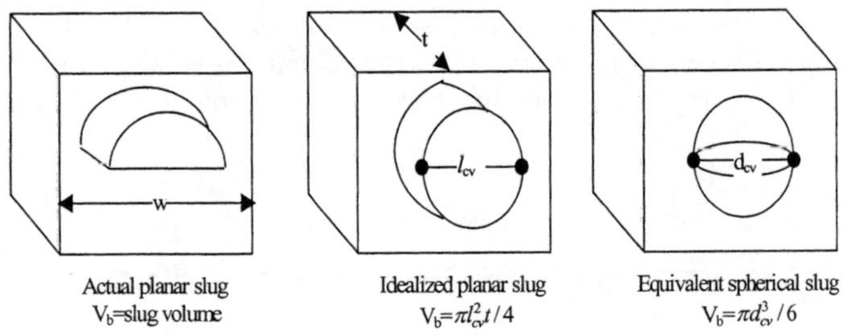

Figure 14: Idealized incipient slug geometry.

3.7.3 Transition drag

An important parameter in the transition regime is the interfacial shear or drag. In a four-field formulation, this drag must be constructed such that the slug flow characteristics of the *cv*-field disappear as the annular flow regime is reached. The drag force is developed exactly as done for the bubbly flow (*cl-dv*) given in eqns. (13)–(16), except here the reference velocity is the drift velocity which remains constant regardless of the bubble size when the incipient slug size is reached. The slug velocity for a rectangular duct is given in eqn. (62). As discussed in section 3.7.2, the bubble volume can be written in terms of a spherically equivalent bubble diameter, d_{cv}. As this volume increases, the projected area will increase from that of the incipient slug to that of the channel cross-sectional area. This idealization is shown in Figure 14. Finally, the drag can be formulated as

$$M_D^{cl-cv} = C_{D_T}(1-\alpha_{cv})^2 \rho_{cl}\alpha_{cv} \frac{wt}{V_b}\frac{u_R^2}{2} + M_{Df}^{cl-cv} \quad (68)$$

As the bubble volume, V_b, increases, the transition drag in the first term drops out, leaving the annular drag term given in eqn. (37).

3.7.4 Liquid bridge breakup

In addition to the entrainment of droplets from roll wave shearing given in section 3.6.2, the droplets are also formed in the transition regime where the liquid bridges break up. Flow instabilities and surface tension cause fragmentation, and the resulting droplets have varying size. A triangular relationship based on the concept of a critical Weber number, breakup time and velocity history was used by Pilch and Erdman [93] to model the bridge breakup. On a local basis, this model may be activated only when the *dv*-, *cv*-, and *cl*-volume fractions are all present in the same node. Thus, the two types of entrainment given in section 3.6.2 and in this section are separable and occur in different regions of the flow, but both contribute to $\Gamma_{cl\text{-}dl}$ in the governing equations. The shift of mass from *cl* to *dl* via the bridge breakup model is assumed to dominate in the transition regime and is given by

$$\begin{aligned} \Gamma_{cl-dl}^{bb} &= C_{bb}\alpha_{dv}\alpha_{cv}\rho_{cl} f \alpha_{cl} \\ f &= F(We)u_{cv}\sqrt{\rho_{cv}/\rho_{cv}}/D_0 \\ D_0 &= 2\sqrt{\sigma/g\Delta\rho} \end{aligned} \quad (69)$$

where f is the bridge breakup frequency and $F(We)$ is a curve-fitted correlation given in [40] based on the frequency measurements made for different We ranges [93]

$$F(We) = 0.07(We^{1/2} - 5.0) \quad (70)$$

This curve-fit of the original data, which contains significant scatter, is not valid for $We < 25$. The constant C_{bb} is taken to be 1.0 as a first approximation; however, it could vary as a function of the film thickness.

3.8 Heat transfer models

3.8.1 Heat partitioning at the wall

At the wall node, the input heat from the wall and initial vapor generation are introduced into the flow field. The models discussed here are based on a wall partition model developed by Kurul [94] for subcooled nucleate boiling, extended by Siebert *et al.* [38] for four fields. The gist of the model is that heat is apportioned to the fluid via various mechanisms such as sensible heating and forced boiling. All mechanisms receive heat simultaneously with the relative distribution being set by heat transfer coefficient models. Using the heat transfer coefficients and the saturation temperatures, a wall temperature is calculated so that the sum of the energy going into each heating mechanism sums to the boundary value.

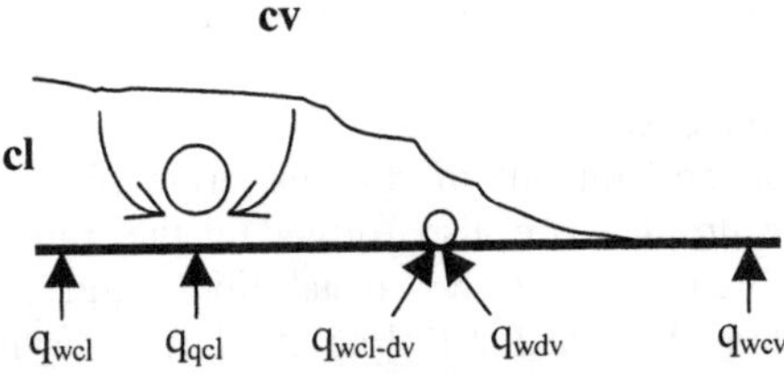

Figure 15: Heat partitioning at the wall based on [38, 94].

Referring to Figure 15, the input heat generation, q_{wtot}, is partitioned as

$$q_{wtot} = q_{wcl-dv} + q_{cl} + q_{wdv} + q_{wcv} \tag{71}$$

where $q_{wcl\text{-}dv}$ is the energy that goes directly into boiling, q_{wdv} is the energy transferred directly to the bubble, and q_{wcv} is the sensible heating of *cv*. The last term would be significant only in post-CHF situations, where the wall dries out. The energy transferred to the liquid, q_{cl} is further partitioned into two separate modes of heat transfer, namely,

$$q_{cl} = q_{qcl} + q_{wcl} \tag{72}$$

where the first term is the quenching mode of heat transfer which is the energy that heats up the cooler liquid that replaces the bubble after it departs the wall or collapses. The second term is the energy transferred to the liquid outside the influence area of the bubbles. In the subcooled boiling region, this energy acts to heat the liquid adjacent to the wall. The wall temperature, T_w can be determined in terms of heat transfer coefficients and ΔT as given in the next equation.

$$\begin{aligned} q_{wtot} = {} & H_{wcl}\left(1 - KN''\frac{\pi d_b^2}{4}\right)A_w'''(T_w - T_{cl}) + H_{qcl}KN''\frac{\pi d_b^2}{4}A_w'''(T_w - T_{cl}) + \\ & H_{dv}F_{dv}d_b A_w'''(T_w - T_{dv}) + fN''\frac{\pi d_b^3}{6}\rho_{dv}h_{fg} + H_{cv}A_w'''(T_w - T_{cv}) \end{aligned} \tag{73}$$

Note that in eqn. (73), Siebert *et al.* [38] took into account only the possible heat transfer paths that contributed the most. The fourth term in eqn. (73) is the fraction of the wall heat generation going directly into boiling. fN'' is the rate at which bubbles form and depart per unit surface area, and $\rho_{dv}h_{fg}$ is the heat content per unit volume. As seen in eqn. (73) and in Table 1, the nucleation site density, N'', is an intimate part of the heat partition model.

Table 1: Heat transfer coefficients in the wall-partitioning model.

Model	Comments
$H_{wcl}=0.023\frac{k_{cl}}{d_h}\left(\frac{\rho_{cl}d_hu_{cl}}{\mu_{cl}}\right)^{0.8}\left(\frac{c_{pcl}\mu_{cl}}{k_{cl}}\right)^{1/3}$	Dittus–Boelter outside bubble area. K in eqn. (73) is taken as 4.0 [95]
$H_{qcl}=2\left(\frac{f\rho_{cl}k_{cl}c_{pcl}}{\pi}\right)^{1/2}; f=(\frac{4}{3d_b}g\frac{\Delta\rho}{\rho_{cl}})^{1/2}$	Quenching model [95]
$H_{dv}=\frac{k_{dv}}{\delta_{dv}}$ where δ_{dv} = effective thermal layer = $\beta_1 d_b$ $F_{dv}=\beta_2 N''\pi d_b^2/4$ in eqn. (72); $(\pi/4)\beta_2/\beta_1=1$ [21, 92]	Energy transferred from wall to vapor. F_{dv} is the fraction of the wall area underneath the bubble where the liquid is totally evaporated.
$H_{cv}=0.023\frac{k_{cv}}{d_h}\left(\frac{\rho_{cv}d_hu_{cv}}{\mu_{cv}}\right)^{0.8}\left(\frac{c_{pcv}\mu_{cv}}{k_{cv}}\right)^{1/3}$	Dittus–Boelter

Following [92, 94], the nucleation site density is correlated as

$$N''=\min\left(\frac{\Delta\rho g}{\sigma}(2374\frac{\Delta T_s}{T_s})^{1.8}(1-\frac{\Delta T_{sonb}}{\Delta T_s})^{1.8}, N''_{\max}\right) \tag{74}$$

where $N''_{\max}$ corresponds to the maximum bubble packing point, the point at which four bubbles occupy a projected surface area of $2/3^{0.5}d_{dv}^{2}$. This correlation does not account for the dependence of nucleation sites on the surface material or surface finish. In eqn. (74), ΔT_s is the difference between the wall and saturation temperatures. ΔT_{sonb} is the nucleation inception superheat given by Davis and Anderson [96] as

$$\Delta T_{sonb}=180\left(\frac{2\sigma T_s\Delta\rho}{h_{fg}\rho_{cl}\rho_{dv}}\right)\frac{1}{k_{cl}}\left[0.023\frac{k_{cl}}{d_h}(\frac{Gd_h}{\mu_{cl}})^{0.8}(\frac{c_{pcl}\mu_{cl}}{k_{cl}})^{1/3}\right] \tag{75}$$

3.8.2 Interfacial heat transfer

The total transfer rate from phase i to phase j is the sum of the wall heat transfer given in section 3.8.1 (which is zero in the interior of the domain) and the interfacial heat transfer. Expressions for heat transfer coefficients on the liquid and vapor sides of the three possible interfaces (*cl-dv, cl-cv, cv-dl*) are given in Table 2.

Table 2: Interfacial heat transfer coefficients.

Model	Comments
$H_{cl-dv}=\frac{k_{cl}}{d_{dv}}\left[2.0+0.6\left(\frac{\rho_{cl}d_b\left\|v_{dv}-v_{cl}\right\|}{\mu_{cl}}\right)^{1/2}\left(\frac{c_{pcl}\mu_{cl}}{k_{cl}}\right)^{1/3}\right]$	Flow around a sphere by Ranz and Marshall [97]
$H_{cl-cv}=0.0045\rho_{cl}c_{pcl}\left\|v_{cl}\right\|\left(\frac{\rho_{cv}\left\|v_{cv}\right\|\mu_{cl}}{\rho_{cl}\left\|v_{cl}\right\|\mu_{cv}}\right)^{1/3}$	Proposed by Bankoff [98] for stratified cocurrent horizontal flows.
$H_{cv-cl}=0.023\frac{k_{cv}}{d_h}\left[\frac{\rho_{cv}d_h\left\|v_{cv}-v_{cl}\right\|}{\mu_{cv}}\right]^{1/2}\left(\frac{c_{pcv}\mu_{cv}}{k_{cv}}\right)^{1/3}$	Dittus–Boelter
$H_{cv-dl}=\frac{k_{cv}}{d_{dl}}\left[2.0+0.6\left(\frac{\rho_{cv}d_d\left\|v_{cv}-v_{cl}\right\|}{\mu_{cv}}\right)^{1/2}\left(\frac{c_{pcv}\mu_{cv}}{k_{cv}}\right)^{1/3}\right]$	Flow around a spherical droplet [97]
$H_{dv-cl}=K\frac{2k_{dv}}{d_{dv}};\quad H_{dl-cv}=K\frac{2k_{dl}}{d_{dl}}$	K is taken to be large to ensure that the droplet and the bubble are saturated and not superheated.

3.9 Turbulence models

Turbulence in two-phase flow in thin passages can significantly affect the lateral void distribution [99]. Therefore, the turbulence structure in *cl* and *cv* needs to be carefully modeled. The dispersed fields, *dv* and *dl*, can be considered to be laminar. Eqns. (22, 23, 55) use the turbulent kinetic energy and the dissipation of turbulent kinetic energy in both *cl* and *cv* fields. These quantities can be calculated from the k-ε transport equations, which are given here as a simple extension of the single-phase model to two-phase flows. Expressions have been developed for the Reynolds stress gradient term in the momentum equation [100]. However, further discussion on these sophisticated models is beyond the scope of this chapter. The eddy viscosity hypothesis is valid and has been used with bubble-induced turbulence as shown in eqn. (22). The steady-state k and ε transport equations are:

$$\nabla.(\rho_i\alpha_i u_i k_i)=\nabla.\{\alpha_i(\mu_i+\frac{\mu_i^T}{\sigma_i})\nabla k_i\}+P_i+G_i-\rho_i\alpha_i\varepsilon_i$$

$$P_i=\alpha_i\mu_{eff}\nabla u.(\nabla u+\nabla u^T);\quad G_i=\alpha_i\frac{\mu_{eff}}{\sigma_T}\beta_i(g.\nabla T) \tag{76}$$

$$\nabla.(\rho_i\alpha_i u_i\varepsilon_i) = \nabla.\{\alpha_i(\mu_i + \frac{\mu_i^T}{\sigma_\varepsilon})\nabla\varepsilon_i\} + C_1\frac{\varepsilon_i}{k_i}(P_i + G_i) - C_2\rho_i\alpha_i\frac{\varepsilon_i^2}{k_i} \quad (77)$$

3.10 Assessment of the multi-field model

Using the 4-field models described in this section, Maneri *et al.* [92] provided integrated comparisons of void fraction with their measurements. Their contour plots given in Figure 16 show the development of subcooled boiling from the inception of bubbly flow transitioning to annular flow. This figure provides a representative picture of the ability of the four-field model to predict the flow regimes without any reliance on flow regime maps.

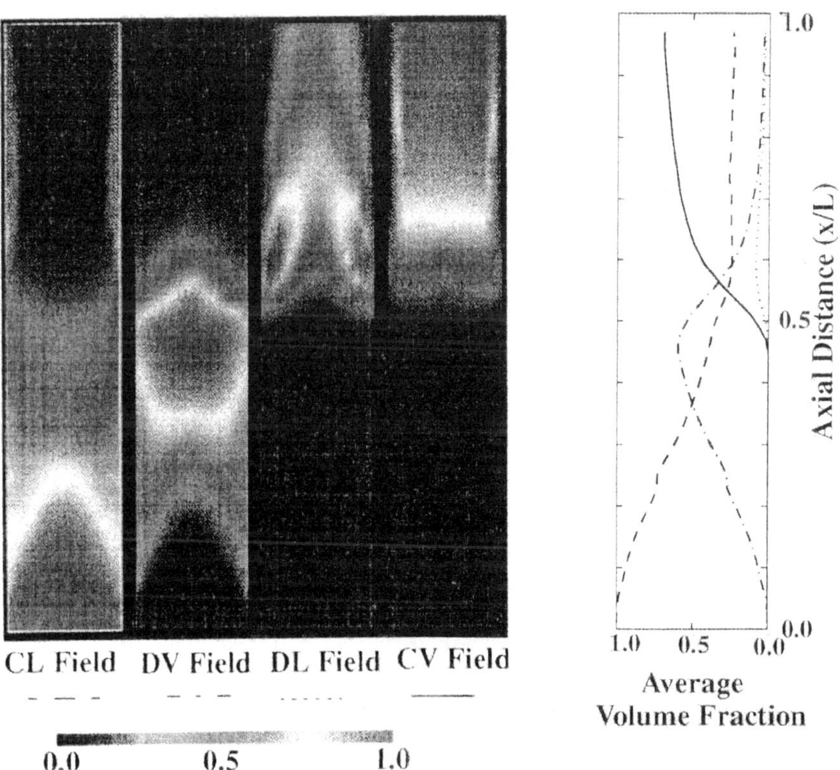

Figure 16: Contour plot of four-field, three-dimensional calculations [92]. Contour planes show the volume fraction variation in the narrow dimension on a center plane for a uniformly heated vertical flow. Individual fields of cross-section averaged volume fraction variation are denoted as: ----, *cl*; -- - --, *dv*; ...., *dl*; _____, *cv*.

The single-phase subcooled flow enters the duct and heats up until sufficient wall superheat is reached to initiate nucleation. The onset of nucleation is indicated by the formation of the *dv* field near the heated walls. The point at which nucleation begins is strongly dependent on the subcooled boiling models employed. Following the onset of nucleation, the *dv*-volume fraction increases as more vapor is formed, and the void peaks in the center. At this point, the bubbles grow by coalescence and interfacial evaporation until they become large enough to be constrained by the walls of the duct. This results in the formation of a new field (*cv*) in the channel. Unlike the *dv*-field, *cv* is forced to the center of the duct, as the flow enters the transition regime. As more *cv* is formed, the coalescence rate increases dramatically, resulting in a rapid transition from a dispersed bubbly flow regime to an annular flow regime. A continuous high velocity vapor core bounded by a thin liquid film on the wall is formed.

Maneri *et al.* [92] showed the significance of the three-dimensional characteristics in a thin duct. Unlike that for a circular geometry, where an axi-symmetric assumption is valid for a majority of the bubbly flow conditions, a two-dimensional assumption for a thin channel could lead to erroneous results. Since the two-dimensional case overpredicts the void fraction, the bubbles are large and the lift force pushes these bubbles away from the wall. The corresponding three-dimensional void fraction profile is considerably different and only slightly overpredicts the measurements (Figure 17).

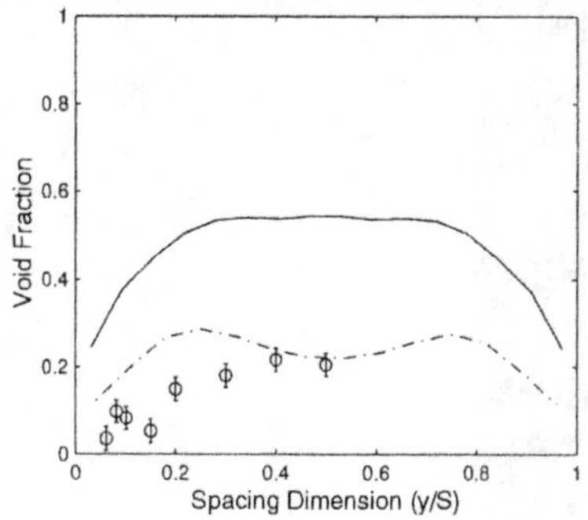

Flow conditions at x/L=0.73:
= 1.35 MPa, G = 528 kg/m^{2}s,
Q_{in} = .6kW

Figure 17: Comparison of two and three dimensional predictions with local void measurements using a hot film anemometer in R-134a [92]: the solid line denotes two-dimensional results, and the dashed line three-dimensional results.

The comparisons between the measurements and predictions in two orientations in a narrow passage given in Figures 18 and 19 demonstrate the ability of the multi-field models to predict the development of the void distribution in the narrow space. Both cases are wall-heated turbulent flow conditions and all the models discussed in this section have been invoked. In the tilted case in Figure 19, the bubbles formed at the lower wall tend to rise and coalesce with the bubbles at the upper wall. However, since the mass flux is high, the bubbles are smaller and the buoyant forces are diminished. In addition, the dispersive forces become more important as the mass velocity is increased. These high-flow effects combine to produce less skewed and more distributed void profiles, which follow the data trends very well at different axial locations. Further validations of the heated annular flow models have not appeared in the literature so far. However, as shown in Figure 16, the four-field formulation is capable of generating an annular vapor core separated from a thin liquid film at the wall. As discussed earlier, adiabatic 3-field modeling (*cl, cv, dl*) in annular flows has yielded good comparisons with available measurements.

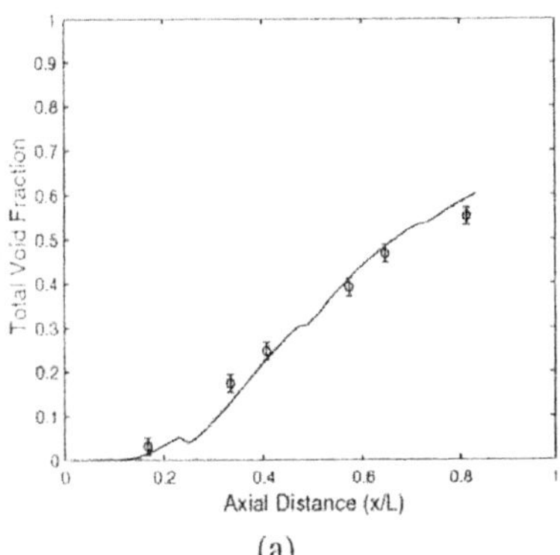

(a)

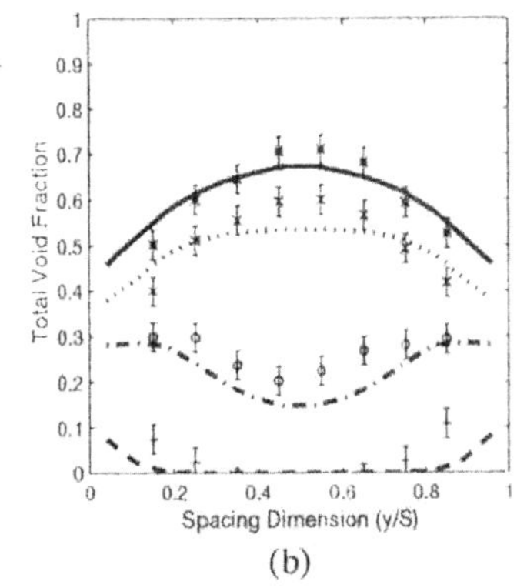

(b)

Figure 18: Comparisons of predictions using a gamma densitometer [92] in R-134a for the vertical case: $P = 2.4$ MPa, $G = 508$ kg/m^{2}s, heat flux = 26.4 kW/m^2: (a) cross-section average, (b) profile in the spacing dimension. Profiles are at locations $x/L = 0.18$, 0.43, 0.68 and 0.86.

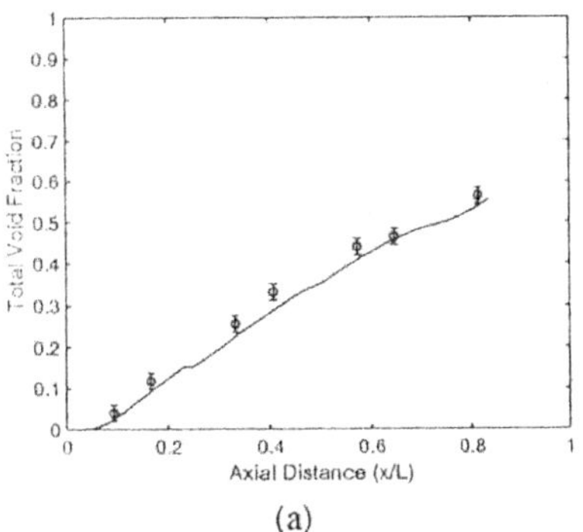

(a)

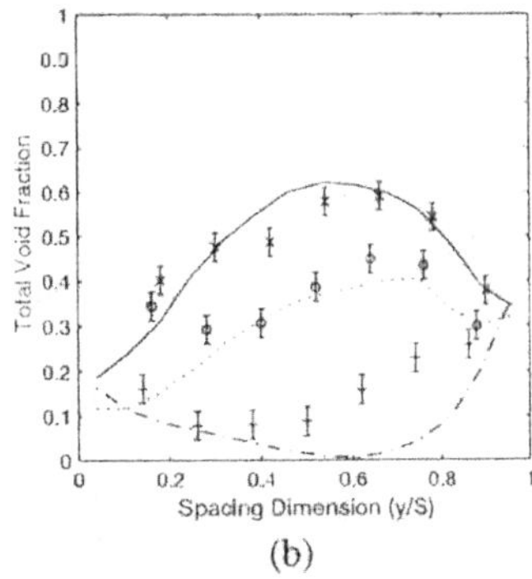

(b)

Figure 19: Predictions in a channel tilted by 45° (rotated about the width) [92]. $P = 2.4$ MPa, $G = 2003$ kg/m^{2}s, heat flux = 70.3 kW/m^2. Profiles are given at locations, $x/L = 0.18$, 0.43 and 0.68.

4 Summary

The performance characteristics of microchannels are vastly different from larger conventional channels. This chapter shows that some progress has been made in obtaining new measurements in microchannels in pressure drop, heat transfer coefficient and critical heat flux. More data are required to develop an understanding of the boiling mechanisms and to enhance the CHF database in microchannels. It was also seen that these global parameters are highly dependent on the channel geometry as well as the flow conditions. The two-phase flow in microchannels has also been visualized to contain transition lines that do not follow the same trends as those of conventional channels. Designers and researchers have opted to develop their own correlations in microchannels because of the geometry effect. Therefore, it is important to understand the dynamic processes and microstructures involved in thin channels for accurate prediction of global quantities.

In the last few years, independent research groups around the world have demonstrated the potential of the multi-dimensional, multi-field approach, which allows for a more mechanistic treatment independent of flow regime maps. In thin geometries, it has been shown that the void gradient in the narrow space principally controls the pressure gradient and the critical heat flux. Seeking a multi-dimensional, multi-field solution is a major step towards improving predictions. However, there are several major challenges facing this evolutionary process.

An optimum number of fields should be used to describe the two-phase flow characteristics. The number of interfaces will increase with the number of fields, along with the number of models. However, modeling itself can be made less complex, since one model does not have to represent multiple interfaces as in the two-field formulation. On the negative side, the closure conditions in the transition and annular flow have been developed based on nonlocal information. New localized models would have to be developed based on new isolated experiments to highlight the individual effects.

Providing robust analytical models that accurately predict the three-dimensional topography is a technical challenge. In the four-field formulation described in this chapter, a number of one-dimensional models developed for conventional channels have been adapted and applied on a local basis in thin channels. This ad hoc approach has introduced empiricism in the modeling coefficients. This is particularly noticeable in the interfacial heat transfer models and the wall-partitioning models.

Proper benchmarking needs to be done since the multi-field formulation is still in a developing stage. This requires new local measurements in various thin geometries to isolate physical mechanisms for model development.

Validation of three-dimensional codes is a time consuming process since the runtime is high. This process is expected to be hastened in the future with the advent of new and faster parallel processors. The design community that relies on empirical correlations needs to embrace the paradigm shift in philosophy.

Nomenclature

A'''	Interfacial area density, 1/mm
A_P	Projected area of a single bubble, m^2
C_D	Drag coefficient
d	Diameter (bubble or droplet), mm
D	Tube diameter, mm
d_o	Bubble diameter corresponding to Levy's model (eqn. (24))
d_h	Hydraulic diameter, mm
E	Entrainment rate
Eo	Eotvos number, $Eo = g\Delta\rho d_{dv}^2 / \sigma$
f	Liquid bridge breakup frequency
f_B	Bubble frequency, 1/s
f_i	Interfacial friction factor
f_{sp}	Single-phase friction factor
F_D	Drag force per unit volume, N/m^3
G	Mass flux, kg/m^2s
h	Enthalpy, kJ/kg
H	Heat transfer coefficient
I	Interfacial momentum terms
k	Thermal conductivity or turbulent kinetic energy or repeated index
L	Length of the duct, m
Mo	Morton number, $Mo = g\mu_{cl}^4 \Delta\rho / \rho_{cl}\sigma^3$
N	Bubble number density, m^{-3}
P	Pressure, MPa
Re	Reynolds number
t	Thickness of the duct, mm
T_S	Saturation temperature, ^oK
$\underline{u}$	Velocity vector
u_R	Relative velocity between phases, m/s
U_B	Single bubble rise velocity, m/s
U_V	Rise velocity in the viscous subregime
U_{DI}	Rise velocity in the distorted-inertial subregime
V_B	Bubble volume, m^3
V_{gj}^B	Drift velocity for a single bubble, m/s
V_{gj}	Drift velocity in a multiple-bubble system, m/s
W	Width of the duct, mm
We	Weber number
X	Flow direction
Y	Lateral or thickness direction
Z	Width direction
α	Volume fraction
$\bar{\alpha}$	Cross-section averaged volume fraction
δ	Liquid film thickness, mm
ε	Turbulent kinetic energy dissipation

Γ Mass transfer rate
ρ Density, kg/m^3
$\Delta\rho$ Density difference between liquid and vapor phases, kg/m^3
ΔT_{sub} $(T_s - T_{cl})$, °K
ΔT_{sonb} Nucleation inception wall superheat, °K
μ Dynamic viscosity, kg/m/s
σ Surface tension, N/m
τ_w Wall shear stress, MPa

Subscripts

cl Continuous liquid
cv Continuous vapor
dl Dispersed liquid
dv Dispersed vapor
f,l Liquid saturation
g Vapor saturation
i Interface
ij Donor phase *i* to receptor phase *j*
lf Liquid film
w Wave

References

[1] Wheeler, D.J., Yi, J.S., Fredley, R., Yang, D., Patterson, T. & VanDine, L., Advancements in fuel cell stack technology at international fuel cells. *J. New Mater. for Electrochemical Sys.*, **4**, pp. 233–238, 2001.
[2] Kandlikar, S.G., Fundamental issues related to flow boiling in minichannels and microchannels. *Exp. Thermal Fluid Science*, **26**, pp. 389–407, 2002.
[3] Kandlikar, S.G. & Grande, W.J., Evolution of microchannel flow passages – thermal hydraulic performance and fabrication technology. *Proc. of the ASME IMECE02*, New Orleans, Nov. 17–22, 2002.
[4] Tran, T.N., Wambsganns, M.W. & France, D.M., Boiling heat transfer with three fluids in small circular and rectangular channels. *Argonne National Lab Report,* ANL-95/9, 1995.
[5] Hosler, E.R., Flow patterns in high pressure two-phase (steam-water) flow with heat addition. *AIChE Symp. Series*, **64**, pp. 54–66, 1968.
[6] Damianides, L.A. & Westwater, J.W., Two-phase flow in a compact heat exchanger and in small tubes. *Proc. of the 2nd UK National Conference on Heat Transfer* **II**, pp. 1257–1268, 1988.
[7] Fukano, T. & Kariyasaki, A., Characteristics of gas–liquid two-phase flow in a capillary tube. *Nucl. Eng. Des.* **141**, pp. 59–68, 1993.
[8] Mishima, K. & Hibiki, T., Some characteristics of air–water two-phase flow in small diameter vertical tubes. *Int. J. Multiphase Flow*, **22**, pp. 703–712. 1996.

[9] Serizawa, A. & Feng, Z.P., Two-phase flow in microchannels. *4th International Conference on Multiphase Flow*, Paper 606, May 27 – June 1, New Orleans, 2001.

[10] Taitel, Y. & Dukler, A.E., A model for predicting flow regime transitions in horizontal and near horizontal gas–liquid flow. *AIChE J.* **22**, pp. 47–55. 1976.

[11] Yang, C.Y. & Shieh, C.C., Flow pattern of air–water and two-phase R-134a in small circular tubes. *Int. J. Multiphase Flow*, **27**, pp. 1163–1177, 2001.

[12] Lowry, B. & Kawaji, M., Adiabatic vertical two-phase flow in narrow flow channels. *AIChE Symp.* **Ser. 84**, 133–139, 1988.

[13] Wambsganss, M.W., Jendrzejczyk & France, D.M., Two-phase flow patterns and transitions in a small horizontal rectangular channel. *Int. J. Multiphase flow*, **17(3)**, pp. 327–342. 1991.

[14] Jones, Jr., O.C. & Zuber, N., The interrelation between void fraction fluctuations and flow patterns in two-phase flow. *Int. J. Multiphase Flow*, **2**, pp. 273–306, 1975.

[15] Mishima, K. & Hibiki, T., Flow regime transition criteria for upward two-phase flow in vertical narrow rectangular ducts. *3rd Int. Conf. on Multiphase flow*, ICMF98, Paper #134, June 8–12, Lyon, 1998.

[16] Troniewski, L. & Ulbrich, R., Two-phase gas–liquid flow in rectangular channels. *Chem. Eng. Sci.*, **39**, pp. 751–765, 1984.

[17] Ali, M. & Kawaji, M., The effect of flow channel orientation on two-phase flow in a narrow passage between flat plates. *ASME/JSME Thermal Eng. Proc.*, **2**, pp. 183–190, 1991.

[18] Mishima, K., Hibiki, T. & Nishihara, H., Some characteristics of gas–liquid flow in narrow rectangular ducts. *Int. J. Multiphase Flow*, **19**, pp. 115–124, 1993.

[19] Wilmarth, T. & Ishii, M., Two phase flow regimes in narrow rectangular vertical and horizontal channels. *Int. J. Heat Mass Transfer*, **37**, pp. 1749–1758. 1994

[20] Griffith, P., The prediction of low quality boiling void. *ASME Trans*. 63-HT-20, 1963.

[21] Kumar, R., Trabold, T.A. & Maneri, C.C., Experiments and modeling in bubbly flows at elevated pressures. *J. Fluids Eng.*, **125**, pp. 469–478, 2003.

[22] Tran, T.N., Chyu, M.C., Wambsganns, M.W. & France, D.M., Two-phase pressure drop of refrigerants during flow boiling in small channels: an experimental investigation and correlation development. *Int. J. Multiphase Flow*, **26**, pp. 1739–1754, 2000.

[23] Eckels, S.J., Doerr, T.M. & Pate, M.B., In-tube heat transfer and pressure drop of R-134a and ester lubricant mixtures in a smooth tube and a micro-fin tube. Part I: Evaporation. *ASHRAE Trans.* **100** (Part 2), 1994.

[24] Cornwell, K. & Kew, P.A., Boiling in small parallel channels. *Energy Efficiency in Process Technology*, ed. P.A. Pilavachi, Elsevier: New York, pp. 624–638, 1993.

[25] Lee, H.J. & Lee, S.Y., Pressure drop and heat transfer characteristics of flow boiling in small rectangular horizontal channels. *4th Int. Conf. on Multiphase Flow*, ICMF2001, Paper #812, New Orleans, May 27 – June 1, 2001.

[26] Moriyama, K. & Inoue, A., The thermohydraulic characteristics of two-phase flow in extremely narrow channels. *Heat Transfer–Japanese Research*, **21(8)**, pp. 838–856, 1992.

[27] Triplett, K.A., Ghiaasiaan, S.M., Abdel-Khlik, S.I., LeMouel, A. & Cord, B.N., Gas–liquid two-phase flow in microchannels. Part II: Void fraction and pressure drop. *Int. J. Multiphase Flow*, **25**, pp. 395–410, 1999.

[28] Peng, X.F., Wang, B.X., Peterson, G.P. & Ma, H.B., Experimental investigation of heat transfer in flat plates with rectangular microchannels. *Int. J. Heat Mass Transfer*, **38(1)**, pp. 127–137, 1995.

[29] Warrier, G.R., Pan, T. & Dhir, V.K., Heat transfer and pressure drop in narrow rectangular channels. *Proc. of the 4th Int. Conf. on Multiphase Flow*, New Orleans, May 27 – June 1, 2000.

[30] Kim, J.M. & Bang, K.H., Evaporation of heat transfer of refrigerant R-22 in small hydraulic diameter tubes. *Proc. 4th Int. Conf. on Multiphase Flow*, Paper #433, New Orleans, May 27 – June 1, 2001.

[31] Vassallo, P.F., Trabold, T.A., Kumar, R. & Considine, D.M., Slug-to-annular regime transitions in R-134a flowing through a vertical duct. *Int. J. Multiphase Flow*, **27**, pp. 119–145, 2001.

[32] Qiu, S., Takahashi, M., Su, G. & Jia, D., Experimental study on heat transfer of single-phase flow and boiling two-phase flow in vertical narrow annuli. *Proc. of the 10th Int. Conf. on Nuclear Engineering*, Arlington, April 14–18, ICONE10-22212, 2002.

[33] Lazarek, G.M. & Black, S.H., Evaporative heat transfer, pressure drop and critical heat flux in a small diameter vertical tube with R-113. *Int. J. Heat Mass Transfer*, **25(7)**, pp. 945–960, 1982.

[34] Zuber, N. & Findlay, J.A., Average volumetric concentration in two-phase flow systems. *J. Heat Transfer*, **87**, pp. 453–468, 1965.

[35] Drew, D.A. & Passman, S.L., *Theory of Multicomponent Fluids*. Springer-Verlag: New York, 1998.

[36] Drew, D.A., Analytical modeling of multiphase flows. *Boiling Heat Transfer*, ed. R.T. Lahey. Elsevier: Amsterdam, pp. 31–84, 1992.

[37] Kelly, J.M., A four-field approach towards developing a mechanistic model for multiphase flow. *PNL-SA-18878*, Pacific Northwest Laboratory, 1993.

[38] Siebert, B.W., Maneri, C.C., Kunz, R.F. & Edwards, D.P., A four-field model and CFD implementation for multi-dimensional, heated two-phase flows. *Proc. of the 2nd Int. Conf. On Multiphase Flow*, Kyoto, 1995.

[39] Antal, S., Kurul, N., Podowski, Z. & Lahey, R.T., The development of multidimensional modeling capabilities for annular flows. *Proc. of the 3rd Int. Conf. On Multiphase Flow*, ICMF98, Lyon, June 8–12, 1998.

[40] Kumar, R. & Trabold, T.A., High pressure annular two-phase flow in a narrow duct: Part II – Three-field modeling. *J. Fluids Eng.*, **122**, pp. 375–38, 2000.

[41] Stosic, Z.V. & Stevanovic, V.D., Schematization of the multi-fluid approach and its application to the CHF prediction. *Proc. of the 4th Int. Conf. on Multiphase Flow, ICMF-2001*, New Orleans, May 27 – June 1, 2001.

[42] Tomiyama, A., Struggle with computational bubble dynamics. *Proc. of the Int. Conf. on Multiphase Flow*, ICMF'98, Lyon, June 8–12, pp. 1–18, 1998.

[43] Mendelson, H.D., The prediction of bubble terminal velocities from wave theory. *AIChE J.*, **13**, pp. 250–253, 1967.

[44] Fan, L.S. & Tsuchiya, K., *Bubble Wake Dynamics in Liquids and Liquid–Solid Suspensions*. Butterworth-Heinemann, Stoneham, MA, 1990.

[45] Lahey, R.T., The prediction of phase distribution and separation phenomena using two-fluid models. In *Boiling Heat Transfer*, ed. R.T. Lahey, pp. 85–122, 1992.

[46] Serizawa, A., Kataoka, I. & Michiyoshi, I., Turbulence structure of air–water bubbly flow, I–III. *Int. J. Multiphase Flow*, **2**, pp. 235–259, 1975.

[47] Zun, I., Transition from wall void peaking to core void peaking in turbulent bubbly flow. *Proc. ICHMT Conference on Transport Phenomena in Multiphase Flow*, Dubrovnik. 1987.

[48] Saffman, P.G., The lift on a small sphere in a slow shear flow. *J. Fluid Mechanics*, **22**, pp. 385–400, 1965.

[49] Drew, D.A. & Lahey, R.T., The virtual mass and lift force on a sphere in rotating and straining inviscid flow. *Int. J. Multiphase Flow*, **13(1)**, pp. 113–121, 1987.

[50] Maneri, C.C. The motion of plane bubbles in inclined ducts. Ph.D. thesis, Polytechnic Institute of Brooklyn, Brooklyn, NY, 1970.

[51] Antal, S.P., Lahey, R.T. & Flaherty, J.E., Analysis of phase distribution in fully developed laminar bubbly two-phase flow. *Int. J. Multiphase Flow*, **17**, pp. 635–652, 1991.

[52] Tomiyama, A., Sou, A., Zun, I., Kanami, N. & Sakaguchi, T., Effects of Eotvos number and dimensionless liquid volumetric flux on lateral motion of a bubble in a laminar duct flow. In: *Advances in Multiphase Flow*, Elsevier: Amsterdam, 1995.

[53] Marie′, J.L., Modeling of the skin friction and heat transfer in turbulent two-component bubbly flow in pipes. *Int. J. Multiphase Flow*, **113**, pp. 309–325, 1987.

[54] Vassallo, P.F., Near wall structure in vertical air/water annular flow. *Int. J. Multiphase Flow*, **25(3)**, pp. 459–476, 1999.

[55] Sato, Y., Sadatomi, M. & Sekoguchi, K., Momentum and heat transfer in two-phase bubble flow, Parts I and II. *Int. J. Multiphase Flow*, **7**, pp. 167–190, 1981.

[56] Lopez de Bertadano, M., Turbulent bubbly two-phase flow in a triangular duct. *Ph.D. Thesis*, Rensselaer Polytechnic Institute, 1992.

[57] Kalkach-Navarro, S., Lahey, R.T. & Drew, D.A., Analysis of the bubbly/slug flow regime transition. *Nucl. Eng. Design*, **151**, pp. 15–39, 1994.

[58] Guido-Lavalle, G., Carrica, P., Clausse, A. & Qazi, M.K., A bubble number density constitutive equation. *Nuc. Eng. Design*, **152**, pp. 213–224, 1994.

[59] Edwards, D.P., Bubble coalescence and breakup modeling in multifield two-fluid predictions. *KAPL-P-000004*, June 1995.

[60] Hibiki, T. & Ishii, M., Development of one-group interfacial area transport equation in bubbly flow systems. *Int. J. Heat and Mass Transfer*, **45**, pp. 2351–2372, 2002.

[61] Levy, S., Forced convection subcooled boiling – prediction of vapor volumetric fractions. *Int. J. Heat Mass Transfer*, **10**, pp. 951–965, 1967.

[62] Hall-Taylor, N.S., Hewitt, G.F. & Lacey, P.M.C., The motion and frequency of large disturbance waves in annular flow in a long vertical tube. *UKAEA Report AERE-R6012*, 1963.

[63] Gottmann, M., Local wall shear stress and interface behavior of adiabatic air–water flows in rectangular ducts. Ph.D. dissertation, University of Arizona, 1997.

[64] Ohba, K., Nakamura, & Naimi, F., A new kind of interfacial wave on liquid film in vertically upward air–water two-phase annular flow. *Proc. 2nd Int. Conf. on Multiphase Flow*, Paper IP1, pp. 27–33, 1995.

[65] Trabold, T.A. & Kumar, R., High pressure annular two-phase flow in a narrow duct: Part 1 – Local measurements in the droplet field. *J. Fluids Eng.*, **122**, pp. 364–374, 2000.

[66] Wallis, G.B., *One-Dimensional Two-Phase Flow*, McGraw-Hill: New York. 1969.

[67] Whalley, P.B. & Hewitt, G.F., The correlation of liquid entrainment fraction and entrainment rate in annular two-phase flow. *AERE-R9187*, 1978.

[68] Sekoguchi, K., Hori, K., Nakazatomi, M. & Nishikawa, K., On ripple of annular two-phase flow: Part 2: Characteristics of wave and interfacial friction factor, *Bull. JSME*, **21(152)**, pp. 279–286, 1978.

[69] Kang, H.C. & Kim, M.H., The relation between the interfacial shear stress and the wave motion in a stratified flow. *Int. J. Multiphase Flow*, **19(1)**, pp. 35–49, 1993.

[70] Henstock, W.H. & Hanratty, T.J., Interfacial drag and height of the wall layer in annular flows, *AIChE J.*, **22(6)**, pp. 990–1000, 1976.

[71] Beus, S.G., Fore, L.B. & Lopez de Bertadano, M.A., Hydrodynamic characteristics of annular two-phase flow. *B-TM-1636*, 2001.

[72] Trabold, T.A., Kumar, R. & Vassallo, P.F., Annular flow of R-134a in a vertical duct: Local void fraction, droplet velocity and droplet size measurements. *Proc. ASME IMECE.,* HTD-Vol. 361–5, **5**, pp. 379–393, Anaheim, 1998.

[73] Kumar, R. & Trabold, T.A., Effect of pressure with wall heating in annular two-phase flow. *J. Fluids Eng.*, **125**, pp. 84–96, 2003.

[74] Kirouac, G.J., Trabold, T.A., Vassallo, P.F., Moore, W.E. & Kumar, R., Instrumentation development in two-phase flow. *Exp. Thermal Science*, **20**, pp. 79–93, 1999.

[75] Chien, S. & Ibele, W., Pressure drop and liquid film thickness of two-phase annular and annular-mist flow,.*J. Heat Transfer*, **86**, pp. 80–96, 1964.

[76] Kumar, R., Gottmann, M. & Sridhar, K.R., Film thickness and wave velocity measurements in a vertical duct. *J. Fluids Eng.*, **124**, pp. 634–642, 2002.

[77] Hewitt, G.F., Disturbance waves in annular two-phase flow. *Proc. Inst. Mech. Eng.*, **184**, pp. 142–150, 1969.

[78] Woodmansee, D.E. & Hanratty, T.J., Mechanism for the removal of droplets from a liquid surface by a parallel air flow. *Chem. Eng. Sci.*, **24**, pp. 299–307, 1969.

[79] Antal, S.P., Edwards, D.P. & Strayer, T.D., Predicting multidimensional annular flows with a locally based two-fluid model. *Proc. of the 3rd Int. Conf. on Multiphase Flow*, ICMF98, Lyon, June 8–12, 1998.

[80] Brackbill, J.U., Kothe, D. & Zemach, C., A continuum method for modeling surface tension. *J. Comp. Physics*, **100**, pp. 335–354, 1992.

[81] Ishii, M. & Mishima, K., Droplet entrainment correlation in annular two-phase flow. *Int. J. Heat Mass Transfer*, **32(10)**, pp.1835–1846, 1989.

[82] Dallman, J.C., Jones, B.G. & Hanratty, T.J., Interpretation of entrainment measurements in annular gas–liquid flows. *Two-phase momentum, heat and mass transfer in chemical process and energy engineering systems*, ed. F. Durst *et al.*, **2**, pp. 681–693, Hemisphere: Washington, DC, 1979.

[83] Lopez de Bertadano, M.A., Jan, C.S. & Beus, S.G., Droplet entrainment correlation for high pressure annular two-phase flow. *ANS Proc., HTC-Vol. 8, ASME National Heat Transfer Conference*, Portland, 1995.

[84] Govan, A.H., Hewitt, G.F., Owen, D.G., & Bott, T.R., An improved CHF modeling code. *2nd UK National Heat Transfer Conf.*, Glasgow, 1988.

[85] Azzopardi, B.J., Drops in annular two-phase flow. *Int. J. Multiphase Flow*, **23**, Suppl., pp. 1–53, 1997.

[86] Ueda, T., Entrainment rate and size of entrained droplets in annular two-phase flow. *Bull. JSME*, **22**, pp. 1258–1265, 1979.

[87] Kataoka, I., Ishii, M. & Mishima, K., Generation and size distribution of droplets in annular two-phase flow. *J. Fluids Eng.*, **105**, pp. 230–238, 1983.

[88] Kocamustafaogullari, G., Smits, S.R. & Razi, J., Maximum and mean droplet sizes in annular two-phase flow. *Int. J. Heat Mass Transfer*, **37**, pp. 955–965, 1994.

[89] Ishii, M. & Zuber, N., Drag coefficient and relative velocity in bubbly, droplet or particulate flows. *AIChE Journal*, **25**, pp. 843–855, 1975.

[90] Ishii, M. & Mishima, K., *Nuc. Eng. and Design*, **82**, pp. 107, 1984.

[91] Prince, M.J. & Blanch, H.W., Bubble coalescence and break-up in air-sparged bubble columns. *AIChE Journal*, **36**, pp. 1485–1499, 1990.

[92] Maneri, C.C., Kumar, R., Strayer, T.D. & Trabold, T.A., Analysis of heated two-phase flows: Four field modeling approach and experimental verification. *KAPL-P-000079*, June 1998.

[93] Pilch, M. & Erdman, C.A., Use of breakup time data and velocity history data to predict the maximum size of stable fragments for acceleration-

induced breakup of a liquid drop. *Int. J. Multiphase Flow*, **13(6)**, pp. 741–757, 1987.

[94] Kurul, N., Multidimensional effects in two-phase flow including phase change. Ph.D. dissertation, Rensselaer Polytechnic Institute, 1990.

[95] Del Valle M.V.H. & Kenning, D.B.R., Subcooled flow boiling at high heat flux. *Int. J. Heat Mass Transfer*, **28**, pp. 1907–1920, 1985.

[96] Davis, E.J. & Anderson, G.H., The incipience of nucleate boiling in forced convection flow. *AIChE J.*, **12**, pp. 774–780, 1966.

[97] Ranz, W.E. & Marshall, W.R., *Chem. Eng. Prog.*, **48**, pp. 141–146, 1952.

[98] Bankoff, S.G., Some condensation studies pertinent to LWR safety. *Int. J. Multiphase Flow*, **6**, pp. 51–67, 1980.

[99] Drew, D.A. & Lahey, R.T., Phase distribution mechanisms in turbulent two-phase flow in a circular pipe. *J. Fluid Mech.*, **117**, 91–106, 1982.

[100] Biesheuvel, A. & van Wijngaarden, L., Two phase flow equations for dilute dispersion of gas bubbles in liquid. *J. Fluid Mech.*, **168**, pp. 301–318, 1988.

CHAPTER 6

Radiative energy transport at the spatial and temporal micro/nanoscales

C.-H. Fan[1] & J.P. Longtin[2]
[1]*Department of Mechanical and Materials Engineering, Florida International University, USA*
[2]*Department of Mechanical Engineering, SUNY – Stony Brook, USA*

Abstract

Radiation energy transport plays a vital role in modern thermal systems. Though radiation energy transport in traditional systems has been well studied for many years, recent developments in micro- and nanoscale engineering provide renewed opportunities to explore the nature and role of radiative heat transport in these systems. In particular, when the feature size approaches, or becomes smaller than, the characteristic wavelength of the radiation, a variety of interesting interactions occur between energy-exchanging bodies. Similarly, when the time scale of the radiation approaches the molecular or material response time, the radiation–material interactions can change dramatically. Finally, when the incident photon density becomes very large, matter responds in new and useful ways. This chapter provides an overview of several current research areas involving radiative energy exchange at micro- and nanoscales. First, a review of radiation preliminaries is put forth regarding radiation and radiation–matter interactions, followed by fundamental issues that arise at the micro- and nanoscales. Next, several contemporary applications are then discussed, including ultrafast laser material processing and novel diagnostics for biological systems. Finally, the chapter is concluded with summary remarks and thoughts for further work.

1 Introduction

The last two decades of the twentieth century witnessed a radical new approach to technological development: engineering at the micro- and nanoscale. The essential idea is that electro-mechanical devices with length scales in the micron

(10^{-6} m) to nanometer (10^{-9} m) length scales can be fabricated into useful devices for sensing, actuation, sampling, analysis, microflow and dispensing, etc. Driven largely by the explosive growth of MEMS (Micro-Electro-Mechanical Systems) and the technologies required to fabricate MEMS structures, a paradigm shift in scientific and engineering research resulted. On the time scale of human innovation, engineers had, virtually overnight, developed a wealth of new tools and techniques to make structures approaching molecular dimensions. Everyday examples include quantum-well lasers used in laser pointers, CD players, and telecommunication equipment, microelectronics with feature sizes approaching less than 200 nm (with smaller sizes on the horizon), and modern medical diagnostics and non-invasive imaging systems. In addition to the technical challenges of *fabricating* such devices, the successful *analysis* and *modeling* of their behavior has required novel new approaches that consider the molecular nature of the materials comprising these devices and structures; traditional, bulk formulations for macroscopic systems often no longer hold.

Though certainly it has received the lion's share of attention from the scientific and lay communities, micro- and nanoscale engineering is but one of several inter-related and rapidly developing fields that require a molecular-based approach for enhanced understanding. The widespread availability of *ultrafast laser systems* with pulse durations approaching 10–100 femtoseconds (1 fs = 10^{-15} s) has resulted in rich and diverse investigations of laser–material interactions in entirely new regimes, resulting in exciting novel processing capabilities, with significant potential for MEMS, medical and bioengineering applications, electronics, sensors, and communication systems. Similarly, in such systems the photon *intensities* associated are enormous, resulting in new and interesting methods for processing materials.

Another such example is the boom in *biotechnology*. The human genome project identifies all of the approximately 30,000 genes in human DNA and the sequences of the 3 billion chemical base pairs that make up human DNA. This information is tremendously valuable for genetics, heritable diseases intervention, and other related technologies in human health applications. Radiative energy transport in biological systems remains central to these discoveries. For example, one technique for DNA sequencing is laser-induced fluorescence (LIF) marking of each DNA with different colors. LIF is also widely used in molecular biology and medical research to discriminate proteins, organelles, and membranes in cellular and even subcellular environments. Recent progress in lasers, optoelectronics, and microscopes makes high-resolution 4-D fluorescence imaging available. In addition, these fluorescence proteins can serve as a physiological indicator to report the status of the live cells and tissues. Investigation of radiation–biomaterial interactions and the associated interdisciplinary topics represents a new trend in microscale technology and development in the beginning of the twenty-first century.

It is the intent of this chapter to provide a brief overview of some of the many recent developments happening in these exciting fields. In the following sections, the impact of microscale length and time scales on radiation phenomena is discussed, along with a description of both linear and nonlinear interactions with matter. Issues related to microfabrication, MEMS, and biotechnology using laser radiation will be introduced as well. Aspects of microscale conduction, convection, phase change, and interface transport are addressed in their respective chapters elsewhere in this text.

2 Fundamentals

Before delving into microscale versus macroscale radiative transport, some preliminary concepts and definitions are discussed.

2.1 Properties of electromagnetic radiation

2.1.1 Wavelength, frequency and energy

Electromagnetic (*EM*) *radiation* is a propagating oscillation of an electric (and magnetic) field. The energy carriers of EM radiation are photons that contain discrete quanta of energy. They propagate in vacuum at the speed of light, $c_0 = 3\times10^8$ m/s and exhibit both wave and particle behavior [1]. Photons have a characteristic frequency of vibration ν that represents the frequency of vibration of the field from which they emanated, and a wavelength λ, which represents the distance over which the photon travels one complete oscillation. The product of vibrational frequency and wavelength results in the propagation speed:

$$c = \lambda \nu \tag{1}$$

The *energy* of a photon depends on its frequency:

$$E = h\nu = hc/\lambda \quad , \tag{2}$$

where h is Plank's constant (h = 6.626 × 10^{-34} J·s). A shorter wavelength corresponds to higher frequency, and accordingly, a higher photon energy as shown in eqn. (2); this is the reason red light ($\lambda \sim 700$ nm) is used in photographic darkrooms: longer wavelengths have less energetic photons and hence cannot initiate photochemical reactions in the light-sensitive film to be developed. In fact, blue light ($\lambda \sim 400$ nm) has nearly twice the photon energy as red light. The photon energy plays a very important role when light interacts with matter at various intensity levels, which will be described in detail in the following sections.

In gases at moderate pressures, the propagation speed of EM radiation is very nearly (99.99%) that in vacuum, however in dense media such as solids and liquids, the propagation speed c is reduced by the real portion of the *refractive index*, n:

$$c = \frac{c_0}{n} \quad . \tag{3}$$

In free space, and weakly absorbing materials, the electrical field $E(z, t)$ as a function of propagation distance z and time t takes the form [1, 2]:

$$E(z,t) = E_0 \exp[i(\omega_0 t - k_0 nz)] \tag{4}$$

with

$$k_0 = \frac{\omega_0}{c}, \tag{5}$$

where E_0 is the peak amplitude, $\omega_0 = 2\pi\nu$ the angular frequency, and k_0 the wavenumber. The *intensity* of the radiation, I, represents the energy delivered by the EM wave crossing a unit area per unit time, and is directly related to the amplitude of the EM wave as follows:

$$I = \varepsilon c \frac{|E|^2}{2}, \tag{6}$$

where ε is the medium permittivity. The *intensity* I (W/cm^2) of the radiation can also be expressed in terms of the total energy of photons crossing a unit area per unit time:

$$I = \frac{N_p h\nu}{At} = \frac{N_p hc}{At\lambda}, \tag{7}$$

where N_p is the number of photons crossing area A within time t.

2.1.2 Propagation speed, spatial extent, and spatial separation

As discussed above, the speed of light in a material is c, and is on the order of 10^8 m/s for virtually all materials. As such, the time it takes radiation to travel length scales of interest in engineering systems is on the order of L/c where L is the characteristic length scale for the system. Assuming a refractive index of unity, it takes radiation roughly 3.3 ns to travel 1 m. In a microscopic system, with dimensions in the micron and nanometer range, this characteristic length scale is reduced accordingly. In many situations, the propagation time is negligible compared with other time scales in the system.

There are, however, situations where the duration of the radiation becomes very short, and the propagation time becomes critical for system analysis. The most notable example is ultrafast laser systems that are finding increasingly widespread application in many areas of science. The spatial extent of such a pulse then scales as $\tau_p c$ where τ_p is the pulse duration. The pulse duration from a commercial ultrafast oscillator can approach 20 fs, for example, with a corresponding spatial extent of only 6 μm! The spatial extent and radiation wavelength form two important length scales that will be delineated below.

2.1.3 Polarization

Another important attribute of EM radiation is its polarization. The *plane of polarization* of an EM wave is defined as the plane in which the electric field or optical disturbance oscillates [2]. If the plane of polarization remains fixed in space, the EM radiation is said to be *linearly polarized.* Other forms of polarization include circular and elliptical polarization, both of which exhibit variations in the orientation of the electric field with time. If the plane of polarization varies rapidly, and in a random way, it is said to be *randomly* polarized. For *thermal* radiation, e.g., radiation from a hot surface or solar radiation, no specific polarization plane exists *per se*, and the radiation is randomly polarized. Lasers, in contrast, generally produce light with high-quality linear or circular polarization.

Circularly and elliptically polarized light can be produced from linearly polarized light by passing it through an optical waveplate of appropriate thickness. The waveplate is made of a birefringent optical material that results in the transmission of light at different velocities along two orthogonal axes. The differences in propagation velocity of these two EM wave components result in a phase retardation as the light passes through. The value of phase retardation is determined by the thickness of the waveplate, which in turn determines the amplitude and orientation of the recombined light. The polarization of light is very critical in many optical technologies, e.g., stress analysis, reflectance at an interface, absorption by an anisotropic media, and feature quality of laser-machined patterns. Randomly polarized light can be "converted" to linearly polarized light by introducing a polarizing element such as a polarizing film or beam splitter into the radiation path. The polarizing element only allows photons with a given orientation to pass through the polarizer. Since photons with the opposite polarization are not passed, a reduction in intensity is observed.

2.1.4 Coherence

As photons exhibit wave-like properties, they can interfere with each other both constructively and destructively. Photons capable of interfering in such a fashion are said to exhibit *coherence.* In particular, one is often interested in the distance from the radiation source in which the photons are capable of exhibiting interference. This distance is called the *coherence length*. Similarly, at a fixed point in space, there is a characteristic time called the *coherence time*, over which the phase relationship of the photons is preserved.

For a given atomic transition with unique energy ΔE_j, eqn. (2) suggests that this transition would require a photon with a single, unique frequency, ν_j, i.e., the *bandwidth,* or frequency variation, of the absorption, $\Delta\nu$ would be zero. In practice, the actual bandwidth is finite due to thermal agitation; the presence of neighboring atoms, electric and magnetic fields, etc. provides slight perturbations to the transition energy ΔE, resulting in a finite frequency bandwidth for the transition. For a transition with bandwidth $\Delta\nu$, the coherence time Δt_c is [1, 2]:

$$\Delta t_c = \frac{1}{\Delta \nu} \tag{8}$$

and the coherence length L_c is

$$L_c = c\,\Delta t_c = \frac{c}{\Delta \nu} \tag{9}$$

Lasers traditionally come to mind when speaking of radiation sources with long coherence lengths. Indeed very narrow-band (highly monochromatic) lasers can have coherence lengths approaching kilometers.

Depending on the material and specific transition, however, even thermal radiation can result in coherence lengths from 1 mm to ~ 10 cm. For a blackbody in a vacuum, the coherence length is [3, 4]:

$$l_c = 0.15ch / k_b T \qquad \text{or} \qquad l_c T \approx 2160\,\mu\text{m} \cdot \text{K} \,. \tag{10}$$

where k_b is Boltzmann's constant and T is the absolute temperature of the radiating surface. Coherence effects can become important in thin films and the interaction of radiation with small-scale structures. A further regime is defined in the vicinity of the coherence length where the light is only partially coherent. Richter and Chen and others discuss *partial coherence* and its implication in thermal radiation–matter interactions [3–8]. The coherence length provides another characteristic length scale for demarcating systems, as discussed below.

2.2 Sources of radiation in thermal engineering

2.2.1 Thermal radiation

The production of EM radiation comes from electron or phonon transitions from higher to lower energy states. The energy difference between states is equal to the energy of the photon emitted, and, therefore, determines the frequency (hence wavelength) of the emission according to eqn. (2). All materials that have a temperature above absolute zero emit radiation due to the random excitation and relaxation of individual molecules as a result of the thermal energy that they possess. This so-called *thermal* radiation results from a large number of discrete molecules whose relaxation and individual photon emission contribute to the net radiative flux. To a good approximation, the excitation and relaxation of the individual molecules occur independently of other molecules. It is for this reason that the radiation is randomly polarized, and the frequency distribution is broad and relatively featureless.

The accurate prediction of the intensity and spectrum of thermal energy was first accomplished by Max Planck in 1902 [9]. Planck proposed the idea – radical at the time – that atoms and molecules could only emit or absorb discrete quanta of energy. Interestingly, it was Planck's attempt to describe so-called blackbody radiation that gave rise to modern quantum mechanics [9].

Planck argued that, for thermal radiation, the transitions are thermally driven and random, which results in a statistical distribution of photon energies in the radiation [9]. For a surface that absorbs (and hence emits) perfectly, i.e., a perfect *blackbody*, the photon intensity depends on wavelength and temperature as follows, and is the so-called *Planck distribution* [9, 10]:

$$I_{b\lambda}(\lambda, T) = \frac{C_1}{\lambda^5\left[\exp\left(C_2/\lambda T\right) - 1\right]} \quad , \tag{11}$$

where $C_1 = 2hc^2 = 11{,}909\ \text{W}\cdot\mu\text{m}^4/\text{cm}^2$, $C_2 = hc/k_b = 14{,}388\ \mu\text{m}\cdot\text{K}$, and T is the temperature. The blackbody distribution for several common temperatures is shown in Figure 1. Note that as the surface temperature increases the intensity increases and the peak shifts to shorter wavelengths.

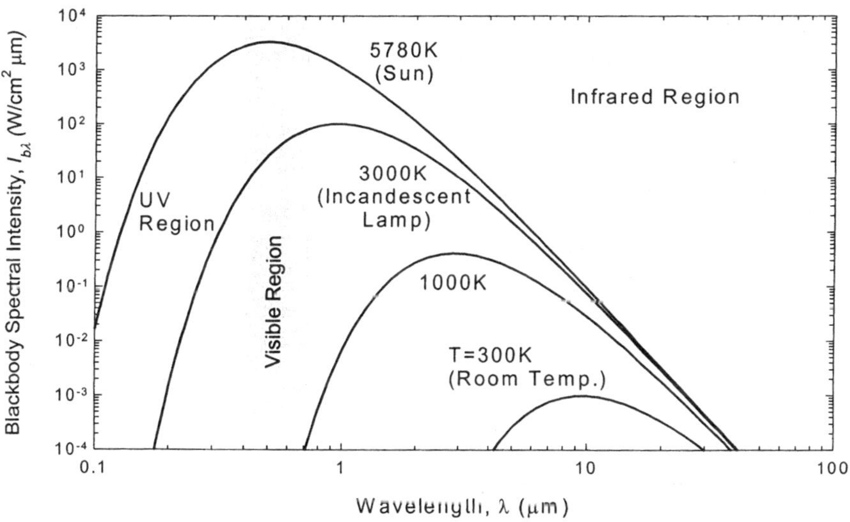

Figure 1: Blackbody radiation versus wavelength at several common temperatures.

2.2.2 Laser radiation

A second important class of radiation for thermal systems is laser radiation. Laser radiation, in contrast to thermal radiation, does not occur naturally, but rather requires a fairly complex sequence of events and precision optical components to produce. Laser radiation results from *stimulated emission* of excited molecules that are forced to transition from higher to lower energy states by the interaction of a pre-existing photon. In doing so, the rather remarkable result is that a second photon, identical in all respects, is produced. Redirecting this stimulated emission back into the excited medium results in a geometric gain or amplification of the radiation until it exits the laser, or all excited states are depleted.

This radiation is essentially monochromatic, highly coherent and directional, and can be produced at *extremely* high intensities and for very short durations. Laser radiation is thus a very useful tool in many diverse technologies, including manufacturing, measurement, communication, information storage, biological and medical areas, and consumer products. Although thermal radiation and laser radiation possess distinctly different qualities, both result from the same mechanism of transitions between different energy states in atoms and molecules.

Lasers can be *continuous-wave* (CW), in which radiation is emitted continuously (unless intentionally modulated by an external source), or *pulsed*, in which photons are produced for some duration of time, called the *pulse duration*, followed by a dwell time in which no radiation is produced [1]. CW lasers find widespread use in measurement, information transmission, diagnostics, and, on occasion, thermal processing, with the notable example being high-power CO_2 lasers. Pulsed lasers produce much higher intensities, but only for the duration of the pulse, with the significant advantage that thermal effects are restricted to the time that thermal energy can diffuse through the material in one pulse duration.

A laser pulse can be imagined as a bullet carrying photon energy. If this light bullet strikes a photodetector, the output signal will follow the temporal profile of the laser pulse (assuming the detector has an infinitely fast response time). Many lasers exhibit *Gaussian* pulse durations, in which the pulse intensity as a function of time can be expressed as:

$$P(t) = P_0 \exp\left[(-4\ln 2)\left(\frac{t}{\tau_p}\right)^2\right], \tag{12}$$

where τ_p is the *full width at half maximum (FWHM)* pulse duration, defined as the time during which the pulse power exceeds one-half of the maximum power at $t = 0$ in eqn. (12) [1]. The factor $-4 \cdot \ln 2$ appears in eqn. (12) so that $P(\pm\tau_p/2) = P_0/2$, according to the FWHM definition. The peak laser *power* P_0 can be related to the measured pulse *energy* E_p, which is straightforward to measure, by the following expression:

$$E_p = P_0 \int_{-\infty}^{\infty} \exp\left[(-4\ln 2)\left(\frac{t}{\tau_p}\right)^2\right] dt = 1.064 P_0 \tau_p \,. \tag{13}$$

In most applications, the laser beam is focused to increase the intensity. Laser beams exhibit unique focusing properties. Many laser beams also are axisymmetric and have a Gaussian distribution in the radial direction about the beam axis. For such Gaussian spatial distributions, the beam radius w as a function of axial position z is characterized by [1]:

$$w(z) = w_0 \left(1 + \frac{z^2}{z_R{}^2}\right)^{\frac{1}{2}}, \tag{14}$$

with

$$z_R = \frac{n \pi w_0{}^2}{\lambda_0}, \tag{15}$$

where w_0 is the beam waist, i.e., the radius of the beam at its smallest, which is taken to be located at $z = 0$, z_R is the Rayleigh length, n is the refractive index of the medium, and λ_0 is the laser wavelength in free space. The *Rayleigh length* of a focused beam is the spatial region over which the beam waist remains less than $\sqrt{2}w_0$ as one moves away from the waist at $z = 0$. The value $2z_R$ is referred to as the *confocal parameter* [1] and represent a symmetric region centered on $z = 0$ for which the beam radius remains less than $\sqrt{2}w_0$.

As a laser pulse propagates in a medium with a speed c, the spatial extent of the laser pulse is related by the pulse length $l_p = c\tau_p$. For a relatively long laser pulse, e.g., with $\tau_p \sim 1$ ns, $\lambda = 800$ nm, and $w_0 = 20$ μm, the pulse length is roughly 30 cm, which is much larger than the focal region characterized by z_R ~1 mm. In this case, the focal region is much smaller than the spatial extent of the pulse, and can be treated as a single point. Therefore the laser power is very nearly uniform within the focal region, and is time dependent *only*, with no spatial variation. Consequently, the laser power can be well described by eqn. (12) with $t = 0$ representing the time when the pulse peak, P_0, reaches the beam focus at $z = 0$. For an ultrashort pulse, e.g., $\tau_p \sim 100$ fs and $l_p = 30$ μm, however, the pulse length is much shorter and cannot overfill the entire focal region. Thus eqn. (12) needs to be modified to take account of the spatial variation.

The ratio of the pulse length to the focal region (Rayleigh range), N_z, is used to determine the size effect of a laser pulse [11]:

$$N_z = \frac{l_p}{z_R} = \frac{c\tau_p}{z_R}. \tag{16}$$

Values of $N_z >> 1$ imply the laser pulse is much larger than the focal region, and the spatial variation in the pulse shape can be ignored. For $N_z << 1$, however, the pulse length is relatively small; hence propagation of the pulse within the focal volume must be taken into account in the expression of a laser pulse, which takes the form [12]:

$$P(t) = P_0 \exp\left[(-4\ln 2)\left(\frac{t - z/c}{\tau_p}\right)^2\right], \tag{17}$$

where z/c terms represents the pulse propagation time in the medium with respect to $z = 0$. The spatial variation in pulse power results in localized interactions with the medium and substantially changes the sequence of energy absorption [13, 14].

2.3 Radiation–matter interactions

2.3.1 Classical absorption

The process of a substance interacting with radiation can be roughly divided into three steps. First is *absorption*, in which an individual particle (electron, atom, molecule, structure, etc.) absorbs the photon and converts the photon energy to thermal energy. Next, the particle will exchange this energy with its neighbors, via collisions with energy carriers and/or other particles. This process is called *thermalization*, and is responsible for transferring the absorbed radiant energy to particles that did not participate directly in the absorption process. Finally, the temperature increase will increase the *emission* of photons, thus increase the emitted intensity. Alternatively, the material may alter the direction of the incident photons without absorbing substantial energy. This phenomenon is called *scattering*. In the following sections, these mechanisms will be discussed in two regimes: (1) macroscopic classical, linear interactions and (2) nonlinear interactions that occur at extreme length, time, and radiation intensity scales.

2.3.2 Reflection and refraction of radiation at an interface

As a wave (light) strikes an interface composed by two different media with refractive indices n_1 and n_2, the disturbances on the interface can spread back to the first medium and, therefore, form a *reflected* wave. The rest of the wave will continue traveling into the second medium through this interface as a *transmitted* wave. When radiation passes from one medium into another at a non-normal angle to the interface, a change in propagation direction is required by the change in velocity of the wave motion (Figure 2). This optical phenomenon is associated with the passage of EM radiation from one medium to another and is known as *refraction*.

For reflection at a smooth interface, the angles of incidence θ_i and reflection θ_r are identical, i.e., $\theta_i = \theta_r$, while the angle of transmission/refraction θ_r is determined by Snell's law, which takes the form [2]:

$$n_1 \sin\theta_i = n_2 \sin\theta_t \quad (18)$$

The incident, reflected, and transmitted waves all lie in the plane of incidence (Figure 2).

For polarized light if the plane of polarization is parallel to the plane of incidence it will be p-polarized; if the plane of polarization is perpendicular to the plane of incidence, it will be s-polarized. The reflectivity or reflectance, which is the ratio of the intensity of reflective light to incident counterparts, depends upon the polarization of incident light, angle of incidence, and the refractive indices of the media, respectively. According to Fresnel's law, the reflectivity of *p*- and *s*-polarized light can be expressed as:

$$R_p = \left[\frac{\tan(\theta_i - \theta_t)}{\tan(\theta_i + \theta_t)}\right]^2, \tag{19}$$

$$R_s = \left[\frac{\sin(\theta_i - \theta_t)}{\sin(\theta_i + \theta_t)}\right]^2, \tag{20}$$

respectively [2]. The angles of incidence θ_i and transmission θ_t follows Snell's law.

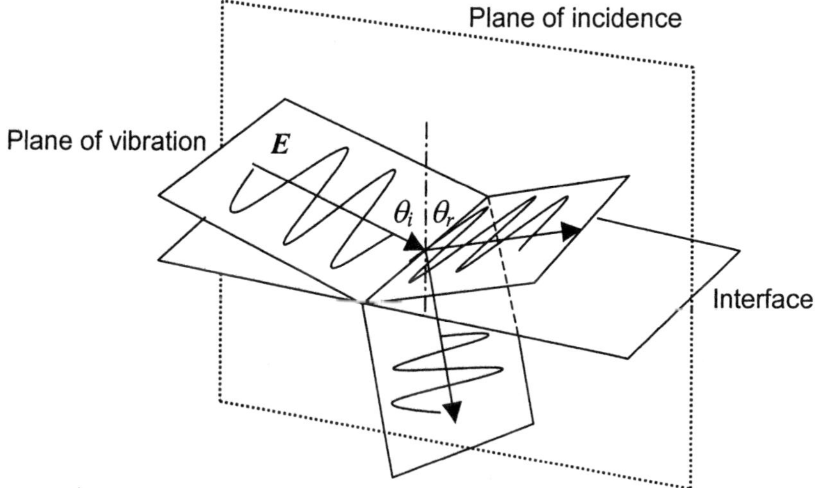

Figure 2: Relationship between *s*- and *p*-polarized incident, reflected, and transmitted waves.

2.3.3 Linear scattering in a medium

The basic concepts of classical interactions of EM waves with bulk materials have been described in the previous section. However, the processes of transmission, reflection, and refraction are microscopic manifestations of scattering occurring on a submicroscopic level [2]. In this section, a deeper exploration of photon–molecule interactions, i.e., linear scattering, which becomes very important for applications in small length and time scale, e.g., biological cells, tissues, and optical fibers, will be addressed and discussed.

Scattering can cause radiant energy losses when the EM wave interacts with particles, which redirect the propagation direction of the wave. *Linear scattering* occurs *without* a frequency (or wavelength) change upon elastic scattering processes: one photon is absorbed and another photon with the same frequency is emitted simultaneously without any delay. Linear scattering includes *Rayleigh scattering* and *Mie scattering*. Rayleigh scattering results from light interacting with medium inhomogeneities that are much smaller than the wavelength of the light [15]. Since most molecules have electronic resonances in the ultraviolet (UV) region, the closer the driving frequency (incoming light) is to the resonant frequency, the more vigorously the oscillator responds. Hence violet light is scattered more strongly than red light. This wavelength-dependent linear scattering was first quantified by Lord Rayleigh in 1871 (long before quantum mechanics). Rayleigh determined that the scattering strength is proportional to $1/\lambda^4$, which successfully explains the blue sky and fiery sunset.

Mie scattering takes place when light interacts with inhomogeneities that are comparable in size to the wavelength [2, 15]. Unlike Rayleigh scattering, Mie scattering is less dependent on wavelength and exhibits a strong angular dependence as the light encounters particles with a size larger than $\lambda/10$. Rayleigh scattering can be treated as a small-size limiting case of Mie scattering [15]. It should be noted that linear scattering will not induce electron excitation in the molecules, i.e., electrons remain in the ground state, which is not the case of nonlinear scattering to be discussed in section 09.

2.3.4 Classical (linear) absorption

Absorption of photons by the medium will attenuate the incident radiation as it passes through. For EM radiation with an initial power P_0, the attenuation of power (or intensity) by absorption can be described by Beer's Law:

$$\frac{P(z)}{P_0} = \frac{I(z)}{I_0} = \exp(-\mu_a z), \tag{21}$$

where μ_a is the absorption coefficient and z is the propagation distance within the medium. As the intensity decreases to 1/e of the original value after the light has advanced a distance d, this travel distance is the so-called *penetration depth* or *skin depth*,

$$\delta = \frac{1}{\mu_a}. \tag{22}$$

For metals and semiconductors, the refractive index is a complex value:

$$\hat{n} = n(1 + i\kappa) \tag{23}$$

where κ is the attenuation index or extinction coefficient. The extinction coefficient determines the absorption coefficient for normal EM radiation absorption, which takes the form:

$$\mu_a = \frac{4\pi\nu}{c} n\kappa = \frac{4\pi}{\lambda_0} n\kappa \quad . \tag{24}$$

Therefore, the penetration depth can be expressed as:

$$d = \frac{1}{\mu_a} = \frac{\lambda_0}{4\pi n\kappa} \quad . \tag{25}$$

A laser pulse propagating in a vacuum or non-absorbing, non-scattering medium is governed by the wave propagation equation:

$$\frac{1}{c}\frac{\partial I}{\partial t} + \frac{\partial I}{\partial z} = 0 \, . \tag{26}$$

For a medium with absorption coefficient μ_a, eqn. (26) can be modified to include the medium absorption:

$$\frac{1}{c}\frac{\partial I}{\partial t} + \partial\frac{\partial I}{\partial z} = -\mu_a I \, . \tag{27}$$

When the time dependent term can be neglected, the solution of eqn. (27) yields the familiar Beer's law in eqn. (21).

2.3.5 Nonlinear optical phenomena on the microscale

At extremely high intensities, or for time durations approaching the characteristic molecular time scales, a host of *nonlinear* radiation-material interactions can occur. Several of these mechanisms are discussed below.

2.3.6 Energy storage in a molecule

In general, a molecule stores energy by three mechanisms: *rotation*, *vibration*, and *electronic transitions*, with each mechanism associated with different energy levels. Each electronic state has a spectrum of vibrational substates, while, in turn, each vibrational state has a series of rotational substates. Due to frequent collisions, the rotation of molecules is severely impeded; hence rotation, which has the smallest energy storage capacity among the three mechanisms, does not contribute considerably to the molecular energy. In general, rotation of a molecule occurs with a typical energy increment of ~0.01 eV, while molecular vibration occurs with an energy difference on the order of ~0.1 eV. For energy transfer via electronic transition, the energy involved is roughly ~1 eV and becomes much higher than the rotation and vibration counterparts. Figure 3 demonstrates the transition between different energy levels, in which thick solid lines, thin solid lines, and dashed lines represent electronic states, vibrational states, and rotational states, respectively. It should be noted that all of the above energy states are quantized; that is energy storage between any two levels is forbidden, although a combination of different states is allowed. For example the molecular energy of a material at room temperature is ~0.25 eV, which is stored in the molecule by a combination of rotational and vibrational excitations.

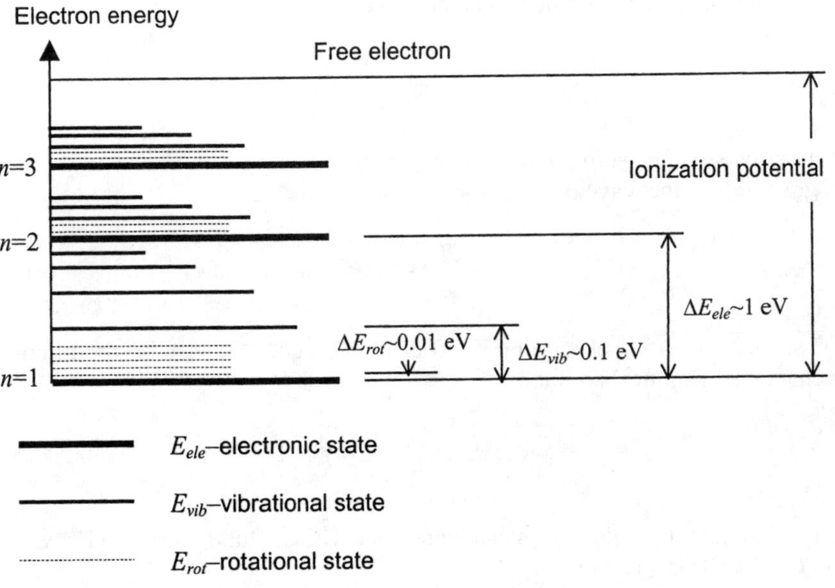

Figure 3: Energy states for a bound electron.

The electrons usually occupy the ground state at room temperature, where the energy level is the lowest and thus the state is stable. In such a case, most of the thermal energy associated with the matter is present in the form of rotational or vibrational excitations. The excited electron states are unstable and can be occupied only for a very short time period, which is typically on the order of ps–ns. Once excited to a higher electron state, there are five possible mechanisms for the molecule to leave the first excited state. (1) A molecule can absorb a second photon while in an excited state, for example from S_1 to S_2 as shown in Figure 4. The molecule will relax back down to S_1 in a very short time (~ 10 fs). Alternatively, the molecule can emit a photon to relax back to the ground state, either (2) spontaneously, which is called *fluorescence decay,* or (3) after being stimulated by a second photon, called *stimulated emission*. Spontaneous emission results in incoherent radiation; stimulated emission results in coherent radiation [1]. (4) The molecule can convert its electronic energy to vibrational and rotational energy. This process is called *non-radiative decay* and represents the primary mechanism for radiative heating from electronic excitations. (5) The molecule can transition from the singlet state S_1 to the lowest triplet state, T_1. Additionally, absorption can occur from the ground state to excited states in the triplet family of states.

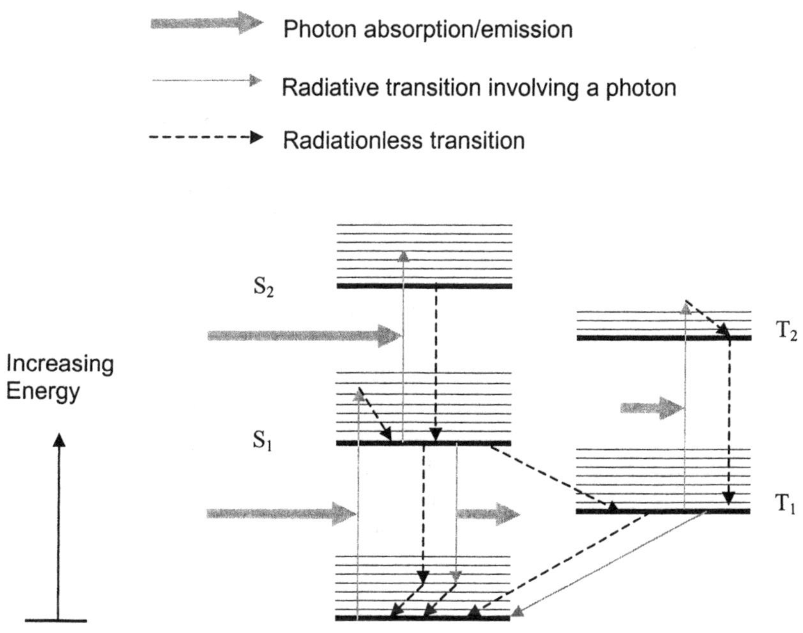

Figure 4: Electronic transitions during radiation-matter interactions.

From the thermal aspect, the occupational probability at different electronic states in dielectrics corresponds to temperature and follows the Boltzmann distribution [9]:

$$\frac{N_n}{N_1} = \exp\left(-\frac{E_n - E_1}{k_B T}\right) , \tag{28}$$

where the subscript n represents the principal quantum number, N is the population of occupied electrons at the corresponding state, and E is the associated electron energy. Here, a hydrogen molecule is used as an example due to its simple structure. The total energy of an electron in the orbit n is composed of potential and kinetic energies, and takes the form [9]:

$$E_n = -\frac{m_e e^4}{8\varepsilon_0{}^2 h^2}\frac{1}{n^2} \quad . \tag{29}$$

At $T = 300$ K, the ratio of occupational probability at an excitation state, e.g., $n = 2$, can be determined using eqns. (28) and (29), which yields $N_2/N_1 \cong 10^{-17}$. For pure water, an even smaller ratio, $N_2/N_1 \cong 10^{-42}$ is obtained. Thus for dielectrics at room temperature, only a very small fraction of electrons can occupy a higher excitation state, and it is reasonable to assume that all the electrons are in the ground state.

When a low-intensity radiation composed of low-energy photons is incident upon a dielectric medium at room temperature, only the rotational and vibrational transitions can contribute to absorbing this electromagnetic radiation as most of the electrons still occupy the ground state, which leads to a slight temperature rise. For radiation with higher photon energy, e.g., violet or ultraviolet radiation, the molecule will absorb a photon and the outermost electron can be excited into a new orbit, raising the energy of the molecule and inducing fluorescence via spontaneous emission.

The molecule does not have to exactly absorb the amount of energy difference between the two electronic ground states. A combination of electronic and vibrational excitations allows for a range of frequencies to be absorbed. This is part of the reason that molecules have broad features in their absorption bands, rather than having a single discrete absorption frequency corresponding to the energy difference between the two ground states. The diagram in Figure 4 illustrates the allowed electronic transitions that a molecule can undergo upon absorbing a photon. Discrete electronic states are denoted by thick horizontal lines, and are labeled S_0, S_1, S_2, T_1, T_2, etc. More energetic states appear vertically above lower states. Most molecules normally have an even number of electrons (the atomic elements, of course, have equal shares of even- and odd electrons). Quantum-mechanically, the molecule will fill up the lowest allowable states with pairs of electrons according to the Pauli exclusion principle, one with spin up (+½) and the other with spin down (–½). Accordingly, the total spin angular momentum Σ is 0, and the multiplicity $M = 2\Sigma + 1 = 1$; these states S_0, S_1, S_2,... are called the *singlet* states. Excitations within the singlet family maintain the molecule's total spin of zero. If, however, the spin of one electron flips, i.e., from +½ to –½ or *vice versa*, the net electron spin becomes $\Sigma = 1$ and $M = 2\Sigma + 1 = 3$; these states are accordingly called *triplet* states, shown on the right of Figure 4 as T_1, T_2, etc.

Ideally, 'pure' singlet and triplet states will never interact; however, actual quantum-mechanical states in the molecule have small admixtures of the opposite type in their wave functions, which give rise to transitions between singlet and triplet states. The probability of these so-called *intersystem crossing (ISC)* transitions is orders of magnitude smaller than transitions within respective families of states, though once a molecule transitions into a triplet state, it can remain in it for a long time. Another result of the electronic configuration of the molecule is that for every excited singlet state, there is a corresponding triplet state of lower energy, as shown in Figure 4, because the spin-parallel electron configuration results in reduced electron–electron repulsion. Triplet formation is therefore thermodynamically favorable.

2.3.7 Multiphoton absorption and ionization of dielectrics

For moderate-intensity laser radiation, the photon density, i.e., number of photons crossing a unit area per unit time, is low; molecules absorb a single photon, then relax back to the ground state before absorbing a second photon. For high-intensity laser radiation, however, e.g., from pulsed lasers, the intensity can reach 10^{10} W/cm^2 and higher, where an extremely large number of photons appear at a focused area during a short time interval. This extremely high photon density makes it possible to absorb several photons by a single molecule, which is called *multiphoton excitation or absorption* [16, 17]. Multiphoton absorption has been used to induce fluorescence to visualize cellular and subcellular structures for biomedical research. Details of this technique will be addressed in a later section.

As the intensity of radiation increases, the photon density grows accordingly. With an even higher photon density, ionization of a molecule can be realized by absorbing several more photons, resulting in *multiphoton ionization*. For a given molecule, the minimum required energy for ionization is determined by its band gap or ionization potential (see Figure 3). For example, the ionization potential for fused silica and water are 9.0 and 6.5 eV, respectively [18–21]. During multiphoton ionization, the number of absorbed photons k depends on the ionization potential of the material ΔE and the photon energy $\hbar\omega$:

$$k = \text{ceil}\left(\frac{\Delta E}{\hbar\omega} + 1\right) , \tag{30}$$

where the ceil(x) function yields the next-larger integer part of x, e.g., ceil(4.5) = 5. The multiphoton absorption process is extremely rapid, typically occurring on the order of a few femtoseconds. It should be noted that multiphoton absorption/ionization is a deterministic process, which will take place once the laser intensity exceeds a threshold value for the specific material. This is in contrast to *optical breakdown*, discussed in the next section, which, in general, is a statistical phenomenon.

2.3.8 Optical breakdown

Free electrons are not bound to individual atoms or molecules and, as such, have a very large number of states available to them, which results in very small energy differences between adjacent states. Such a condition is sometimes referred to as a *quasi-continuum*. As a result, free electrons exhibit a much stronger capability to absorb radiation than electrons strongly bound to individual ions and molecules. *Laser-induced ionization* provides an efficient approach to deposit photon energy into the bulk material via producing a large number of free electrons. One application that takes advantages of this strong interaction between matter and radiation is laser processing and micromachining of a variety of materials, especially high-bandgap dielectric materials, including diamond, glasses, plastics, oxide ceramics, and liquids.

To efficiently deposit photon energy into the material for processing, a high electron density, i.e., a *plasma*, is required to be generated within a certain volume; this process is called *optical breakdown*. During optical breakdown, a free electron density on the order of 10^{18}–10^{20} electrons/cm^3 can be reached [11, 18]. Two distinct mechanisms are responsible for laser-induced optical breakdown: *multiphoton ionization* and *avalanche ionization*, with the latter also called cascade or impact ionization. Multiphoton ionization occurs when a molecule simultaneously absorbs enough photons to ionize the molecule [11, 22]. Avalanche ionization is initiated when free electrons already existing in the material or produced by multiphoton ionization continuously absorb the laser energy through inverse bremsstrahlung absorption [11, 22]. During inverse bremsstrahlung absorption, seed electrons absorb laser photons by favorable collisions with heavier particles, i.e., molecules and ions. If they sustain enough favorable collisions, they will eventually gain sufficient energy to impact ionize other molecules, freeing new electrons to repeat the process, which results in a geometric increase in the free electron density. Due to the collision nature of avalanche ionization, it generally takes 10^{-9} – 10^{-12} s to induce sufficient electrons for breakdown. Therefore, for long pulses, e.g., in the nanosecond regime, breakdown is primarily caused by avalanche ionization, while multiphoton ionization dominates breakdown in the ultrafast (< 1 ps) regime, e.g., $\tau_p < 40$ fs for water [23].

2.3.9 Nonlinear scattering

Nonlinear scattering is accompanied by a frequency shift between the incident and scattered photons, and generally occurs in the presence of a strong electromagnetic field. Scattered light can be directed forward, backward, or sideways, depending upon the nature of the interaction. Nonlinear scattering includes *Raman scattering* and *Brillouin scattering*.

Nonlinear scattering is accompanied with a wavelength shift, which is not the case for linear scattering (section 2.3.3), and is called a *Stokes* or *anti-Stokes* wavelength shift. Stokes shifts occur when an electron transitions to a higher electronic state and emits a photon with a *longer* wavelength λ (lower frequency υ) than the excitation wavelength [24, 25]. This phenomenon is referred to as *fluorescence*. Referring to Figure 5, a photon from the light source raises the molecule from the electronic and vibrational ground state S_{0G} to a vibrationally excited state of the first electronic state S_1. The molecule, however, very quickly (typically < 100 fs) relaxes to the vibrational ground state of the first excited state, S_{1G}, with the energy difference between states S_1 and S_{1G} being distributed over the vibrational modes of the molecules as internal energy, i.e., the molecule is heated. Furthermore, the molecule undergoes a radiative transition from the state S_{1G} to a vibrationally excited mode in the electronic ground state S_0 (Figure 5). The molecule then relaxes very quickly back to the vibrational ground state S_{0G}, again with the energy difference between S_0 and S_{0G} being transferred to the molecule as heat. The emitted photon has less energy than the absorbed one; hence its wavelength is longer, which follows the law of energy conservation:

$$h\nu_i = h\nu_f + h\nu_{phonon} \quad , \tag{31}$$

where υ_i υ_f and υ_{phonon} are the frequencies of the incident excitation photon, fluorescence photon, and the phonon associated with the thermal vibrational energy, respectively.

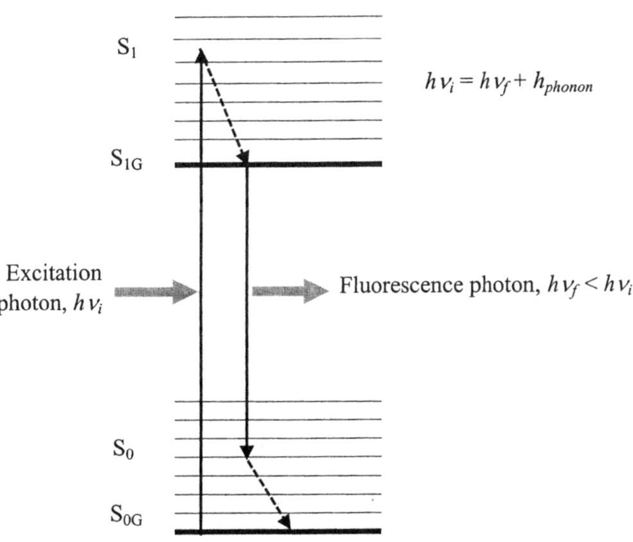

Figure 5: Stokes shift during fluorescence.

Alternatively, a molecule can produce a photon with a *shorter* wavelength (higher frequency) after absorbing a photon. This phenomenon is called the *anti-Stokes shift*, where a fraction of the thermal energy will be converted into radiation,

$$h\nu_i + h\nu_{phonon} = h\nu_f \quad . \tag{32}$$

Both Raman and Brillouin scattering follow the rule of the Stokes shift to produce lower frequency photons and vibrational energy in the molecules, i.e., phonons.

For molecules with high vibrational energy, anti-Stokes Raman scattering can occur and emit photons with higher frequencies; interestingly, the electrons do not necessarily have to be excited to an electron state for this to occur. The phonons produced have relatively low frequencies, i.e., they are *acoustic* phonons with υ_{phonon}~30 GHz; this optical scattering is classified as Raman scattering. Brillouin scattering produces higher frequency phonons in the optical branch, which is ~135 GHz in fused silica [26]. It should be noted that Raman scattering actually always occurs as light passes through a material, though it is too weak to be noticed in most situations.

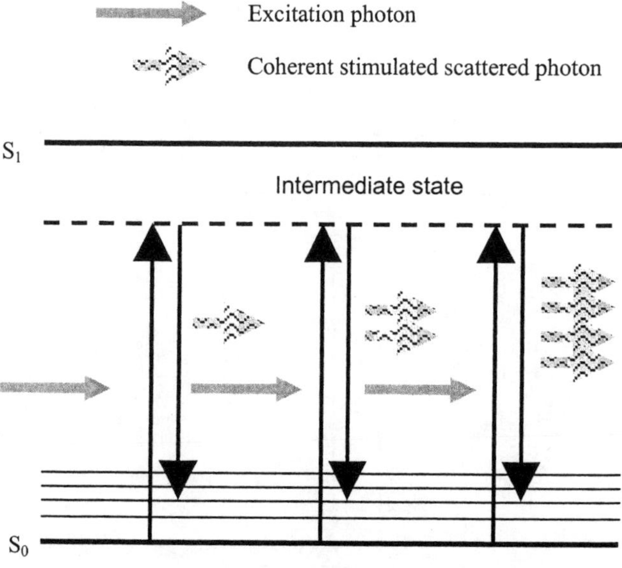

Figure 6: Stimulated Raman scattering.

As the excitation radiation increases its intensity above a specific threshold, *coherent* scattered photons can be generated, which is called *Stimulated Raman scattering*. For Raman scattering, scattered photons with energy $h\upsilon_f$ are incoherent and emitted in random directions. For high-intensity excitation radiation, the excitation and previously scattered photons are simultaneously incident on a molecule, resulting in *two* stimulated photon emission with energy of $h\upsilon_f$. This concept is schematically shown in Figure 6. Since one more photon is produced through each scattering process, the emission is subsequently amplified. Raman amplifiers are still of laboratory interest [26, 27], although the efficiency of the erbium-doped fiber amplifiers has made these latter amplifiers dominant in commercial use. Stimulated Raman scattering can occur in solids, liquids, and even dense gas media interacting with strong radiation, with scattered emission over a variety of wavelengths. Therefore, this nonlinear scattering mechanism can potentially provide a whole new range of high flux-density coherent sources from infrared to ultraviolet [2].

2.3.10 Kerr effect and self-focusing

This section introduces another interesting nonlinear optical phenomenon, *self-focusing*, which also becomes important in the presence of very high-intensity radiation, e.g., femtosecond laser pulses. For most cases, the refractive index of a homogeneous and isotropic optical medium can be treated as a constant, which refers to the linear part of the refractive index. In the presence of a strong electrical field, however, an intensity-dependent nonlinear refractive index can be induced in the neutral medium. This phenomenon, called the *optical Kerr effect*, can be described as [28, 29]:

$$n(I) = n_0 + n_2 I \ , \tag{33}$$

where n_0, n_2, and I are the linear part or the nominal refractive index, the optical Kerr coefficient, and the optical intensity (W/cm^2), respectively. For most condensed media, n_2 is on the order of 10^{-16} cm^2/W, which is very small, hence the nonlinear contribution to the refractive index is negligible under low and even moderate intensity situations, and thus n can be taken as the constant value, n_0. However, for ultrashort laser pulses with peak intensities as high as 10^{19}–10^{20} W/cm^2 [30], the nonlinear refractive index becomes comparable to the linear counterpart, and the total refractive index becomes correspondingly intensity-dependent.

For a laser beam, the Gaussian intensity distribution $I(r)$ as a function of the beam radius r is given by:

$$I(r) = I_0 \exp\left(-2\left(\frac{r}{r_0}\right)^2\right) \ , \tag{34}$$

where I_0 is the maximum intensity at the beam center and r_0 is the beam radius measured to the $1/e^2$ intensity points. Substituting eqn. (34) into eqn. (33), the corresponding distribution of refractive index as a function of the beam radius is obtained, with the beam center having the maximum value of the refractive index that decreases axisymmetrically to the beam edge.

As a high intensity Gaussian laser beam propagates in a medium, the nonlinear refractive index will be induced and a spatial gradient in refractive index is formed accordingly, with the maximum refractive index along the beam propagation axis. As different parts of the beam encounter different refractive indicies, non-uniform phase retardation is introduced along different regions of the wavefront: the high intensity region encounters a higher refractive index and is delayed more than the outer regions that encounter lower refractive indices, with the central part of the beam suffering the maximum delay. The total delay accumulates as the light propagates through the medium, and results in a spatially distorted wavefront at the medium exit. The resulting beam distortion is similar to what is imposed on the beam by a converging focusing lens. However, this distortion is induced by the high intensity laser beam itself; therefore, it is called *self-focusing*, where the beam appears to focus by itself [31–35].

2.3.11 Plasma effect and defocusing

Another mechanism that can cause variations in refractive index through the interaction between high-intensity radiation and the medium is due to the generation of free electrons, i.e., plasma production via laser-induced ionization. Once ionization is induced by multiphoton and/or cascade ionization, free electrons are generated, resulting in plasma formation around the focal region and variations in refractive index. For a gas medium, ionization can be induced by high intensity laser pulses either via multiphoton or tunneling ionization [29], although cascade ionization may supplement multiphoton ionization in heavy inert gases under high pressure conditions [36]. In regions where plasma is generated, the refractive index becomes [29, 37]:

$$n = n_0 - \frac{\omega_p^2}{2\omega^2}, \tag{35}$$

where ω_p and ω are the plasma frequency and laser frequency respectively, and n_0 is the nominal refractive index under low-intensity radiation. Note that, in contrast to the Kerr effect in eqn. (34), the refractive index *decreases* with the plasma effect. The plasma frequency ω_p is defined by [29, 37]:

$$\omega_p = \left(\frac{e^2 N_e}{\varepsilon_0 m}\right)^{1/2}, \tag{36}$$

where e is the electron charge, N_e the plasma (electron) density, m the electron mass, and ε_0 the permitivity of free space. A critical plasma density N_{cr} is obtained when the plasma frequency equals the laser frequency [38]:

$$N_{cr} = \frac{\omega^2 \varepsilon_0 m}{e^2}. \tag{37}$$

Using eqns. (36) and (37), eqn. (35) can be rewritten as [38]:

$$n = n_0 - \frac{N_e}{2N_{cr}}. \tag{38}$$

Equation (38) suggests that the presence of plasma decreases the refractive index, and the variation in refractive index depends on the plasma density. Due to the strong nonlinear dependence on laser intensity of the ionization process, temporal and spatial variations in the beam intensity give rise to both time- and space-dependent plasma density, which is a maximum on the beam axis and decreases rapidly in the radial direction. Therefore, the refractive index is smallest on the beam axis and the beam is *defocused* by plasma, which acts as a diverging lens.

2.3.12 Self-phase modulation

Once ionization is induced, the density of free electrons increases as the plasma grows, resulting in a temporal gradient in plasma density. While the spatial gradient in plasma density results in beam defocusing and wavefront distortion, rapid *temporal* variations in plasma density can cause another nonlinear optical effect: *self-phase modulation* [29, 39–41], which shifts the laser beam wavelength from its nominal value.

Referring to eqn. (4), the instantaneous phase $\phi(t)$ of the EM wave in the exponential term is:

$$\Phi(t) = \omega_0 t - k_0 nz \ . \tag{39}$$

The derivative of the phase with respect to time gives the instantaneous frequency $\omega(t)$, which is:

$$\omega(t) = \frac{d\Phi(t)}{dt} = \omega_0 - k_0 z \frac{dn}{dt} \ . \tag{40}$$

The change in refractive index with time, dn/dt, can be determined by differentiating eqn. (38) as well:

$$\frac{dn}{dt} = -\frac{1}{2N_{cr}} \frac{dN_e(t)}{dt} \ . \tag{41}$$

Equation (40) suggests that temporal variations in refractive index, dn/dt, modulate the frequency, or the phase, of the field associated with a laser beam. If the variations in refractive index are induced by the laser beam itself, then it is called *self-phase modulation.*

Since the wavelength, λ is inversely proportional to the frequency, ω, i.e.,

$$\lambda = \frac{2\pi c}{\omega} \ , \tag{42}$$

$dN_e(t)/dt > 0$, i.e., $dn/dt < 0$, during the laser pulse, which will shift the spectrum to shorter wavelength, and is called *blue-shifting* [36]. Many experimental investigations have demonstrated that plasma-induced self-phase modulation shifts components of the laser pulse to higher frequencies, resulting in a broadening of the spectral distribution in the pulse as shorter-wavelength components are introduced. This mechanism provides a potential way to generate a *supercontinuum* or a *white light* laser.

2.4 Characteristic length, time, and structure regimes for radiation–material interactions

There are several intrinsic length and time scales associated with radiation–material interactions, and, depending on the values of the individual parameters, different regimes are defined. These so-called *microscale regimes* provide a framework and the corresponding analytical tools that need to be used on the analysis of a given system. The definitions below come from the literature, and are discussed in detail elsewhere. The interested reader is referred to Ref. [41] and the references therein. A summary of the salient results is presented below.

2.4.1 Microlength scales and regimes

The following length scales are associated with radiation–material interactions:

- Wavelength, λ
- Rayleigh range, z_R
- Skin depth, δ
- Coherence length, l_c
- Medium spatial dimension, L
- Energy carrier mean-free-path, Λ
- Energy carrier wavelength, λ_c

Here L is the characteristic length of the device or structure, e.g., the thickness of a thin film coating, the width of a patterned line on a MEMS structure or grating, or the diameter of the fibers in a fibrous insulating material. The carrier mean free path(s) (MFP) depends on the type of energy carrier(s) in the material, and may include electrons, phonons, holes, or actual molecules. From the uncertainty principle, every energy carrier has a wavelength, or, equivalently, an uncertainty in its position at any given time.

First microscale regime

The first microscale is defined as:

$$L/l_c \precsim O(1) \ . \tag{43}$$

Thus, when the characteristic dimension of the medium L becomes comparable to or less than the coherence length l_c, the photons can interact coherently, resulting in interference phenomena.

Second microscale regime

The second microscale regime is defined as:

$$L/\Lambda \precsim O(1) \tag{44a}$$

$$\delta/\Lambda \precsim O(1) \tag{44b}$$

$$L/\lambda_c \succsim O(1) \, . \tag{44c}$$

Equation (44a) quantifies the *size effect*, in which the characteristic dimension L is smaller than the carrier mean free path, Λ. Values less than one imply that the majority of carriers strike a boundary rather than each other when they experience a collision, which imposes an upper limit on the maximum MFP and also makes the MFP a function of sample dimension. Equation (44b) states that the skin depth is less than the carrier MFP. When the skin depth is less than the carrier MFP, carriers near the surface experience a *non-uniform* electric field between collisions, and thus interact differently with the incident radiation than if the electric field were uniform, as is assumed in the classical model. This is the *anomalous skin effect* [1], and can result in optical properties that differ from the classical case in which $\delta/\Lambda > O(1)$. Equation (44c) states that the characteristic material dimension L exceeds the heat carrier wavelength λ_c. For the second microscale regime, the optical properties of the material will be altered.

Third microscale regime
The third microscale regime is characterized as:

$$L/\lambda_c \lesssim O(1) \ . \tag{45}$$

In this case, the characteristic material dimension is less than the heat carrier wavelength. Quantum mechanics must be used to correctly characterize the transport in such cases. The unique optical properties of a quantum well or a nanoparticle, for example, result from the condition in eqn. (45) being satisfied [42]. Though the carrier wavelengths typically fall into the 10–100 nm range, such effects are becoming increasingly more important with the tremendous interest of late in nanotechnology, i.e., the fabrication and engineering of devices and structures with nanometer-sized dimensions.

2.4.2 Radiation phenomena on the temporal microscale

The following time scales are important for radiation–material interactions:

- Laser pulse duration, τ_p,
- Thermal diffusion time, τ_d,
- Radiation propagation time, τ_c,
- Relaxation/thermalization time, τ_r.

Though in the following τ_p represents the pulse duration for laser radiation, the arguments hold for *all* types of short-duration radiation, e.g., radiation from a spark discharge, fast combustion, or a light-producing chemical reaction. In such cases, τ_p is taken as the duration of the radiation. The laser pulse duration is an *imposed* time scale and the temporal microscale regimes are determined based on the magnitude of the laser pulse duration compared to the other time scales.

First temporal microscale regime
The first temporal microscale regime is defined as:

$$\tau_p \lesssim \tau_d = L^2/\alpha \ , \tag{46}$$

where α is the *thermal diffusivity* (m^2/s). Equation ((46) states that the radiation time scale τ_p is comparable or shorter than the time required for thermal energy to diffuse through the medium, τ_d. In this case, the thermal energy is deposited in the material before it can be conducted away, and the initial temperature profile will approximate the EM source distribution inside the material.

Second temporal microscale regime
The second temporal microscale regime is defined by:

$$\tau_p \lesssim \tau_c \ . \tag{47}$$

In such situations, the laser pulse duration τ_p is on the order of or shorter than the light propagation time τ_c across the spatial extent of the medium. As a consequence, the radiation energy or intensity will vary as a function of position, and must be taken into account if (1) there are intensity-dependent effects exhibited by the medium, such as intensity dependent absorption or scattering, or (2) the short-time-scale nature of the laser pulse propagation is of interest, for example, in diagnostic measurements that rely on the delay time of the laser pulse to obtain information about the medium it passes through. The propagation of the radiation must be then modeled as a function of position and time.

Third temporal microscale regime
The third microscale time regime involves phenomena that are limited by the relaxation time of the excited electrons, atoms and/or molecules. In classical absorption molecules are assumed to relax instantaneously after absorbing radiant energy. In reality, molecules have finite relaxation times. Thus, after absorbing a photon the molecule remains in an excited energy state for a period of time before returning (relaxing) back to its original state, where it exchanges the absorbed energy with its neighbors.

2.4.3 Microstructural scales and regimes
A third class of microscale radiation phenomena involves radiation interactions with microstructures. *Microstructure* in this sense refers to a *collection* of individual elements, including particles, thin film layers, molecular clusters, grooves or channels, cells, bubbles, and combinations of these elements. In such cases, the proximity of neighboring elements must be accounted for in terms of both the radiation interaction and energy transfer.

Two length-scale criteria for microstructures can be established based on the characteristic separation distance l_s between adjacent elements. When either criterion is met, the interaction of the incident radiation with the microstructure will differ from the interaction of radiation with a single, isolated element. The first criterion is

$$l_s \lesssim \lambda \ , \tag{48}$$

which states that the separation distance l_s is smaller than the wavelength of the incident radiation. In this case, the neighboring elements perturb the EM field near the element of interest, rendering the EM solution for an isolated element inapplicable.

The second criterion,

$$l_s \lesssim l_c \quad , \tag{49}$$

indicates that the separation distance is smaller than the coherence length. In this case, interference effects due to emission from adjacent structures modify the radiation interaction with the microstructure.

3 Applications

3.1 Ultrafast laser materials processing

The advent of the laser some 40 years ago has made an unprecedented impact on manufacturing technology and capability. Indeed, lasers are routinely used today for cutting, drilling, welding, marking, measurement, etc. on scales ranging from microns to meters in nearly all branches of manufacturing, including the electronics, biomedical, aerospace, automotive, appliance, and food industries. As lasers continue to enjoy innovative new applications, however, relentless demands for faster, smaller, and cheaper processing are being made by laser users. In particular, materials processing at very small length scales (micron and submicron resolution), termed *laser micromachining*, is a rapidly growing research and application area [23, 31, 43–47]. Fueled by the explosive growth in nanotechnology, micro-electro-mechanical systems (MEMS), and biomedical fields, laser micromachining provides many advantages over other technologies, e.g., silicon lithography, due to versatility in materials processed, ability to generate high aspect ratios, flexibility in making process changes on the fly, and conformal processing of non-planar surfaces. Example applications include: fabrication [48–54] and repair [55, 56] of micro-electro-mechanical systems (MEMS); bioengineering (drug delivery, *in situ* sensors and diagnostics) and biomedicine (precision surgery, e.g., laser eye surgery for improved vision) [57, 58]; micro- and nano- surface texturing [59, 60]; and for small-scale fluid processing, including analysis and diagnostics using minimal fluid amounts (lab-on-a-chip), e.g., crime analysis using trace evidence [61, 62].

Most traditional laser processing relies on thermal-based processing, in which surface temperatures are increased to the material melting and/or boiling points for material removal. The large temperature increase in the workpiece and the conduction of heat to adjacent structures, however, make thermal-based machining of microstructures difficult or impossible.

An exciting and promising recent development in laser technology is high-power *ultrafast* laser systems, in which the laser pulse duration is measured in femto- or picoseconds. Ultrafast systems have many advantages over their thermal-based counterparts, including [16, 23, 63]:

Table 1: Advantages of ultrafast laser processing.

Features	Thermal Processing	Ultrafast Processing
Negligible heat diffusion	No	Yes
Sub-diffraction-limited features	No	Yes
Processing of many materials with same system	No	Yes
Micron-size features	Difficult	Straightforward

- *Minimal temperature rise and thermal damage in processed material*
- *Very wide range of applicable materials*
- *Precision machining capability*
- *Subsurface (3-D) machining*
- *High-aspect-ratio processing*

The salient differences between thermal-based and ultrafast laser processing are summarized in Table 1. Though used in research laboratories since the 1960s, two significant developments in the early 1990s made ultrafast systems practical for far more researchers: 1) the development of titanium-doped sapphire (Ti:sapphire) as the lasing medium, and 2) chirped-pulse amplification (CPA) to produce femtosecond laser pulses with millijoule energy levels [23, 64]. These developments opened the door to significant research endeavors into ultrafast laser-materials processing.

3.1.1 Thermal-based laser processing

The vast majority of laser processing operations relies on a *photothermal* approach: laser light striking the workpiece is absorbed at and near the surface, which raises the material temperature in the irradiated region to its melting or vaporization point, after which material removal occurs [65,66]. Thermal-based processing, however, has several disadvantages that are particularly problematic for next-generation laser processing and micromachining:

- *Thermal damage to adjacent structures.* Thermal energy that diffuses away from the initial deposition region elevates the material temperature in nearby regions. Such temperature excursions can result in permanent damage to the material, thermal stresses, warpage and feature distortion, and surface property alteration. Also, the feature size is determined not by the minimum laser spot size, but the larger region, in which the material is heated above the melting point. Thermal-based material removal makes fabricating small-scale, high-quality features difficult or impossible.

- *Difficulty in processing composite and multi-layer materials.* Thermal-based processing of composite and multi-layer materials is particularly challenging, due to the high temperatures required for processing as well as difficulties in dealing with materials with large disparities in their thermal and optical properties. Examples include plastic–metal, plastic–ceramic systems, biological systems such as bone and tissue, and multi-layer materials.
- *Restricted to classical absorption.* Most laser processing to date relies on classical absorption of the incident light in which laser light is absorbed near the surface of a material, with the intensity decaying exponentially with depth underneath the surface. The absorption coefficient, α (cm^{-1}), characterizes absorption in a material, and the intensity, I, as a function of depth, z, becomes [64]:

$$I(z) = I_0 \exp(-\alpha z) \,, \tag{50}$$

where I_0 is the initial intensity at the material surface (after correcting for reflection losses). The initial temperature distribution in the material is directly related to the intensity profile in eqn. (50); that is, the maximum temperature will initially occur near the surface ($z = 0$), and decrease with distance z into the material. Classical absorption is limited to this temperature profile, and all thermal-based processing must accommodate this energy distribution.

3.1.2 Ultrafast laser processing

To minimize the disadvantages of thermal-based processing and obtain finer feature size and resolution, two common approaches have been used: 1) use of a short-pulse laser, and/or 2) use of a laser that employs a non-thermal material removal mechanism. *Short pulse durations* result in less heat diffusion, and more precise control of the processing region because the laser light is only present on the material surface for the duration of the laser pulse. Even for pulse durations as short as 1 ns, however, many such lasers still rely on thermal mechanisms for material removal, resulting in large local temperature excursions, and a feature size determined by the extent of the heat-affected zone. *Non-thermal* or *photochemical* material removal has been employed by using lasers that break chemical bonds in the material directly with the laser light – rather than relying on heating – for material removal. Typically these lasers operate at wavelengths in the ultraviolet (UV) range, where photons, due to their shorter wavelength, have sufficient energy to break chemical bonds directly, which allows the removal of material without bringing the irradiated region to the melting and vaporization temperature. Examples include the ubiquitous excimer laser and frequency-upshifted Nd:YAG systems.

So-called *ultrafast* lasers embrace the best features of both: very short laser pulses *and* a non-thermal material removal mechanism. If the pulse duration is reduced to the pico- and femtosecond regime, the pulse intensity becomes high enough to initiate optical breakdown in the material, which results in non-thermal energy deposition and material removal.

Commercial ultrafast lasers have a pulse duration τ_p that is extremely short: typically 100 fs – 1 ps , with pulse energy, E_p ~1–10 mJ. The *intensity, I,* (energy delivered per unit time and area), scales as

$$I \sim \frac{E_p}{A_p \tau_p} \quad , \tag{51}$$

where A_p is the area of the laser beam. For very short focused pulses, the intensity becomes enormous.

For metals, the laser pulse is absorbed by the large number of conduction electrons, which is the same mechanism for absorption of light at low intensities. Dielectric materials, including glass, ceramics, plastics, and liquids, have an insignificant number of free electrons, and thus do not absorb appreciably at low laser intensities. At high intensities, however, the laser intensity becomes large enough to remove electrons from the lattice atoms, resulting in ionization, plasma formation, and ejection of material.

As discussed in previous sections, there are two primary mechanisms for absorption of laser light at high intensities in dielectric materials. The first is multiphoton ionization, in which several photons are absorbed simultaneously to provide enough energy to ionize the molecule. Multiphoton ionization scales as I^k, where k is the number of photons required for ionization (typically k is 2–6 for visible light and determined by eqn. 30) [18, 66, 67]. The second mechanism is avalanche (or cascade) ionization, in which a few seed electrons absorb laser energy via inverse bremsstrahlung absorption [11, 23, 68]. Some electrons collide with heavier particles (atoms and ions), and occasionally undergo favorable collisions that increase their translational energy. If the electron gains enough energy to impact ionize a neutral atom, an additional electron is released. The process repeats geometrically (hence the name) and results in high electron densities in the material. Both mechanisms require a threshold intensity, I_{th}, to be reached before absorption occurs, and also exhibit a dependence on the pulse duration. At very short pulse durations, multiphoton absorption dominates the initial breakdown, and provides seed electrons for subsequent avalanche ionization [19, 67–70].

In dielectrics, multiphoton absorption and avalanche ionization form the basis for materials processing using ultrafast lasers. These mechanisms – coupled with the ultrashort duration of the laser pulse – provide for several unique advantages:

- *Negligible thermal damage.* Thermal damage is negligible in ultrafast processing for two reasons. First, bond breaking and material removal arise due to the extremely high laser intensity, rather than raising the material temperature to the melting or boiling point. Secondly, the distance thermal energy travels away from the deposition site, i.e., the thermal penetration depth, l, is on the order of 10 nm or less, which, for most applications is insignificant. Such small thermal penetration is ideally suited for laser micromachining.

- *Extremely wide range of material processing capability*. Since multiphoton ionization and avalanche ionization are nonlinear optical breakdown mechanisms, they do not depend on the material's classical physical and optical properties; since virtually all dielectric materials exhibit breakdown at ultrafast laser intensities, the same system can process many materials with no change in laser setup or optics.
- *3-D micromachining (subsurface machining)*. A very promising, novel aspect of ultrafast laser micromachining is the nonlinear absorption of laser light, which can be used for subsurface or 3-D micromachining.

3.1.3 3-D volume micromachining of dielectric materials

An intriguing aspect of nonlinear energy deposition during ultrafast laser processing is the possibility for *subsurface* or *volume* micromachining in dielectric materials. Both avalanche ionization and multiphoton ionization require a threshold intensity I_{th} before significant energy deposition occurs in the material. This threshold-based absorption mechanism allows the beam to be focused *underneath* the material surface to perform processing, while leaving the surface undisturbed. Subsurface machining using ultrafast lasers would be very well suited for microfluidics and MEMS applications, and is particularly exciting because it may provide the possibility for *true* three-dimensional fluid networks at the microscale.

To achieve the intensities required for breakdown, the beam is focused using one or more optical components. Most laser beams can be described using Gaussian optics, which results in a well-defined beam shape as a function of position in the optical system. For Gaussian beams, the beam spot size w after the focusing optics is a function of axial position z, and is characterized by the beam waist w_0 at the focal point, and Rayleigh range or focal region z_R, as discussed in section 2.2.2:

$$w(z) = w_0\left(1+\frac{z^2}{z_R^2}\right)^{1/2}, \tag{14}$$

$$z_R = \frac{n\pi w_0^2}{\lambda_0}. \tag{15}$$

Here $w(z)$ is the beam radius as a function of position z, λ the wavelength, and n the index of refraction of the medium. Note that the minimum beam diameter is w_0, occurs at $z = 0$, and is determined by the wavelength and focusing optics. The term z_R is called the *Rayleigh range* and represents the distance from $z = 0$ over which the beam waist remains within $\sqrt{2}w_0$.

The intensity of the beam, I, depends on the laser pulse energy, E_p, pulse duration, τ_p, and the area of the beam, $A_p = \pi w(z)^2$:

$$I \sim \frac{E_p}{\tau_p A_p} \sim \frac{E_p}{\tau_p w_z^2} \quad . \tag{52}$$

In eqn. (52) the intensity increases with decreasing beam diameter. The beam profile has an hourglass shape, with the beam waist at the center. The maximum intensity occurs at the beam waist, where the beam is at its smallest diameter, and this is the region where breakdown is initiated. If the beam waist is located underneath the surface of a transparent material, and the laser pulse energy is selected such that only the intensity near the beam waist is sufficient to induce breakdown, then the beam will pass undisturbed through the surface and the bulk material above the beam waist until the threshold intensity is reached, at which point optical breakdown occurs. Thus nonlinear absorption mechanisms provide the ability to generate features and perform processing underneath the surface [71]. This phenomenon has received relatively little attention to date, despite the large potential for innovative micromachining possibilities, though some researchers have used the technique to fabricate waveguides in glass [72–74].

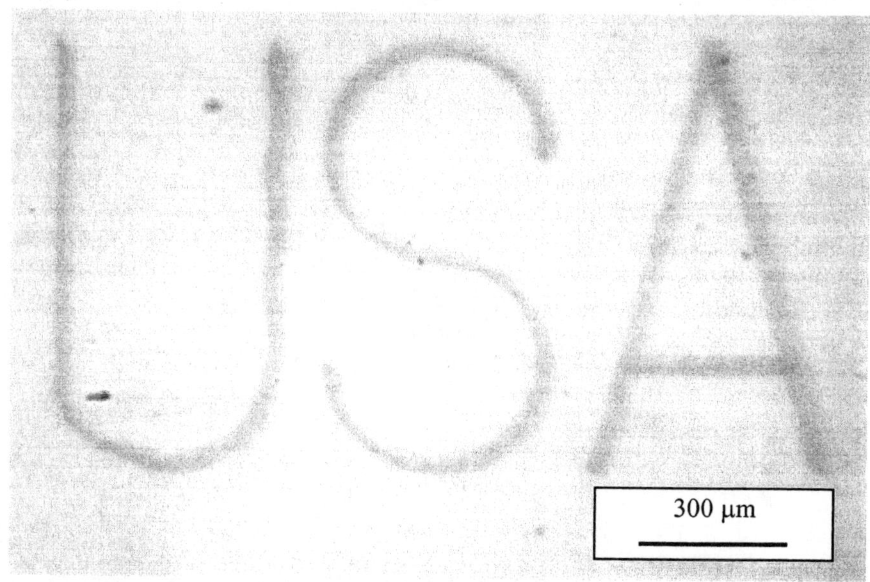

Figure 7: 3-D micromachining inside 1 mm-thick glass slide.
Top and bottom surfaces of slide undisturbed.

Furthermore, by translating the focal region within the material in the x, y and z directions, a three-dimensional region can be fabricated. To illustrate this capability, subsurface machining was performed using a 150 fs Ti:sapphire laser and an ordinary glass slide used for optical microscopy, as shown in Figure 7. The slide is 75 mm long, 25 mm wide and 2 mm thick. The beam was focused using a three-lens optical system resulting in a small beam waist and a small Rayleigh range (eqns. (14) and (15), respectively), resulting in breakdown inside the material. Both front and back surfaces of the slide, however, are optically smooth and have not been disturbed during the micromachining process. The width of the feature in the material is roughly 40–50 μm.

Ultrafast laser processing is poised to become a significant and viable technology for precision micromachining for a vast array of applications. At this time the costs of ultrafast laser systems and their frequent maintenance needs have prevented their widespread industrial acceptance. These technological hurdles will be overcome, however, and future systems, which will be less expensive, more reliable, and more powerful, will pave the way for a new era in ultrafast laser materials processing.

3.2 Laser scanning microscopy for biological systems

The advent of lasers opens up new possibilities for visualizing the distribution and dynamics of biological tissues, cells, and subcellular organelles via a scanning head coupled with a regular bright-field microscope, which is the so-called *Confocal laser scanning microscopy* (*CLSM*) and was first successfully presented by White *et al.* in 1987 [75–77]. Soon after publication, the demand for this user-friendly CLSM increased dramatically, and it quickly became a powerful facility in biological as well as medical research laboratories. With the rapid progress in laser technology, femtosecond lasers were brought into laser scanning microscopy to replace the traditional CW laser source used in CLSM, which gave birth to a new generation of the technique: *Multiphoton laser scanning microscopy* (*MPLSM*). The following sections will introduce this powerful tool as an application example for laser radiation in biomedical engineering and sciences based upon the fundamental optical principles addressed in the previous sections.

3.2.1 Confocal laser scanning microscopy

Confocal laser scanning microscopy (*CLSM*) is a technique to image a stained specimen by collecting the fluorescence intensity from each scanned pixel. Fluorescence can be induced by external light sources, e.g., lasers. Irradiated by a laser beam with sufficient photon energy E_p, the electrons of the stained fluorophores can be excited to higher electron states, through single-photon excitation. Fluorescence emission occurs afterwards when the excited electrons release energy radiatively (details are described in section 2.3.6 and shown in Figure 5).

The proper laser wavelength for fluorescence microscopy depends on the fluorophores stained or tagged in the specimens. In biomedical research, fluorophores include both chemical fluorescent dyes and fluorescent proteins. *Fluorescent dyes* are used for fixed-slice specimens, while *fluorescent proteins*, such as CFP (cyan fluorescent protein), GFP (green), and YFP (yellow), can tag specific proteins and generate light emission from live specimens. The fluorescence emission from the fluorophores exhibits a longer wavelength than that from the excitation radiation due to the Stokes shift (section 2.3.9). For example, the laser wavelength to excite the fluorescent dye, *cy3*, is 554 nm, whereas the nominal fluorescence wavelength is 566 nm. Similarly, the fluorescent protein *GFP* exhibits excitation and emission wavelengths of 498 and 515 nm, respectively.

The basic setup of fluorescence microscopy is shown in Figure 8a. The excitation laser beam is directed by a dichroic mirror, which reflects the shorter

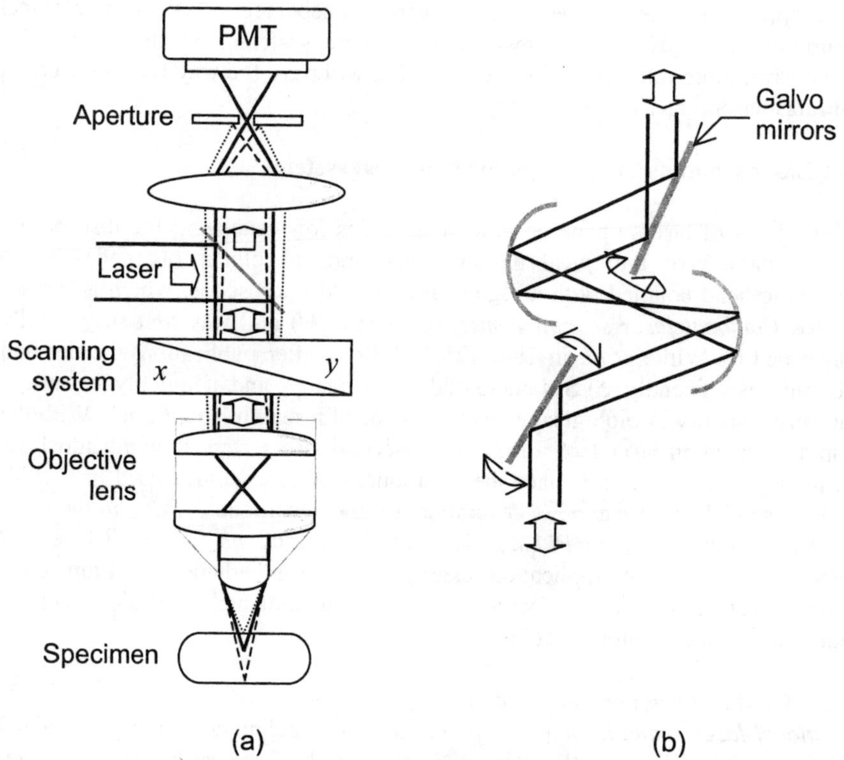

Figure 8: (a) Optical path and depth discrimination in a confocal laser scanning microscope. (b) Optical path and geometry of a x-y scanning head.

wavelength component (excitation laser beam) and allows the longer wavelength counterpart (fluorescence emission) to pass. Such a design can minimize the contribution from the scattered excitation photons to the detector. Since the fluorescence is induced by a laser through single photon excitation and collected by the identical objective lens, the emission occurs along the optical path within the specimen, which blurs the optical information at the focal point (Figure 8a). One solution is to set up a pinhole or so-called *confocal aperture* at the back focal plane of the objective lens. This pinhole allows the emission from the focal plane to completely pass through, and effectively blocks the emission from other areas along the optical path; therefore a clear optical signal is able to reach the photomultiplier tube (PMT) behind the dichroic mirror.

In practice, instead of pointwise information, two-dimensional sectioning images are generally required, which are realized in modern scanning microscopes by using a pair of motorized galvo mirrors to scan the specimens, as shown in Figure 8b. Each galvo mirror has its own axis of rotation to achieve the *x-y* scan. This design maximizes the detection sensitivity and uniformity during the scanning cycle. The PMT detects the fluorescence intensity associated with each pixel and sends the optical information to a computer. The computer synchronizes the optical information with the beam location from the scanning system to form a digital 2-D image. Brightness variation is displayed in the images according to the fluorescent intensity at each scanned pixel. By multi-staining or labeling the specimens, images based upon different emission wavelength (colors) can be resolved, while providing more histological information for biomedical research.

The smallest separation between two point objects to be resolved is referred to as the optical *resolution* of an imaging system. The resolution of the microscope depends on two critical parameters: the nominal wavelength of the laser λ_0 and the numerical aperture (NA) of the objective lens. The NA of an objective lens is the product of the sine of the half-angle θ of the light cone and the refractive index of the imbibing medium n [27,78]:

$$NA = n \sin \theta \ . \tag{53}$$

This same definition of the numerical aperture appears in other optical systems, e.g., fiber optics and telescope design.

In general, two resolution capabilities are of interest: *lateral* and *axial* resolution. For lateral resolution, the *Rayleigh criterion* – which states that two images are said to be resolved if the smallest distance of two equally bright points in the specimen plane is equal to or larger than the radius of the Airy disk r_{Airy} [78] – provides a lower resolution bound:

$$r_{Airy} = 0.61 \frac{\lambda_0}{NA_{obj}} \ . \tag{54}$$

As the size of the confocal pinhole is decreased in Figure 8a, the field of view becomes negligible, resulting in an improvement of lateral resolution by a factor of $\sqrt{2}$. The confocal lateral resolution thus can be expressed as:

$$R_{lat} = \frac{r_{Airy}}{\sqrt{2}} = 0.43 \frac{\lambda_0}{NA_{obj}} \ . \tag{55}$$

Thus, using a blue laser for illumination ($\lambda \sim 400$ nm) and a high-*NA* objective lens, e.g., an oil-immersion lens with $NA = 1.4$, a lateral image resolution of 120 nm can be achieved.

The determination of the axial resolution is, unfortunately, not as easy as lateral resolution, especially when dealing with highly scattering or reflecting specimens. One commonly used expression takes the form [78]:

$$R_{axial} = 1.77 \frac{\lambda_0}{(NA_{obj})^2}, \tag{56}$$

resulting in an axial resolution of ~100 nm–1 μm in sectioning depth. A recent variation of these ideas called *4π confocal microscopy* employs two opposing objective lenses for illumination and/or detection, and has broken the diffraction barrier with a reported $\lambda/23$ focal spot and 33 nm axial resolution [79]. Interested readers can find the details in references [78–81].

3.2.2 Multiphoton microscopy

Despite the impressive capabilities of traditional microscopy, including confocal microscopy and its variations, all of these techniques rely on *linear* interactions between light and matter: doubling the photon density, for example, results in a signal increase that also roughly doubles. A clever extension to traditional microscopy is the use of *nonlinear* radiation–matter interaction mechanisms to enhance signal-to-noise ratios, provide increased contrast, and, with care, exceed the diffraction-limited spot size resolution.

Electron excitation to higher states can be realized not only by energetic photons via single-photon excitation but also by multiphoton absorption at high photon densities (section 2.3.7). The CW laser used in CLSM discussed above can be replaced by a very-high-intensity pulsed laser source to produce multiphoton excitation in the sample [82]. The technique is called *multiphoton laser scanning microscopy* (*MPLSM*). For two-photon excitation, the wavelength to excite fluorescent emission should be roughly twice that for single-photon excitation. Similarly, three-photon excitation should employ a laser source with three times the single-photon wavelength. For example, the wavelength used for two-photon excitation of GFP is 996 nm, which arises from the 498 nm wavelength for single-photon excitation.

One of the immediate benefits of a longer wavelength is reduced scattering: as the wavelength of light increases, the Rayleigh scattering effect decreases as λ^4 (section 2.3.3). If the wavelength is doubled, the scattering is 16 times weaker than that of the original wavelength, over an order of magnitude reduction! This unique feature makes it possible to induce deeper fluorescence sectioning in a highly scattering biological environment, e.g., lipids in cytoplasm. Thus the MPLSM significantly improves the sectioning capability of a laser scanning microscope [83].

Another unique, and significant, feature of multiphoton absorption is that, because it is an intensity-dependent phenomenon, multiphoton absorption *only* takes place in regions with a high photon density; with appropriate adjustment of the incident laser power, this region can be confined only to the focal region, with *no* fluorescent emission occurring at any other place in the material except at the beam focus. The confocal pinhole used to block fluorescence out of the focal region in CLSM can be removed, while still maintaining good axial resolution. Due to the smaller interaction region, MPLSM also minimizes phototoxicity and photobleaching during the scanning process, which makes it valuable for biomedical studies *in vivo*.

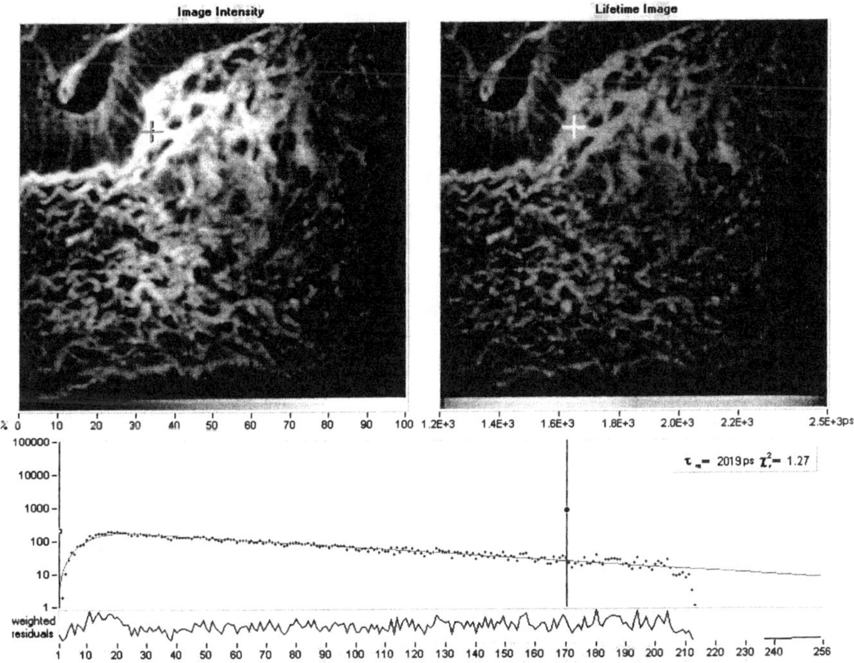

Figure 9: Multiphoton and lifetime images of a uterus slide.

When the laser beam interacts with the fluorophores in the cells, an image can be achieved by collecting the fluorescence information from each scanned pixel as done for traditional CLSM, which is the so-called multiphoton image. The black and white picture in the upper-left of Figure 9 shows an example of a multiphoton image of a uterus slice, stained by acridine orange (AO) (Figure 9 will be further discussed in the following section). The AO molecules are excited by Nd:YLF fs laser pulses (λ = 1047 nm) to induce fluorescence emission at a nominal wavelength of 550 nm via two-photon excitation. For a living system, fluorescence proteins can be tagged with a specific protein in a cell, and this protein movement in a living cell can be tracked. Current biochemical technologies allow the cells and tissues to be stained with several fluorescence dyes or tag with different fluorescence proteins; therefore, a variety of biological information, e.g., tissue histogram and protein movement, become available by observing the multi-label specimens. Moreover, since deeper sectioning images are obtainable with MPLSM, a 3-D biological architecture and image can be reconstructed [84–87]. If the MPLSM records the 3-D images as a time series, a "movie" can be produced that depicts the biological system evolution over the time scale of the measurements [88–90], providing biological information in both *spatial* and *temporal* domains.

3.2.3 Lifetime and spectral imaging

Although MPLSM has been used to image and trace labeled proteins, observe thick sections, and visualize three-dimensional structures, due to the overlapping of florescence spectra in a multi-label specimen, e.g., YFP and GFP, optical filters and specific computer algorithms are generally required to enhance image discrimination if only the florescence wavelength is monitored, since a photon emitted in a wavelength region shared by two (or more) labeling proteins has no history of which particular protein it came from. However, an additional piece of information that can be extracted from the fluorescence emission is the *spectral and lifetime shift* that occurs when the histological stains and the fluorescence proteins are bound to different components of the cells and tissues. Such information can be significant for imaging and medical diagnosis.

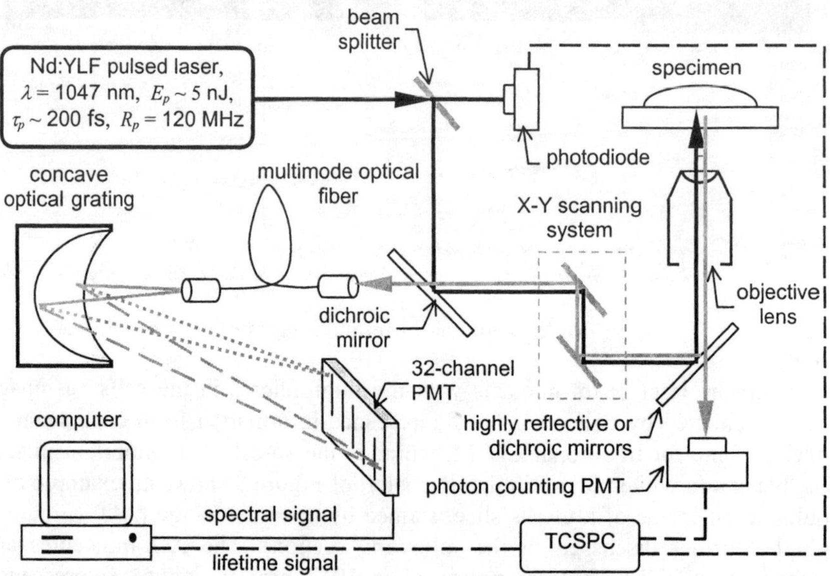

Figure 10: Experimental setup for spectral and lifetime imaging.

For both CLSM and MPLSM, the images are formed by collecting fluorescence intensity at each scanned pixel. The fluorescence is composed of different wavelengths as determined by the particular choice of protein label. The spectral information from the florescence can be made available by sending the fluorescence to dispersive optics, e.g., prisms or gratings to resolve the spectrum. One example to realize spectral images through regular laser scanning microscopy is shown in Figure 10. In this figure, the 200-fs laser serves as an excitation source to induce fluorescence via a scanning head. The fluorescence emission is then collected by the same objective lens and directed back to a multimode optical fiber coupled with an aberration-corrected concave optical grating. The grating separates different wavelength components from the fluorescence and focuses each component to a linear multi-channel PMT, which is used to produce 32 separate images for each optical section with a spectral resolution of ~10 nm. By calibrating each channel using a light source with a known wavelength, spectral images are formed associated with the specific channel.

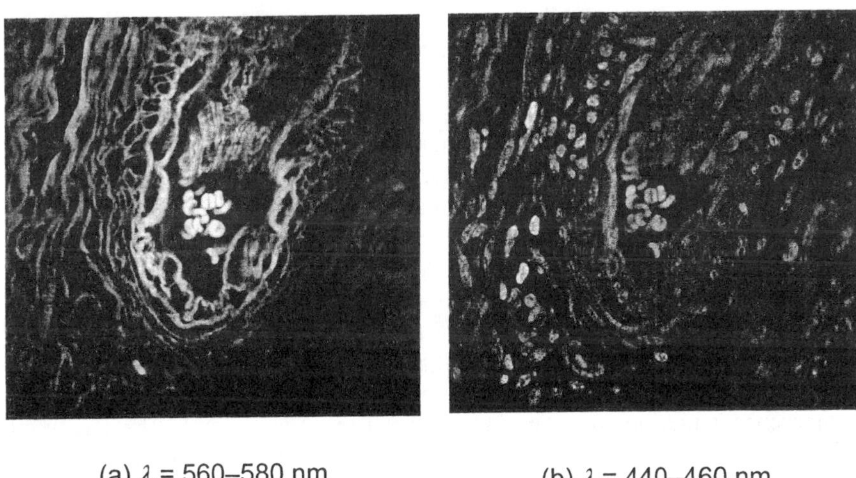

(a) λ = 560–580 nm (b) λ = 440–460 nm

Figure 11: Spectral images of uterus slide at different wavelength (stained with AO).

One example of spectral images is shown in Figure 11, in which the specimen is a uterus slice stained by AO and displayed in two spectral ranges. In Figure 11a, the image at $\lambda = 560$–580 nm is shown, while Figure 11b is the image at $\lambda = 440$–460 nm. As can be seen, the tissue histogram and cell membranes clearly appear in the longer-wavelength image, however, the nuclei show up in the shorter wavelength counterpart, which provides additional information of biomedical significance based upon the original laser scanning microscope.

Another important piece of information extracted from fluorescence emission is the fluorescence lifetime. As the electrons of the fluorophore are excited to a higher energy state S_1, it takes a finite time to return to the ground state S_0. If one monitors the fluorescent emission with time, an exponential decay of the intensity I can be observed, which takes the form:

$$I_f = I_{f0} \exp\left(-{}^{t}/_{\tau_0}\right), \tag{57}$$

where I_{f0} is the peak intensity of fluorescence emission, t is time, and □$_0$ is the fluorescence or excited electron lifetime. When the time t equals the fluorescence lifetime τ_0, 63% of the excited electrons have returned to the ground state. Instead of monitoring the intensity decay, current technology employs a fast, high-sensitivity PMT and a time-correlated single-photon counting system to efficiently measure the fluorescence lifetime [91]. The experimental setup is shown in Figure 10 as well. To collect fluorescence lifetime using the lifetime detector, a dichroic mirror is used to replace the fully reflective mirror for spectral images. The correlated lifetime signals can be synchronized by the fast photodiode located at the laser exit. Since this approach requires a high-frequency excitation cycle, mode-locked blue diode lasers for CLSM or femtosecond laser oscillators for MPLSM with a repetition rate of 80–120 MHz are appropriate excitation light sources for the measurement.

As the fluorophore molecules bind with specimens, their fluorescence lifetime becomes very sensitive to the microenvironments within the cells, e.g., the proteins they bind. As a consequence, classifying these measured lifetime values to form lifetime images could provide valuable information for diagnostics as well as for the identification of subcellular structures. For lifetime images, one commercial time-correlated single photon counting system is used for the measurement (Becker and Hickl SPC 730). The photo detector has 256 time channels for each pixel, which can be fitted to the exponential decay of the fluorescence emission to determine the lifetime. According to the measured values, different lifetimes of the fluorophore in the specimens are displayed in pseudocolors. The same uterus slice sample shown for multiphoton and spectral imaging is employed again for lifetime imaging and shown in Figure 9.

In this figure, the color figure (upper right) is the lifetime image, while the black and white counterpart is the multiphoton image and only represents the fluorescence intensity. The measured lifetime ranges from 1.2 (red) to 2.5 ns (blue) shown as a continuous spectrum. As can be seen, a shorter lifetime of fluorophore appears when binding with the cells around the vesicle at the top left corner in the image, making them easier to be distinguished from the surrounding cellular structures. With existing analysis software a histogram of photon counting can be obtained as shown in the bottom of Figure 9. The blue dots represent the measured photon counts at each time bin and the red solid line is the curve fit to determine its lifetime. In addition to the studies with fixed slides, encouraging lifetime results have also been obtained from a live specimen, e.g., a GFP tagged with *C. elegans* and drosophila embryos.

The above results suggest that the lifetime imaging technique can successfully avoid the drawbacks of spectrum overlapping of multi-stained specimens and can be applied to a variety of living specimens to increase image contrast and discrimination

3.2.4 Future research in cancer and pre-cancer diagnosis

Fluorescence spectroscopy and imaging in the ultraviolet–visible wavelength spectrum has been an exciting new modality for detecting human epithelial pre-cancers and early cancers. Noninvasive and fast detection of epithelial pre-cancers and early cancers through the use of fluorescence techniques could have significant impact on cancer screening and diagnostic programs. The unique spectral signatures of neoplastic cells can be observed from fluorescence excitation-emission matrices, which is an intensity contour diagram of the excitation and fluorescence emission. Recent research results for breast cancer study have shown that the largest difference in spectral intensity appears at 340 nm excitation [92], at which squamous cancer cells exhibit higher intensity and a narrower spectrum than normal immortalized keratinocytes. Thus, an excitation at 340 nm maximally exploits the endogenous fluorophores present in the epithelial tissues, and is promising for auto-fluorescence diagnosis of human epithelial pre-cancers and cancers.

To date, most studies have exploited the spectral properties of tissue endogenous fluorescence and presented their spectral signatures, however, far fewer studies of fluorescence lifetime and imaging of the cancerous human pre-cancer and cancer have been conducted. In conjunction with the unique signature of neoplastic cells (or other forms of cells), a spectral and lifetime imaging technique (section 0) capable of visualizing biological histograms and subcellular structures is possible to effectively detect human neoplasia *in vivo*. This represents an exciting, active area of research for many years to come.

4 Future directions and concluding remarks

Radiation on the microscale will become even more important and integral to small-scale system design, very fast, high-speed devices, telecommunications, optical MEMS, biological systems, including both tools, e.g., for diagnostics, as well as *in vivo* functional devices. As photons have no mass and can be appropriately delivered, optical technologies have an impact on a system ranging from negligible to catastrophic. This chapter reviews recent interest and application areas of radiation at small length, time, and structural scales, and discusses application where such phenomena occur, and have even been used to advantage. The field is dynamic and constantly evolving with many exciting, challenging, and important problems remaining to be addressed. It is hoped that the discussion in this chapter will give the reader an overview of the current state of the art, the important issues that must be addressed, and a set of criteria to determine the appropriate microscale regime for the radiation transport problem at hand.

Acknowledgment

The authors appreciate the microscopy facility provided by Laboratory for Optical and Computational Instrumentation (LOCI) at the University of Wisconsin–Madison for biological imaging. The authors also thank Prof. John G. White (LOCI director) and Mr. Kevin W. Eliceiri (LOCI manager) for their assistance in bioimaging project funded by NIH under Grant No. RR00570–31.

References

[1] Siegman, A.E., *Lasers*, University Science Books: Sausalito, CA, 1986.
[2] Hecht, E., *Optics,* 3rd edn, Addison Wesley Longman: New York, 1998.
[3] Richter, K., Chen, G. & Tien, C.L., Partial coherence theory of multiplayer thin-film optical properties. *Optical Engineering*, **32(8)**, pp. 1897–1903, 1993.
[4] Chen, G. & Tien, C.L., Partial coherence theory of thin-film radiative properties. *Journal of Heat Transfer*, **114(3)**, pp. 636–643, 1992.
[5] Chen, G., Wave effects on radiative transfer in absorbing and emitting thin-film media. *Microscale Thermophysical Engineering*, **1(3)**, pp. 215–224, 1997.
[6] Anderson, C.F. & Bayazitoglu, Y., Radiative properties of films using partial coherence theory. *Journal of Thermophysics and Heat Transfer*, **10(1)**, pp. 26–32, 1996.
[7] Wong, P.Y., Hess, C.K. & Miaoulis, I.N., Coherent thermal-radiation effects on temperature-dependent emissivity of thin-film structures on optically thick substrates. *Optical Engineering*, **34(6)**, pp. 1776–1781, 1995.
[8] Grossman, E.N. & McDonald, D.G., Partially coherent transmittance of dielectric lamellae. *Optical Engineering*, **34(5)**, pp. 1289–1295, 1995.
[9] Tien, C.L. & Lienhard, J.H., *Statistical Thermodynamics*, Hemisphere Publishing: New York, 1979.
[10] Modest, M.F., *Radiative Heat Transfer*, McGraw-Hill: New York, 1993.
[11] Kennedy, P.K., Hammer, D.X. & Rockwell, B.A., Laser-induced breakdown in aqueous media. *Progress in Quantum Electronics*, **21(3)**, pp. 155–248, 1997.
[12] Fan, C.H. & Longtin, J.P., Modeling optical breakdown in dielectrics during ultrafast laser processing. *Applied Optics*, **40(18)**, pp. 3124–3131, 2001.
[13] Fan, C.H., Sun, J. & Longtin, J.P., Plasma absorption of femtosecond laser pulses in dielectrics. *Journal of Heat Transfer*, **124(2)**, pp. 275–283, 2002.
[14] Fan, C.H., Sun, J. & Longtin, J.P., Breakdown threshold and localized

electron density in water induced by ultrashort laser pulses. *Journal of Applied Physics*, **91(4)**, pp. 2530–2536, 2002.

[15] Bohren, C.F. & Huffman, D.R., *Absorption and Scattering of Light by Small Particles*, Wiley: New York, 1983.

[16] Shen, Y.R., *The Principles of Nonlinear Optics*, Wiley: New York, 1984.

[17] Denk, W., Strickler, J.H. & Webb, W.W., Two-photon laser scanning fluorescence microscopy. *Science*, **248**, pp. 73–76, 1990.

[18] Noack, J. & Vogel, A., Laser-induced plasma formation in water at nanosecond to femtosecond time scales: calculation of thresholds, absorption coefficients, and energy density. *IEEE Journal of Quantum Electronics*, **35(8)**, pp. 1156–1167, 1999.

[19] Kennedy, P.K., A first-order model for computation of laser-induced breakdown thresholds in ocular and aqueous media: 1. Theory. *IEEE Journal of Quantum Electronics*, **31(12)**, pp. 2241–2249, 1995.

[20] Longtin, J.P. & Tien, C.L., Efficient laser heating of transparent liquids using multiphoton absorption. *International Journal of Heat and Mass Transfer*, **40(4)**, pp. 951–959, 1997.

[21] Williams, F., Varma, S.P. & Hillenius, S., Liquid water as a lone-pair amorphous-semiconductor. *Journal of Chemical Physics*, **64(4)**, pp. 1549–1554, 1976.

[22] Liu, X., Du, D. & Mourou, G., Laser ablation and micromachining with ultrashort laser pulses. *IEEE Journal of Quantum Electronics*, **33(10)**, pp. 1706–1716, 1997.

[23] Feng, Q., Moloney, J.V., Newell, A.C., Wright, E.M., Cook, K., Kennedy, P.K., Hammer, D.X., Rockwell, B.A. & Thompson, C.R., Theory and simulation on the threshold of water breakdown induced by focused ultrashort laser pulses. *IEEE Journal of Quantum Electronics*, **33(2)**, pp. 127–137, 1997.

[24] Stuke, M, *Dye Lasers: 25 Years*, Springer-Verlag: Berlin, 1992.

[25] Seilmeier, A. & Kaiser, W., Ultrashort intramolecular and intermolecular vibrational energy transfer to polyatomic molecules in liquids. *Ultrashort Laser Pulses: Generation and Applications,* 2nd edn, ed. W. Kaiser, Springer-Verlag: Berlin, 1993.

[26] Powers, J., *An Introduction to Fiber Optic Systems,* 2nd edn, McGraw-Hill Higher Education: New York, 1997.

[27] Agrawal, G.P., *Nonlinear Fiber Optics,* 2nd edn. Academic Press: San Diego, 1995.

[28] Gross, B. & Manassah, J.T., Propagation of femtosecond pulses in focusing nonlinear Kerr planar waveguides. *Optics Communications*, **129(1-2)**, pp. 143–151, 1996.

[29] Chin, S.L., Brodeur, A., Petit, S., Kosareva, O.G. & Kandidov, V.P., Filamentation and supercontinuum generation during the propagation of powerful ultrashort laser pulses in optical media (white light laser). *Journal of Nonlinear Optical Physics and Materials*, **8(1)** pp. 121–146, 1999.

[30] Mourou, G., The ultrahigh-peak-power laser: present and future. *Applied*

Physics B, **65(2)**, pp. 205–211, 1997.
[31] Ameer-Beg, S., Perrie, W., Rathbone, S., Wright, J., Weaver, W. & Champoux, H., Femtosecond laser microstructuring of materials. *Applied Surface Science*, **129**, pp. 875–880, May 1998.
[32] Sarkisov, G.S., Bychenkov, V.Y., Novikov, V.N., Tikhonchuk, V.T., Maksimchuk, A., Chen, S.Y., Wagner, R., Mourou, G. & Umstadter, D., Self-focusing, channel formation, and high-energy ion generation in interaction of an intense short laser pulse with a He jet. *Physical Review E*, **59(6)**, pp. 7042–7054, 1999.
[33] Infeld, E. & Zycki, P., Instability of self-focused optical beams in plasmas – numerical analysis. *Physical Review A*, **46(4)**, pp. 2173–2175, 1992.
[34] Ashkenasi, D., Varel, H., Rosenfeld, A., Henz, S., Herrmann, J. & Cambell, E.E.B., Application of self-focusing of ps laser pulses for three-dimensional microstructuring of transparent materials. *Applied Physics Letters*, **72(12)**, pp. 1442–1444, 1998.
[35] Strickland, D. & Corkum, P.B., Short pulse self-focusing. *SPIE Vol. 1413: Short-Pulse High-Intensity Lasers and Applications*, pp. 54–58, 1991.
[36] Penetrante, B.M., Bardsley, J.N., Wood, W.M., Siders, C.W. & Downer, M.C., Ionization-induced frequency-shifts in intense femtosecond laser-pulses. *Journal of the Optical Society of America B*, **9(11)**, pp. 2032–2040, 1992.
[37] Sprangle, P., Esarey, E. & Krall, J., Self-guiding and stability of intense optical beams in gases undergoing ionization. *Physical Review E*, **54(4)**, pp. 4211–4232, 1996.
[38] Rae, S.C., Spectral blueshifting and spatial defocusing of intense laser pulses in dense gases. *Optics Communications*, **104(4–6)**, pp. 330–335, 1994.
[39] Rae, S.C. & Burnett, K, Detailed simulations of plasma-induced spectral blueshifting. *Physical Review A*, **46(2)**, pp. 1084–1090, 1992.
[40] Yablonovitch, E., Self-phase modulation of light in a laser-breakdown plasma. *Physical Review Letters*, **32(20)**, pp. 1101–1104, 1974.
[41] Tien, C.L., Majumdar, A. & Gerner, F.M. (eds). *Microscale Energy Transport*, Taylor and Francis: Washington DC, 1997.
[42] Yariv, A., *Quantum Electronics,* 3rd edn. Wiley: New York, 1989.
[43] Ashkenasi, D., Varel, H., Rosenfeld, A., Noack, F. & Campbell, E.E.B., Pulse-width influence on laser structuring of dielectrics. *Nuclear Instruments and Methods in Physics Research Section B*, **122(3)**, pp. 359–363, 1997.
[44] Bor, Z., Racz, B., Szabo, G., Xenakis, D., Kalpouzos, C. & Fotakis, C., Femtosecond transient reflection from polymer surfaces during femtosecond UV photoablation, *Applied Physics A*, **60(4)**, pp. 365–368, 1995.
[45] Bloembergen, N., Laser-induced electric breakdown in solids, *IEEE Journal of Quantum Electronics*, **QE10(3)**, pp. 375–386, 1974.
[46] Varel, H., Ashkenasi, D., Rosenfeld, A., Wahmer, M. & Campbell, E.E.B.,

Micromachining of quartz with ultrashort laser pulses. *Applied Physics A*, **65(4–5)**, pp. 367–373, 1997.

[47] Tonshoff, H.K., Momma, C., Ostendorf, A., Nolte, S. & Kamlage, G., Microdrilling of metals with ultrashort laser pulses. *Journal of Laser Applications*, **12(1)**, pp. 23–27, 2000.

[48] Weichenhain, R., Horn, A. & Kreutz, E.W., Three dimensional microfabrication in ceramics by solid state lasers. *Applied Physics A*, **69(Suppl.)**, S855–S858, 1999.

[49] Papakonstantinou, P., Vainos, N.A. & Fotakis, C., Microfabrication by UV femtosecond laser ablation of Pt, Cr and Indium oxide thin films. *Applied Surface Science*, **151(3–4)**, pp. 159–170, 1999.

[50] Mailis, S., Zergioti, I., Koundourakis, G., Ikiades, A., Patentalaki, A., Papakonstantinou, P., Vainos, N.A. & Fotakis, C., Etching and printing of diffractive optical microstructures by a femtosecond excimer laser. *Applied Optics*, **38(11)**, pp. 2301–2308, 1999.

[51] Horiyama, M., Sun, H.B., Miwa, M., Matsuo, S. & Misawa, H., Three-dimensional microstructures created by laser microfabrication technology. *Japanese Journal of Applied Physics Part 2 – Letters*, **38(2B)**, L212–L215, 1999.

[52] Zergioti, I., Mailis, S., Vainos, N.A., Ikiades, A., Grigoropoulos, C.P. & Fotakis, C., Microprinting and microetching of diffractive structures using ultrashort laser pulses. *Applied Surface Science*, **139**, pp. 82–86, Jan. 1999.

[53] Jandeleit, J., Horn, A., Weichenhain, R., Kreutz, E.W. & Poprawe, R., Fundamental investigations of micromachining by nano- and picosecond laser radiation. *Applied Surface Science*, **129**, pp. 885–891, May 1998.

[54] Maboudian, R., Surface processes in MEMS technology. *Surface Science Reports*, **30(6–8)**, pp. 209–270, 1998.

[55] Phinney, L.M., Fushinobu, K. & Tien, C.L., Subpicosecond laser processing of polycrystalline silicon microstructures. *Microscale Thermophysical Engineering*, **4(1)**, pp. 61–75, 2000.

[56] Fushinobu, K., Phinney, L.M., Kurosaki, Y. & Tien, C.L., Optimization of laser parameters for ultrashort-pulse laser recovery of stiction-failed microstructures. *Numerical Heat Transfer Part A –Applications*, **36(4)**, pp. 345–357, 1999.

[57] Oraevsky, A.A., Dasilva, L.B., Rubenchik, A.M., Feit, M.D., Glinsky, M.E., Perry, M.D., Mammini, B.M., Small, W. & Stuart, B.C., Plasma mediated ablation of biological tissues with nanosecond to femtosecond laser pulses: relative role of linear and nonlinear absorption. *IEEE Journal of Selected Topics in Quantum Electronics*, **2(4)**, pp. 801–809, 1996.

[58] Docchio, F., Spatial and temporal dynamics of light attenuation and transmission by plasmas induced in liquids by nanosecond Nd-YAG laser-pulses, *Nuovo Cimento Della Societa Italiana Di Fisica D*, **13(1)**, pp. 87–98, 1991.

[59] Wang, C.Z., Ho, K.M., Shirk, M.D. & Molian, P.A., Laser-induced graphitization on a diamond (111) surface. *Physical Review Letters*,

85(19), pp. 4092–4095, 2000.
[60] Ozkan, A.M., Malshe, A.P., Railkar, T.A., Brown, W.D., Shirk, M.D. & Molian, P.A., Femtosecond laser-induced periodic structure writing on diamond crystals and microclusters. *Applied Physics Letters*, **75(23)**, pp. 3716–3718, 1999.
[61] Ho, C.M. & Tai, Y.C., Micro-electro-mechanical systems (MEMS) and fluid flows. *Annual Review of Fluid Mechanics*, **30**, pp. 579–612, 1998.
[62] Ho, C.M. & Tai, Y.C., Review: MEMS and its applications for flow control. *Journal of Fluids Engineering*, **118(3)**, pp. 437–447, 1996.
[63] Campbell, E.E.B., Ashkenasi, D. & Rosenfeld, A., Ultra-short-pulse laser irradiation and ablation of dielectrics. *Materials Science Forum*, **301**, pp. 123–144, 1999.
[64] Ganesh, R.K., Faghri, A., & Hahn, Y., A generalized thermal modeling for laser drilling process 1: mathematical modeling and numerical methodology. *International Journal of Heat and Mass Transfer*, **40(14)**, pp. 3351–3360, 1997.
[65] Ganesh, R.K., Faghri, A. & Hahn, Y., A generalized thermal modeling for laser drilling process 2: numerical simulation and results. *International Journal of Heat and Mass Transfer*, **40(14)**, pp. 3361–3373, 1997.
[66] Girardeau, M.D., Kim, K.G. & Widmayer, C.C., Theory of atomic excitation and ionization by ultrashort laser pulses. *Physical Review A*, **46(9)**, pp. 5932–5937, 1992.
[67] Perry, M.D., Stuart, B.C., Banks, P.S., Feit, M.D., Yanovsky, V. & Rubenchik, A.M., Ultrashort pulse laser machining of dielectric materials. *Journal of Applied Physics*, **85(9)**, pp. 6803–6810, 1999.
[68] Du, D., Liu, X. & Mourou, G., Reduction of multi-photon ionization in dielectrics due to collisions. *Applied Physics B*, **63(6)**, pp. 617–621, 1996.
[69] Sparks, M., Mills, D.L., Warren, R., Holstein, T., Maradudin, A.A., Sham, L.J., Loh, E. & King, D.F., Theory of electron-avalanche breakdown in solids. *Physical Review B*, **24(6)**, pp. 3519–3536, 1981.
[70] Pronko, P.P., Vanrompay, P.A., Horvath, C., Loesel, F., Juhasz, T., Liu, X. & Mourou, G., Avalanche ionization and dielectric breakdown in silicon with ultrafast laser pulses. *Physical Review B*, **58(5)**, pp. 2387–2390, 1998.
[71] Ashkenasi, D., Varel, H., Rosenfeld, A., Henz, S., Herrmann, J. & Cambell, E.E.B., Application of self-focusing of ps laser pulses for three-dimensional microstructuring of transparent materials. *Applied Physics Letters*, **72(12)**, pp. 1442–1444, 1998.
[72] Korte, F., Adams, S., Egbert, A., Fallnich, C. & Ostendorf, A., Sub-diffraction limited structuring of solid targets with femtosecond laser pulses. *Optics Express*, **7(2)**, pp. 41–49, 2000.
[73] Kirby, K.W. & Jankiewicz, A.T., Laser machining of glass forming ceramics. *Journal of Laser Applications*, **10(1)**, pp. 1–10, 1998.
[74] Lenzner, M., Kruger, J., Sartania, S., Cheng, Z., Spielmann, C., Mourou, G., Kautek, W. & Krausz, F., Femtosecond optical breakdown in dielectrics. *Physical Review Letters*, **80(18)**, pp. 4076–4079, 1998.

[75] White, J.G. & Amos, W.B., Confocal microscopy comes of age. *Nature*, **328(6126)**, pp. 183–184, 1987.

[76] White, J.G., Amos, W.B. & Fordham, M., An evaluation of confocal versus conventional imaging of biological structures by fluorescence light microscopy. *Journal of Cell Biology*, **105(1)**, pp. 41–48, 1987.

[77] Amos, W.B., White, J.G. & Fordham, M., Use of confocal imaging in the study of biological structures. *Applied Optics*, **26(16)**, pp. 3239–3243, 1987.

[78] Pawley, J.B. (ed). *Handbook of Biological Confocal Microscopy*, Kluwer Academic Publishers: New York, 1995.

[79] Dyba, M. & Hell, S.W., Focal spots of size lambda/23 open up far-field florescence microscopy at 33 nm axial resolution. *Physical Review Letters*, **88(16)**, Art. No. 163901, 2002.

[80] Klar, T.A., Engel, E. & Hell, S.W., Breaking Abbe's diffraction resolution limit in fluorescence microscopy with stimulated emission depletion beams of various shapes. *Physical Review E*, **64(6)**, Art. No. 066613, 2001.

[81] Klar, T.A., Jakobs, S., Dyba, M., Egner, A. & Hell, S.W., Fluorescence microscopy with diffraction resolution barrier broken by stimulated emission. *Proceedings of the National Academy of Sciences of the United States of America*, **97(15)**, pp. 8206–8210, 2000.

[82] Denk, W., Strickler, J.H. & Webb, W.W., 2-photon laser scanning fluorescence microscopy. *Science*, **248(4951)**, pp. 73–76, 1990.

[83] Centonze, V.E., Wokosin, D.L. & White, J.G., Improved deep optical sectioning capabilities rendered by 2-photon excitation imaging. *Molecular Biology of the Cell*, **6(Suppl.)**, pp. 657–657, 1995.

[84] Clem, C.J. & Rigaut, J.P., Computer simulation modeling and visualization of 3D architecture of biological tissues: simulation of the evolution of normal, metaplastic and dysplastic states of the nasal epithelium. *Acta Biotheoretica*, **43(4)**, pp. 425–442, 1995.

[85] Cogswell, C.J., Larkin, K.G. & Arnison, M.R., Powerful tools for 3D microscopy image analysis, processing and visualization based on 2D and 3D Fourier transforms. *Zoological Studies*, **34(Suppl.)**, pp. 146–147, 1995.

[86] Tekola, P., Baak, J.P.A., Vanginkel Hahm, Belien, J.A.M., Vandiest, P.J., Broeckaert, M.A.M. & Schuurmans, L.T., Three-dimensional confocal laser scanning DNA ploidy cytometry in thick histological sections. *Journal of Pathology*, **180(2)**, pp. 214–222, 1996.

[87] Hell, S.W., Schrader, M. & Vandervoort, H.T.M., Far-field fluorescence microscopy with three-dimensional resolution in the 100-nm range. *Journal of Microscopy – Oxford*, **187**, pp. 1–7, 1997.

[88] Mohler, W.A. & White, J.G., Stereo 4-D reconstruction and animation from living fluorescent specimens. *Biotechniques*, **24(6)**, pp. 1006–1012, 1998.

[89] Mohler, W.A. & White, J.G., Multiphoton laser scanning microscopy for four-dimensional analysis of *Caenorhabditis elegans* embryonic development. *Optics Express*, **3(9)**, pp. 325–331, 1998.

[90] Thomas, C., Devries, P., Hardin, J. & White, J., Four-dimensional imaging: computer visualization of 3D movements in living specimens. *Science*, **273(5275)**, pp. 603–607, 1996.
[91] O'Connor, D.V., *Time-Correlated Single Photon Counting*, Academic Press: New York, 1984.
[92] Gill, E.M., Palmer, G. & Ramanujam, N., Steady state, fluorescence imaging of neoplasia. *Methods of Enzymology*, ed. G. Marriot, Vol. 361–*Biophotonics*, 2002.

CHAPTER 7

Direct simulation Monte Carlo of gaseous flow and heat transfer in a microchannel

R. Sayegh[1], M. Faghri[2], Y. Asako[3] & B. Sundén[4]
[1]*Naval Undersea Warfare Center, USA*
[2]*University of Rhode Island, USA*
[3]*Tokyo Metropolitan University, Japan*
[4]*Lund Institute of Technology, Sweden*

Abstract

This chapter presents a methodology and numerical results for two-dimensional flow and heat transfer characteristics of gaseous flow and heat transfer in a microchannel. The numerical methodology is based on direct simulation Monte Carlo (DSMC). In this method, the molecules are sampled by using a probability density function and a random number generator. The model is validated for the case of low Knudsen number (*Kn*) where exact analytical results are available. It is used to obtain velocity and temperature profiles, pressure distribution, friction factor, slip velocity and temperature and Nusselt number for a range of *Kn* in the slip flow and transition regimes. The simulation results for helium gas under the constant wall temperature condition showed that the flow and heat transfer characteristics were changed significantly as *Kn* was increased. Specifically, shear stress, friction coefficient, slip temperature and Nusselt number were decreased and slip velocity was increased as the gas became more rarefied.

1 Introduction

DSMC was originally developed by Bird [1] to solve problems related to flow around space vehicles at high altitude. The flow was modeled at the microscopic level due to the noncontinuum nature of the gas at high altitude, where the gas has low density and large mean free path. Nanbu [2] presented a theoretical basis for the DSMC methodology based on the solution of the Boltzmann equation. The methodology has been widely used for problems in rarefied gas dynamics in the past, e.g., [3–5].

Recent developments in fabrication and utilization of microelectromechanical systems (MEMS) opened a new frontier in the application of DSMC in modeling fluid flow and energy transport in microchannels. The gas inside these channels can be considered to be rarefied when the channel height, H, is small compared to the mean free path, λ, of the molecules. This is best described by the Knudsen number ($Kn = \lambda/H$), which characterizes the degree of rarefaction. The traditional momentum equation, called the "Navier–Stokes" equation, has failed to provide accurate results based on the continuum assumption, which does not exist for flows in microchannels. The mathematical model that describes the flow at the microscopic level is the Boltzmann equation, which is very difficult to solve. Therefore, DSMC is an attractive tool for solving fluid flow and energy transport problems involving noncontinuum flows. Four flow regimes (continuum, slip flow, transition, and free molecular) can be specified based on the degree of rarefaction of a gas, and each flow regime has a unique mathematical model. Kn can range between zero and infinity. For $0 \leq Kn < 0.01$ the flow regime is considered to be continuum, and the Navier–Stokes equations are valid. For $0.01 \leq Kn < 0.1$ which is known as the slip flow regime, the Navier–Stokes equations are valid only with the application of the slip boundary condition. The transition regime is present for $0.1 \leq Kn < 3$ when the flow is too rarefied (low density) to be solved by the Navier–Stokes equations and therefore the Boltzmann equation or direct simulation Monte Carlo must be used. For $Kn > 3$ the flow regime is considered to be in the free molecular regime where collisions between molecules are neglected and the collisionless Boltzmann equation can be used.

Direct simulation Monte Carlo (DSMC) can be described as a probability/statistical simulation method that utilizes a sequence of random numbers to perform the simulation. Unlike computational fluid dynamics (CFD), DSMC simulates the physical process directly without performing a mathematical discretization of the partial differential equations that describe the mathematical system. However, the DSMC requires that the system be described by a probability density function. Once this requirement is achieved, the simulation is performed by a random sampling of this function. The random sampling also requires generating a random number, Rf, which is uniformly distributed between 0 and 1. The output of the simulation is then accumulated at each sampling and averaged over the number of samples to arrive at the solution of the physical problem. The utilization of DSMC in gas dynamics has been very useful particularly for dilute gases. G. Bird [6] developed a DSMC code to perform molecular simulation. In this simulation, the gas is modeled by a large number of simulated molecules, which ranges in millions using current computer capabilities. Each simulated molecule represents a large number of real molecules. The information about the simulated molecules (position in space, velocity components, internal state) is saved and updated with time after representative collisions and boundary interactions in the simulated physical space. The simulation follows the trajectory of a very large number of simulated molecules.

There is no convergence criterion for the DSMC method nor is there a requirement for an initial approximation of the flow and temperature. However, there are numerical criteria for DSMC that must be satisfied. These criteria are: the ratio of the time step to the local mean collision time (CTR) and the ratio of the mean separation between collision partners to the local mean free path (CSR) which must be very small compared to unity over the entire flow field. In addition, the cell size should be chosen such that its linear length is small in comparison with the local mean free path of the molecules.

Several studies have been conducted on rarefied gas dynamics for microchannels using DSMC concepts. Oh *et al.* [7] reported the effects of *Kn* (0.07, 0.14, and 0.19) for high-speed flows by varying the channel height. The study was conducted for a rectangular channel, 6 μm long by 1.2 μm high filled with helium gas. The computations were performed using the DSMC method combined with a monotonic Lagrangian grid (MLG). DSMC-MLG produces a method for automatic grid adaptation to account for variation of flow properties. In addition, the results were presented for pressure, temperature jump and slip velocity as a function of position along the wall using the variable hard sphere (VHS) model.

Marvriplis *et al.* [8] investigated the applicability of the DSMC method to the fluid and thermal analysis of a microchannel. The simulations included supersonic, subsonic, and pressure-driven flows in a microchannel. The results presented the effects of varying channel aspect ratio along with *Kn* on the pressure and temperature distribution for a range of continuum to transitional regime.

Piekos *et al.* [9] reported the effect of increasing rarefaction on the fluid flow behavior. This study considered both microchannel and micronozzle and the results were presented for the pressure and velocity distribution. For the case of microchannel, two regimes were investigated: the slip and transition regimes. The channel size and the pressure ratio for the slip and the transition regime were 900 × 30λ with a pressure ratio of 2.4 and 120 × 2λ with a pressure ratio of 4.2, respectively. The code written for this chapter used an unstructured grid and a trajectory tracing particle movement scheme. The VHS model was used for the calculation of the collision cross-sections. In addition, the code was validated by comparing the flow characteristics for low *Kn* with the analytical results.

A study performed by Kavehpour *et al.* [10] investigated two-dimensional fluid flow and heat transfer characteristics in microchannels numerically by a control volume finite difference scheme for a slip-flow regime. The microchannel was modeled as a parallel plate duct with uniform wall temperature and uniform wall heat flux. The conservation equations of mass, momentum, and energy were solved using slip velocity and temperature jump boundary conditions. It was found that the friction coefficient and Nusselt number, for both constant wall temperature and constant wall heat flux, were reduced in comparison with the continuum results.

Most of the available DSMC simulation results are limited to velocity, pressure, and temperature distribution without reporting the friction factor and

Nusselt number. The effects of Knudsen number on the wall shear stress, Nusselt number, and friction factor for microchannels were first examined by Sayegh *et al.* [11, 12] and will be presented in this chapter. The effects of the Knudsen number on the heat transfer were also recently examined by Liou *et al.* [13]. The same authors [14] examined heat transfer and the fluid characteristics of subsonic gas flows through microchannels. Other recent contributions are by Piekos and Breuer [15], Beskok and Karniadakis [16] and by Hadjiconstantinou and Simek [17].

2 Description of the DSMC method

2.1 Grids

For the rectangular microchannel the flow-field is divided into a number of four-sided cells of constant area. Each cell is divided into a number of four-sided subcells of constant area. DSMC uses the cells to sample the macroscopic properties and the formation of subcells causes all collisions to be between near neighbors. The DSMC procedure includes an optimally efficient indexing scheme to allow pairs within a particular cell to be selected from an unsorted array of molecules. The linear dimension of the cells is assumed to be small in comparison with the scale length of the macroscopic flow gradient in the direction in which the dimension is measured. In regions with large macroscopic gradients, this means that the cell dimension should be of the order of one third of the local mean free path of the molecules.

2.2 Molecular approximation

Molecular approximation is essential due to memory constraints as well as CPU processing time considerations. It is impractical to use the actual number of real molecules of a gas, therefore, the real molecules are represented by a lesser number of molecules called "simulated molecules". Each simulated molecule represents a number of real molecules and it is recommended that a minimum of 10–20 simulated molecules are contained within a cell. The ratio of the real molecules to the simulated molecule, $R_{r,s}$, is entered into the simulation program and the number of simulated molecules in each cell, N_{smc}, can be calculated from:

$$N_{smc} = \frac{n \times V_{cell}}{R_{r,s}} \tag{1}$$

where n is the number density and V_{cell} is the cell volume. The results become more accurate as the number of simulated molecules is increased. The initial distribution of the simulated molecules within the flow domain is random, and the initial coordinates (P_x, P_y) of a simulated molecule within a cell is calculated using the following equations:

$$P_x = x_1 + Rf(x_2 - x_1) \tag{2}$$

$$P_y = y_1 + Rf(y_2 - y_1) \tag{3}$$

where (x_1, x_2) and (y_1, y_2) are the cell coordinates in the x and y directions respectively.

2.3 Time step

The time increment, dt, is utilized in the simulation to uncouple the movement and collision of the molecules. It should be much less than the mean collision time and is selected such that the fastest molecule moves about one third of the cell dimension. For a square cell this is given as:

$$dt = \frac{x_2 - x_1}{3 \times U_{mp}} \tag{4}$$

where U_{mp} is the most probable speed of the simulated molecules. The simulation accuracy is improved with decreasing time step and cell size.

2.4 Molecular models

The DSMC method utilizes a molecular model to approximate the unknown intermolecular interaction. The model specifies the collision cross-section and the post-collision scattering law. In addition, it enables the computations to reproduce the coefficient of viscosity and its temperature dependence. There are several molecular models that have been developed (e.g., inverse power law (IPL), hard sphere (HS), variable soft sphere (VSS), and variable hard sphere (VHS)). In this study, the VHS model was used to obtain the effective diameter for real molecules and to define the mean free path in the numerical studies.

2.5 Molecular movement

Molecules are allowed to move along their trajectory over a time interval for a distance depending on their calculated speed. When a molecule is moved, it either passes on to an adjacent cell, reflects from the wall boundary, collides with another molecule, or leaves the calculation domain at the outflow boundary. The molecular movement is uncoupled from its collision phase through the use of some fraction of its mean time between collisions. However, the collision of molecules with the boundaries is still performed in the molecular movement phase.

Let's consider the physical movement of one simulated molecule Z. The initial location of molecule Z can be found using eqns. (2) and (3). When molecule Z is moved, the new (P_x, P_y) coordinate of the molecule is determined from the following equations:

$$P_{x\_new} = P_x + dx$$
$$P_{y\_new} = P_y + dy \tag{5}$$

where (dx, dy) is the distance traveled in the Cartesian coordinate system. Given the speed of molecule Z and the total time step, the distance traveled is calculated from the following:

$$dx = u \times dt_f$$
$$dy = v \times dt_f \tag{6}$$

where (u, v) are the velocity components of the molecule and dt_f is the fraction of the time step for the molecular movement which is calculated from:

$$dt_f = Rf \times dt \tag{7}$$

It is important to note that even though the microchannels under study are two-dimensional, the molecules are still considered to have three-dimensional velocity components (u, v, w) corresponding to the (x, y, z) coordinates. The velocity components for each molecule are given as:

$$u = u' + U_x$$
$$v = v' + U_y \tag{8}$$
$$w = w'$$

where u', v', and w' are the thermal velocity components, and U_x, U_y are the stream velocity components.

After calculating the new location of molecule Z, it is determined whether the molecule collides with a surface, leaves the flow domain, collides with another molecule, or moves to a different cell. The decision for which of the above conditions apply is described as follows:

a) Collision with surface: A collision with a surface boundary occurs if the new coordinate (P_{x_new}, P_{y_new}) of molecule Z is outside the flow region and its path has crossed the surface boundary. Then, the location of collision between the molecule Z and the boundary is calculated as follows:

$$xc_{SB} = P_x + (y_{SB} - P_y) \times \frac{dx}{dy} \tag{9}$$

where y_{SB} is the location of the surface boundary. If, on the other hand, the boundary is considered to be a plane of symmetry, then it causes the molecule to reflect specularly such that the normal velocity component is reversed in direction and the parallel component remains unchanged.

b) Leaving the flow boundary: The molecule Z is removed from the calculation domain when its new coordinate (P_{x_new}, P_{y_new}) is outside the flow domain and its path crosses the stream boundary.

c) Moving to a different cell: When the new coordinate of molecule Z is outside the boundary of its current cell, the molecule is moved to the next cell and the new cell number, m, is based on the following equation:

$$m = 1 + \ln\{1 - (1 - r) \cdot \hat{x}/a\}/\ln(r) \tag{10}$$

where:

$$a = (1 - r)/(1 - r^{n_c})$$

$$r = C_w^{1/(n_c - 1)}$$

$n_c =$ cell number

$C_w =$ cell width

$$\hat{x} = (x - x_1)/(x_2 - x_1)$$

The probability of collision of the molecule Z with another molecule is discussed in the following section.

2.6 Collisions between molecules

The current DSMC code utilizes the "no-time-counter" (NTC) method developed by G. Bird [18]. The NTC method avoids the problems in computational efficiency associated with the "time-counter" (TC) method that was introduced earlier by Bird [1]. The TC method was developed in order to obtain a computation time directly proportional to the number of molecules. This involves the calculation of representative collisions and at each collision, a time parameter in each cell is advanced by the amount appropriate to that collision. The process is repeated until the cell time reaches the flow time. However, in extreme nonequilibrium situations such as the leading edge of strong shock waves, there are problems due to correlations between the probability and the spacing of the collisions in time. The occasional acceptance of a collision pair with a small collision probability could result in the cell time being advanced by a large interval that no other collisions could be computed for a number of time steps. However, the NTC method determines the number of pair selections in advance rather than within the collision subroutine when the cell time reaches the flow time. The NTC method gives the number of pair selections for possible collisions per time step as follow:

$$NP = \frac{1}{2} N_{smc} \overline{N}_{smc} R_{r,s} (\sigma_T c_r)_{\max} \frac{dt}{V_{cell}} \tag{11}$$

where σ_T is the total collision cross-section and c_r is the relative speed. The collision is accepted with a probability of

$$\Psi = \frac{\sigma_T c_r}{(\sigma_T c_r)_{\max}} \tag{12}$$

The value of $(\sigma_T c_r)_{\max}$ for each cell is saved and updated if a larger value is encountered during sampling. To better understand the use of the above equations, consider two molecules (*h, q*) that are chosen at random within a cell. The relative speed of these molecules is calculated from:

$$c_r = \left\{u_r^{\,2} + v_r^{\,2} + w_r^{\,2}\right\}^{1/2} \tag{13}$$

where,

$$u_r = u_h - u_q$$

$$v_r = v_h - v_q$$

$$w_r = w_h - w_q$$

The collision cross-section is based on

$$c_r \sigma_T = \left[\frac{\{2kT_{ref}/(m_r c_r^2)\}^{\omega-1/2}}{\Gamma(5/2-\omega)}\right]^{0.5} \tag{14}$$

where m_r is the reduced mass that is equal to *m/2* for a simple gas; ω is an empirical constant that is related to the VHS model; *k* is the Boltzmann constant; and T_{ref} is the reference temperature. The value of $(\sigma_T c_r)_{\max}$ of the cell is compared with the new calculated $(c_r \sigma_T)_{h,q}$ and if the condition $(c_r \sigma_T)_{h,q} > (\sigma_T c_r)_{\max}$ is true, the value of $(\sigma_T c_r)_{\max}$ is replaced by $(c_r \sigma_T)_{h,q}$. The probability of collision is accepted if the condition

$$Rf < \frac{(c_r \sigma_T)_{h,q}}{(c_r \sigma_T)_{\max}} \tag{15}$$

is true. The sum of the collision pair separation (cps) is calculated by

$$cps = \left\{(u_h - u_q)^2 + (u_h - u_q)^2\right\}^{1/2} \tag{16}$$

If the collision is accepted, the post-collision velocity components (u^*, v^*, w^*) for the molecules (*h,q*) are calculated using the VHS model:

$$u_h^* = u_c + u_r \frac{m_r}{m_h} \qquad u_q^* = u_c - u_r \frac{m_r}{m_q}$$

$$v_h^* = v_c + v_r \frac{m_r}{m_h} \qquad v_q^* = v_c - v_r \frac{m_r}{m_q} \tag{17}$$

$$w_h^* = w_c + w_r \frac{m_r}{m_h} \qquad w_q^* = w_c - w_r \frac{m_r}{m_q}$$

where,

$$u_c = u_h \frac{m_r}{m_h} + u_q \frac{m_r}{m_q}$$

$$v_c = v_h \frac{m_r}{m_h} + v_q \frac{m_r}{m_q} \tag{18}$$

$$w_c = w_h \frac{m_r}{m_h} + w_q \frac{m_r}{m_q}$$

and the relative velocity components are:

$$u_r = (2Rf - 1) \cdot c_r$$

$$v_r = \left\{1 - (2Rf - 1)^2\right\}^{1/2} \cos(2\pi Rf) \cdot c_r$$

$$w_r = \left\{1 - (2Rf - 1)^2\right\}^{1/2} \sin(2\pi Rf) \cdot c_r \tag{19}$$

2.7 Boundary interaction

When the simulated molecules strike a surface boundary, some reflect off the surface without transferring any of their streamwise momentum and others transfer all of their momentum. To simulate the energy exchange between molecules and a surface, several gas–surface interaction models such as diffuse, specular, and Cercignani–Lampis–Lord (CCL) models, have been utilized in the past for many aerospace engineering problems. Fortunately, experiments with engineering surfaces interacting with a rarefied gas indicate that the molecular reflection process can be approximated by a diffuse reflection model. The diffuse surface causes the molecules to reflect off the surface at a random angle without regards to their incoming direction. The velocities of the reflected molecules are distributed in accordance with the Maxwell distribution. Therefore, the reflected velocity component $U_{y,ref}$ is calculated from:

$$U_{y,ref} = \{-\ln(Rf)\}^{1/2} / \beta \tag{20}$$

where β is the reciprocal of the most probable speed. The other two tangential velocities are obtained by first generating thermal velocities through a distribution function and then adding them to the stream velocity as was shown in eqn. (8).

Generation of molecular thermal velocity components (u', v') with the most probable speed is often required by the DSMC simulation during initialization of molecular velocities, molecular reflection off the surface boundary, and introduction of new molecules into the calculation domain. The procedure to generate these velocities is straightforward. First, the distribution function fu' for the thermal velocity component in an equilibrium gas is obtained as:

$$fu' = (\beta / \pi^{1/2}) \cdot \exp(-\beta^2 \cdot u'^2) \tag{21}$$

Then, from the normal distribution, fu', the thermal velocity component (u', v') can be generated as:

$$u' = r \cdot \cos(\theta)$$

$$v' = r \cdot \sin(\theta) \tag{22}$$

where

$$\theta = 2 \cdot \pi \cdot Rf$$

$$r = \frac{\partial(u', v')}{\partial(r, \theta)} = \{-\ln(Rf)\}^{1/2} / \beta$$

3 DSMC simulation of microchannel

3.1 Computational methodology

The DSMC algorithm consists of four primary processes: molecular movements, indexing and cross-referencing the molecules, simulating the collisions, and sampling the flow field. Figure 1 illustrates the DSMC processes for steady state flow simulation. Prior to simulation, the domain is divided into a number of cells that are utilized to sample the macroscopic properties. The cells are further discretized into subcells that are used to select possible collision pairs. In addition, the number of real molecules in the gas is approximated by a lesser number of molecules called simulated molecules because of memory constraints as well as CPU processing time considerations. During the simulation, these four processes are uncoupled using a time step that is much less than the mean collision time.

The molecular movement process is modeled deterministically and the simulation follows on a fixed spatial grid that defines the spatial cells. Modeling molecule–surface interactions requires applying the conservation laws to individual molecules. The indexing and tracking process are the prerequisite for modeling collisions and sampling of the flow field. Molecular collision is a probabilistic process that sets DSMC apart from deterministic simulation methods such as molecular dynamics. Several collision-modeling techniques have been applied to DSMC and the current preferred model is the no-time-counter technique.

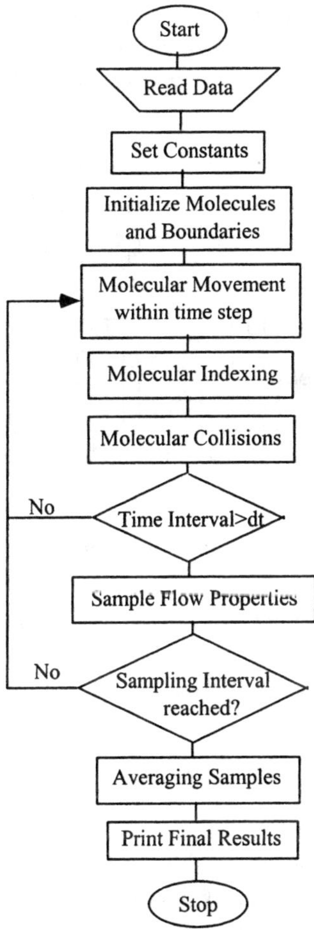

Figure 1: Flow chart of direct simulation Monte Carlo.

The final process is sampling the macroscopic flow properties. The spatial coordinates and velocity components of molecules in a particular cell are used to calculate macroscopic quantities at the geometric center of the cell. The pressure and the shear stress at the walls are calculated from the momentum exchange between the simulated molecules and the boundaries according to a surface model (diffuse, specular, or a combination of both). More details of the DSMC methodology are documented by Bird [18] and by Sayegh [11, 12].

3.2 Description of the problem

The DSMC simulation is applied to helium gas flowing inside a two-dimensional microchannel (parallel plate), 0.02m long by 0.01m high. Geometrical symmetry was used to reduce the model size and computational time. The solid wall was modeled as a diffuse surface and the boundary of symmetry was modeled as a specular surface. The inlet and outlet boundaries were modeled as stream boundaries. The flow domain, as shown in Figure 2, was divided into a number of cells with dimensions that satisfy the DSMC computational requirements.

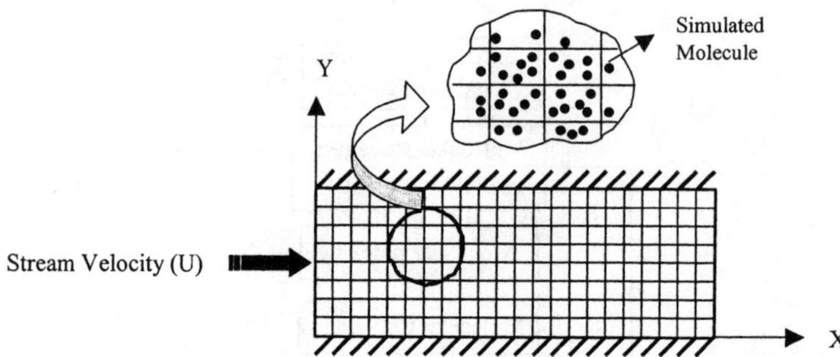

Figure 2: Schematic of microchannel set-up for DSMC simulation.

The variation of *Kn* was achieved by varying the fluid number density, while keeping the channel size constant. The relationship between the number density and the mean free path is as follows:

$$n = \frac{1}{\sqrt{2}\pi \bar{d}^2 \lambda} \tag{23}$$

where d is the molecular diameter.

Utilizing the above equation, Figure 3 was plotted for helium gas where ρ_0 is the density at a pressure of 1 atm and a temperature of 0 °C. This figure was then utilized to provide an approximation of the fluid density for a specific value of *Kn* for the simulation.

A total of seven simulations were conducted. Four simulations were used to study the fluid flow and the remaining three simulations to study heat transfer characteristics.

3.2.1 Fluid flow simulation in the microchannel

The first DSMC simulation (case 1) was used to validate the DSMC code based on the value of the friction coefficient for the near continuum flow. Then, higher values of Knudsen number were investigated in the slip flow and transition regime. For the slip flow regime (cases 2 and 3), the Navier–Stokes equation was also used with the slip boundary condition and was compared with the output of the simulation. Case 4 was used to simulate the transition regime. In these simulated cases, the wall and the fluid inlet temperatures were set at 300 K. The calculated flow parameters are listed in Table 1.

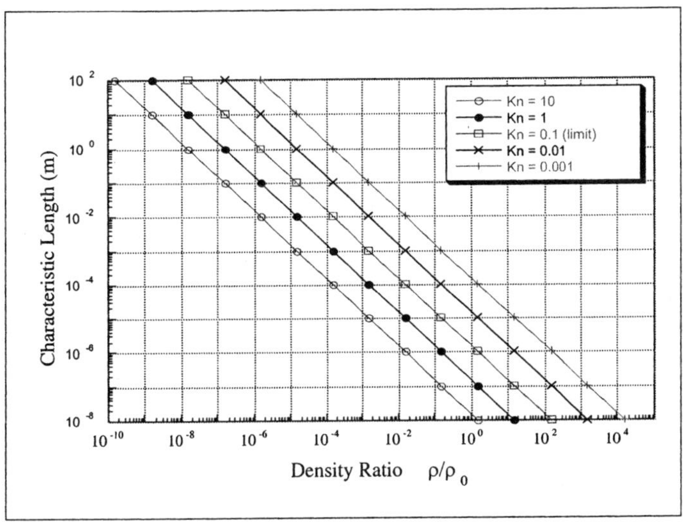

Figure 3: Relationship of Kn to channel size and fluid density for helium gas.

Table 1: Calculated parameters for DSMC flow simulations.

Flow regime	Continuum	Slip flow		Transition
Cases	**1**	**2**	**3**	**4**
Kn	0.009	0.017	0.03	0.11
Re	60.0	30.1	16.2	4.2
n	4.6066E+22	2.4388E+22	1.3820E+22	4.1459E+21
U (m/s)	200	200	200	200
dt (s)	8.9597E–09	1.69239E–08	2.98657E–08	9.95522E–08
$R_{r,s}$	3.455E+12	6.526E+12	1.1517E+13	3.8388E+13
T (K)	300	300	300	300
Number of cells (Y×X)	167 × 667	88 × 353	50 × 200	15 × 60
Channel Size (H/2×L)	0.005 × 0.02	0.005 × 0.02	0.005 × 0.02	02

3.2.2 Thermal simulation in the microchannel

The DSMC simulation (case 5) was used to validate the DSMC code except in this case, the validation was based on the value of the Nusselt number for the near continuum flow. For higher values of the Knudsen number, the simulations were also conducted using the same approach outlined in section 3.2.1. That is, in the slip flow regime (cases 6), the simulation output was compared with the slip-boundary/Navier–Stokes equation results, and case 7 was used to study the transition regime. The wall and fluid inlet temperatures for these cases are set at 400 K and 300 K, respectively, and other calculated parameters for thermal simulations are listed in Table 2.

Table 2: Calculated parameters for DSMC thermal simulations.

Flow regime	Continuum	Slip flow	Transition
Cases	**5**	**6**	**7**
Kn	0.009	0.045	0.11
Wall temperature (K)	400	400	400
Fluid inlet temperature (K)	300	300	300
n (molecules/m^3)	4.6066E+22	9.2132E+21	4.14594E+21
U (m/s)	200	200	200
dt (s)	8.9597E–09	4.47985E–08	9.95522E–08
Number of cells ($Y \times X$)	167×667	33×133	15×60
Channel size ($H/2 \times L$)	0.005×0.01	0.005×0.01	0.005×0.01

3.3 Methodology of calculating important parameters

a) *Pressure distribution*: Bird [18] has shown that the ideal gas equation shown below, is valid for dilute gas even in nonequilibrium flow.

$$p = nkT \tag{24}$$

where k is the Boltzmann constant. The number density (n) and temperature (T) distribution was obtained from the simulation and the pressure distribution was calculated over the entire flow-field.

b) *Slip velocity*: The velocity profile at several locations was obtained from the simulation and fitted into a 9th-order polynomial. The slip velocity distribution was obtained by two different methods. The first method involved evaluating the polynomials at the wall ($y = 0$). The second method involved evaluating the first derivative at the wall, then using the following equation to obtain the slip velocity:

$$u_{slip} = \frac{2-F}{F}\lambda\frac{du}{dy}|_{wall} \quad (25)$$

where F is the specular reflection coefficient and was assumed to have a value of one.

c) *Friction coefficient*: The local friction coefficient was calculated using the following equation:

$$C_f = \frac{2\tau}{\rho u_{avg}^2} \quad (26)$$

where τ is the local shear stress value; ρ is the fluid density; and u_{avg} is the local average velocity which is calculated as follows:

$$u_{avg} = \frac{\int_{y_1}^{y_2} u(y)dy}{H} \quad (27)$$

Again, the friction coefficient was obtained by two methods. The first method utilizes the local shear stress calculated during the simulation. The second method obtains the local shear stress from the following equation:

$$\tau = \mu\frac{du}{dy} \quad (28)$$

d) *Slip temperature*: The slip temperature was obtained similar to the slip velocity by evaluating the first derivative of the temperature profile at the wall and by using the following equation:

$$T_{slip} = \frac{2-\alpha}{\alpha}\frac{2\gamma}{\gamma+1}\frac{\lambda}{\Pr}\frac{dT}{dy}|_{wall} \quad (29)$$

where Pr is the Prandtl number, γ is the specific heat ratio of the fluid, and α is the fraction of the impinging molecules that become accommodated to the wall temperature.

e) *Nusselt number*: The Nusselt number was determined by evaluating the variation of the heat transfer coefficient, h, in the axial direction as follows:

$$Nu(x) = \frac{hH}{k} \quad (30)$$

where k is the thermal conductivity of the fluid and the heat transfer coefficient is given by:

$$h = \frac{k \frac{dT}{dy}|_{wall}}{T_{wall} - T_b(x)} \tag{31}$$

where $T_b(x)$ is the local mean bulk temperature of the fluid and is calculated by:

$$T_b(x) = \frac{\int_0^H uTdy}{u_m H} \tag{32}$$

where u_m is the mean velocity, and H is the channel height.

The methodology for obtaining the parameters in Tables 1 and 2 is as follows:

1. Set the dimensions of the channel size.
2. Set the temperature of the fluid and the channel wall.
3. Set the desired value of Kn.
4. Calculate the mean free path from the definition of Kn.
5. Calculate the number density using eqn. (23).
6. Choose the cell dimension to be one-third of the mean free path and calculate the total number of cells in the channel.
7. Calculate the time step using eqn. (4).
8. Obtain the ratio of the real molecules to the simulated molecules using eqn. (1).
9. Set the mean velocity, U, of the flow inside the domain.

4 Results and discussion

4.1 Numerical criteria

Table 3: DSMC numerical criteria results.

Case	Kn	CSR	CTR	N_{smc}
1	0.009	< 0.058	< 0.14	18 → 14
2	0.017	< 0.050	< 0.13	21 → 19
3	0.03	< 0.060	< 0.13	13 → 11
4	0.11	< 0.060	< 0.14	13 → 11
5	0.009	<.055	<.134	17→14
6	0.045	<.056	<.136	12→10
7	0.11	<.057	<.138	12→10

As mentioned earlier, meeting the numerical criteria is important in order to obtain valid results. Table 3 summarizes the DSMC numerical criteria results for the fluid flow simulations (cases 1–4) and for the heat transfer simulations (cases 5–7). The values of CTR and CSR are well below unity over the entire flow-field. In addition, the number of simulated molecules per cell was also in the recommended range, which is between 10 and 20.

4.2 Effects of cell size

The effects of cell size were studied to determine if the numerical criteria discussed are met prior to considering any outputs from the simulation. A typical rule for the DSMC technique is that the cell size should be smaller than one-third of the mean free path of the gas. However, it is difficult to satisfy the cell size condition for very low Knudsen number flow. The cell size effects were tested to determine the outcome if the cell size condition is violated. Five cases were tested for the flow of $Kn = 0.002$. In these cases, at least 1000 cells were deployed in the channel width direction. The cell-Knudsen number for the cases 1–5 was varied about 30 to 2.

As seen in Table 4, the computed C_fRe values were 84–96% of the analytical value even though the cell size condition is violated. However, the dynamic viscosity of the gas computed from the simulation increased with increasing cell-Knudsen number. In case 5 where the cell-Kundsen number is 2, the computed value of the viscosity is 50% higher than the analytical value. This suggests that the effects of cell size is a strong function of computed viscosity of the gas.

Table 4: Effects of cell size in DSMC simulation ($Kn = 0.002$).

Case	Channel size ($H \times L$)	number of cells	Helium viscosity	C_fRe	Slip velocity (m/s)
1	0.01×0.04	64×16	8.3E–5	20.03	13.00
2	0.01×0.04	128×32	4.5E–5	22.10	7.50
3	0.01×0.04	128×64	3.5E–5	23.03	2.75
4	0.01×0.04	64×128	3.6E–5	21.98	6.58
5	0.01×0.02	32×256	2.7E–5	21.50	9.40
Analytical	–	–	1.865E–5	24.00	2.00

4.3 Pressure distribution

Figure 4 shows a plot of the pressure distribution along the channel length at the wall (y = 0) and at the centerline (y = 0.005). The data points with circular and square symbols refer to the pressure at the wall and at the centerline, respectively.

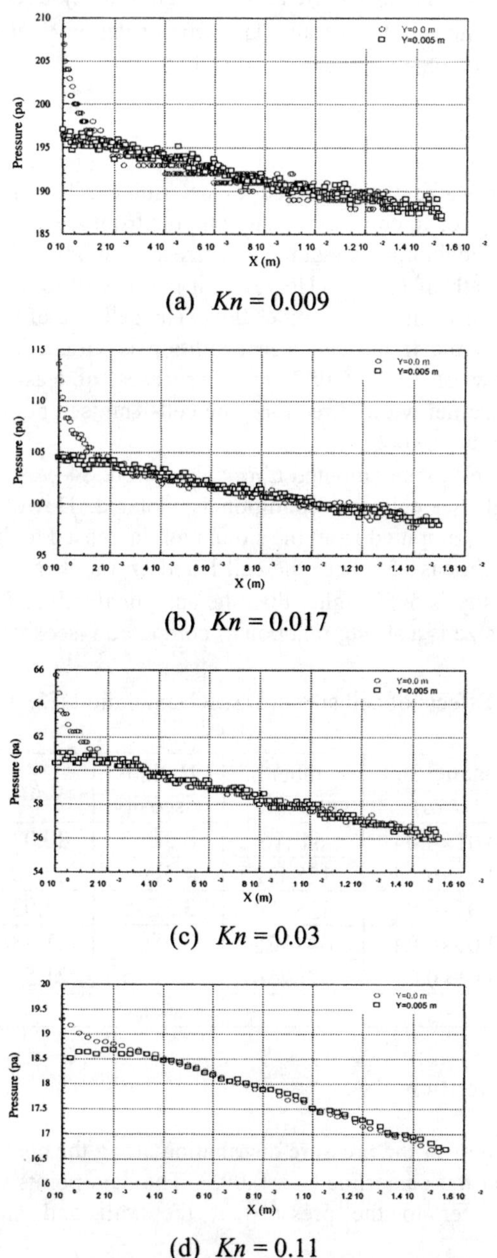

(a) $Kn = 0.009$

(b) $Kn = 0.017$

(c) $Kn = 0.03$

(d) $Kn = 0.11$

Figure 4: Pressure distribution along the channel length.

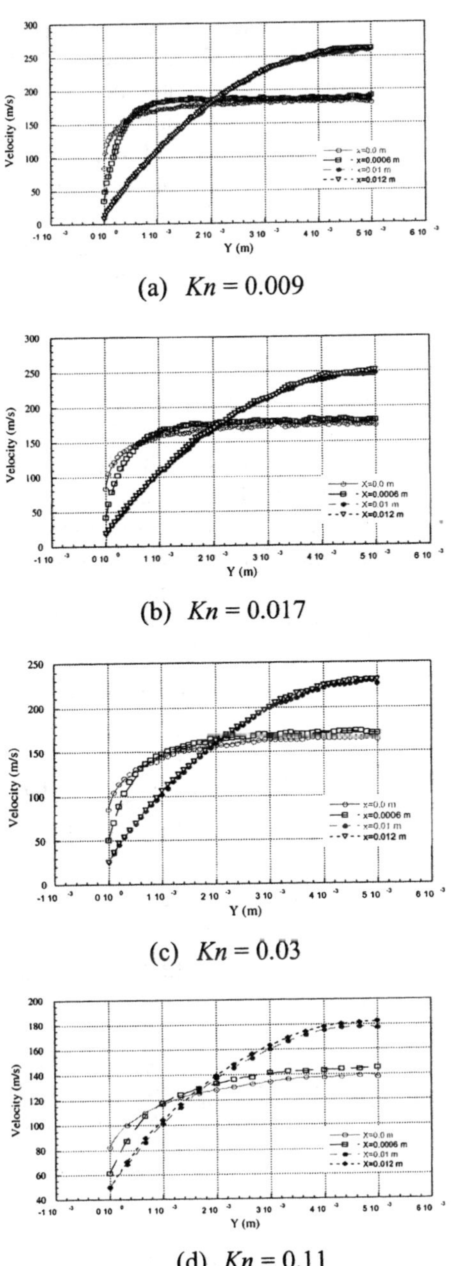

(a) $Kn = 0.009$

(b) $Kn = 0.017$

(c) $Kn = 0.03$

(d) $Kn = 0.11$

Figure 5: Velocity profile at $x = 0.0$, 0.0006, 0.01, and 0.012 m.

These data were obtained from eqn. (24) as discussed earlier. For cases 1 through 3, the pressure distribution at the centerline shows a linear variation along the channel. However, the pressure at the wall behaves slightly differently near the entrance. It decreases exponentially for about 10% of the channel length then decreases linearly similar to the centerline pressure. For case 4, the centerline pressure at the inlet increases due to the increase of the boundary layer thickness. The same trend was reported by Mavriplis *et al.* [8]. The pressure decreases linearly due to the decrease in the number density. Unlike the previous cases, the pressure distribution at the wall shows a linear variation for the entire channel length. The linearity of pressure distribution is due to the absence of compressibility effects since the channel aspect ratio is small.

4.4 Velocity profile

Figure 5 presents the velocity profiles obtained from the simulation at locations x=0.0, 0.0006, 0.01, and 0.012. For cases 1 through 3, the velocity profile is flat at the entrance for all cases, and becomes a fully developed parabolic profile at the exit. For case 4, the velocity profile at the entrance seemed to have a sharp gradient near the wall but not as steep as the previous cases. Away from the walls, the velocity profile is almost flat as expected since the boundary layer has not been developed yet. For all cases, at approximately $x = 0.01$ m, the velocity profile has reached the fully developed condition.

Table 5: Output of DSMC simulation.

Case			1	2	3	4
Kn			0.009	0.017	0.03	0.11
$C_f Re$	DSMC	τ eqn. (28)	23.9	23.3	22.0	–
		τ (DSMC)	24.3	22.7	21.2	14.0
	Analytical $24/(1+3Kn)$		23.4	22.8	21.2	14.0
U Slip	DSMC		12.8	19.2	27.0	48.0
	Analytical		11.2	17.2	26.9	–
μ	DSMC		1.991E–5	1.844E–5	1.801E–5	1.801E–5
	Analytical		1.865E–5	1.865E–5	1.865E–5	–

4.5 Slip velocity

Figure 6 presents the slip velocity distribution along the channel for $Kn = 0.009$, 0.017, 0.03, and 0.11. The data in this figure represent the actual slip velocity obtained from the simulation. As seen from this figure, the slip velocity is high at the channel entrance and decreases exponentially and then levels off for most of the channel length as the flow becomes fully developed. Away from the entrance, the slip velocity values were averaged and compared with the analytical value as shown in Table 5. For a continuum flow (Kn = 0), the velocity of the fluid at the wall is zero. A slight increase in the degree of rarefaction would enable the molecules at the wall to increase their tangential velocity (slip velocity).

Case 1 represents a near continuum condition and since Kn is not exactly zero, the slip velocity does not disappear completely where its value is about 6% of the free stream velocity. In addition, this figure shows that away from the entrance, the slip velocity is essentially a function of Kn. That is, a slip velocity increase by a factor of about 2 was matched by an increase *of* Kn from 0.03 to 0.11. A similar trend was observed by Oh *et al.* [7].

4.6 Shear stress

Figure 7 shows the shear stress distribution obtained by the simulation for Kn = 0.009, 0.017, 0.03 and 0.11. As seen in this figure, the value of the shear stress at the wall is high near the entrance of the channel and decreases exponentially with the presence of statistical fluctuation until it levels off to a steady value. This trend is consistent with the boundary layer theory for internal flows. For the transition case, the trend is similar to cases 1 through 3; however, near the entrance the stress does not exhibit a sharp decrease. In addition, the stress value does not vary significantly as compared with previous cases.

4.7 Friction coefficient

Figure 8 presents the distribution of the friction coefficient. For cases 1 through 3, the friction coefficient was calculated by the two methods as discussed earlier. In addition, the dashed line represents the analytical value of C_fRe that is equal to 24 for the continuum flow regime. However, with the presence of gas rarefaction this value is reduced and approximated by $24/(1+3Kn)$.

In cases 1 through 3 and for both calculation methods, the local friction values at the channel entrance are high due to high local shear stresses and then fall to a constant value as the distance from the entrance increases. It is noteworthy that there is a departure of the local C_fRe value between the two methods and it increases with an increase of Kn. However, far away from the entrance, both methods yield a friction coefficient value that is very close to the analytical value as shown previously in Table 5. Case 4 indicates that the friction value has decreased by a factor of 1.5 from Kn = 0.009 to 0.11. The C_fRe value for this case is 14.0 and cannot be compared with the analytical solution since the Kn falls in the transition regime. However, to prove this, the analytical equation was evaluated at Kn = 0.11 and it was found that the difference in C_fRe values between the simulation and the analytical was much larger than that for the slip flow regime. This indicates that the error in the continuum equation is increased as the Kn falls above 0.1.

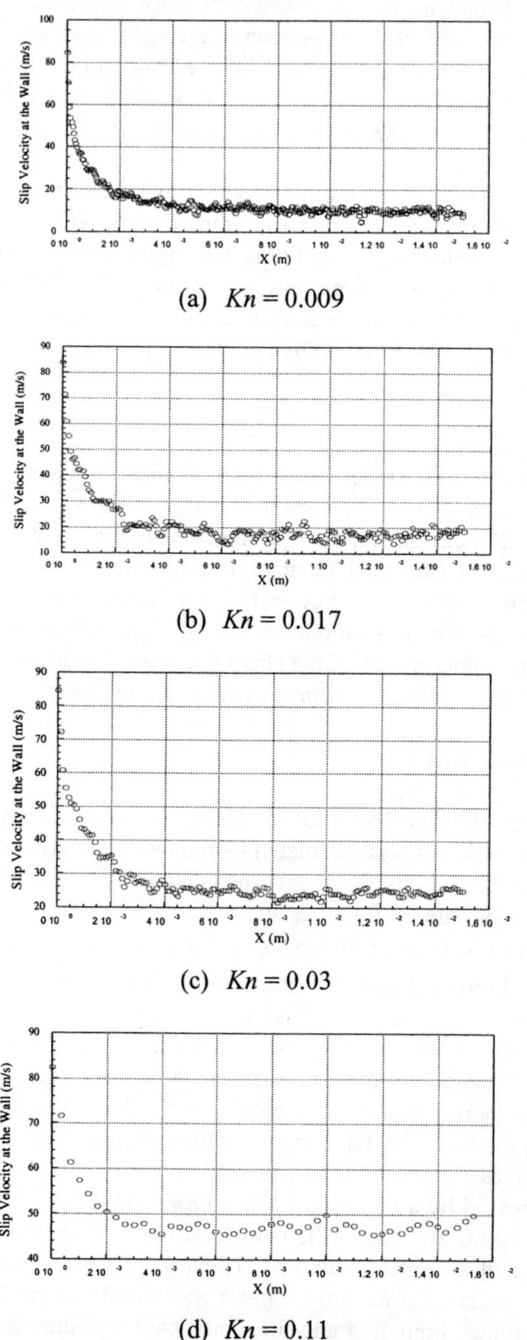

(a) $Kn = 0.009$

(b) $Kn = 0.017$

(c) $Kn = 0.03$

(d) $Kn = 0.11$

Figure 6: Slip velocity distribution along the channel length.

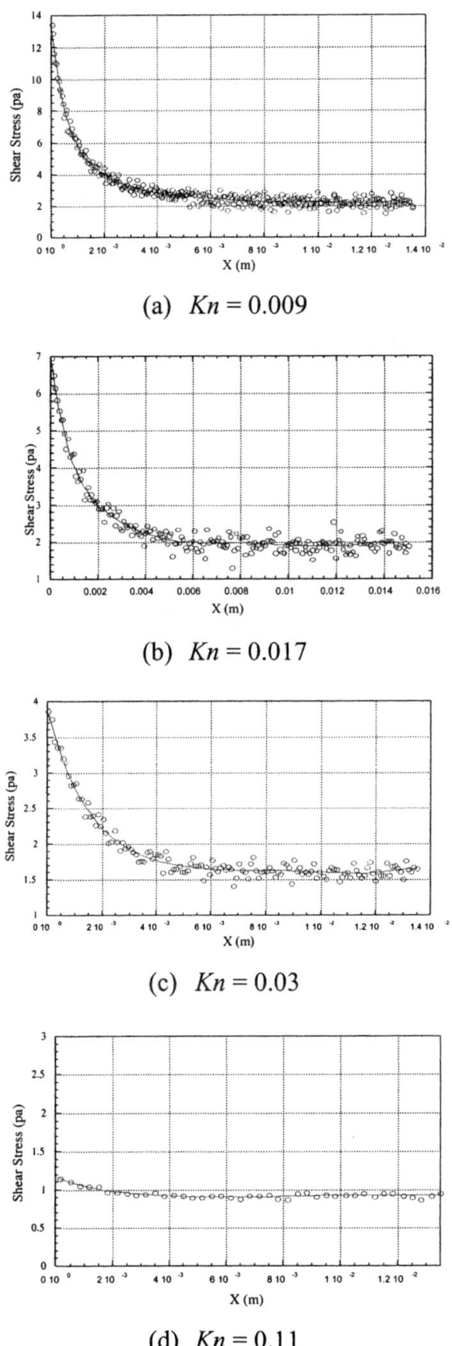

Figure 7: Shear stress distribution at the wall.

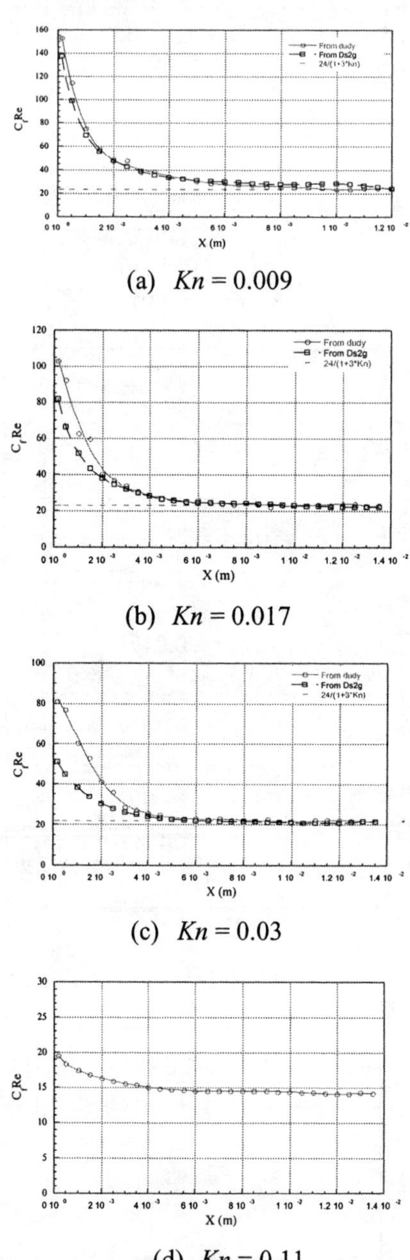

Figure 8: Distribution of friction coefficient at the wall.

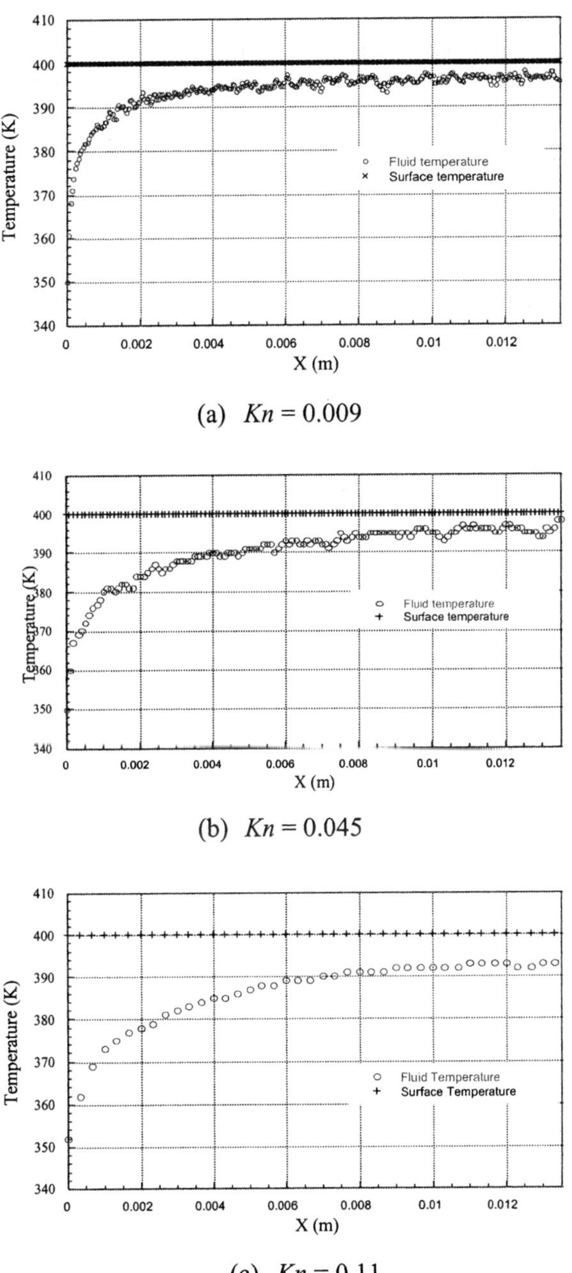

(a) $Kn = 0.009$

(b) $Kn = 0.045$

(c) $Kn = 0.11$

Figure 9: Slip temperature distribution along the channel length.

4.8 Slip temperature

Figure 9 shows the slip temperature distribution along the channel length and for Kn = 0.009, 0.045, and 0.11. The fluid temperature increases as a result of energy exchange between the molecules and the channel wall as the wall is hotter than the fluid. At the entrance, the slip temperature increased by 2% as *Kn* increased from 0.009 to 0.11 and there is a sharp decrease in the slip temperature that levels off as the flow becomes fully developed. Away from the channel entrance, the slip velocity data were averaged and then compared with the available analytical solution as shown in Table 6. Good agreement between the simulated and analytical results for the slip temperature can be seen.

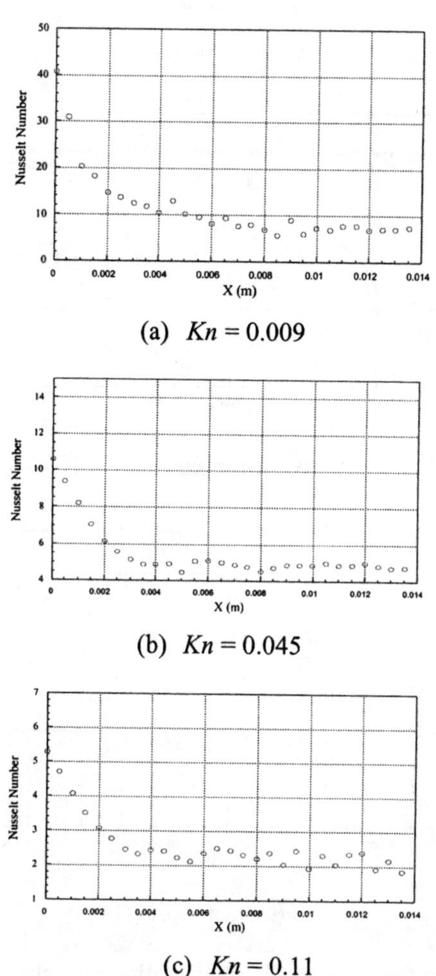

Figure 10: Nusselt number distribution along the channel.

4.9 Nusselt number

Figure 10 shows the Nusselt number distribution along the channel length for *Kn* = 0.009, 0.045, 0.11. The *Nu* decreases exponentially in the downstream direction with the presence of statistical fluctuations, and then levels off as the flow becomes fully developed. The overall simulation results show a reduction in *Nu* as the gas becomes more rarefied. For flows between parallel plates, the value of *Nu* in the continuum regime is 7.54 and for higher *Kn* the Nusselt number is as listed in Table 6.

Table 6: Output of DSMC simulation.

CASE		1	2	3
Kn		0.009	0.045	0.11
Nu		7.27	4.82	2.19
Slip velocity (m/s)	DSMC	13.6	35.2	47.5
	Analytical	11.8	32.7	–
Slip temperature (K)	DSMC	398.0	395.3	393.0
	Analytical	396.4	391.9	–

5 Conclusions

The DSMC methodology was introduced and was utilized to study the effects of Knudsen number on fluid flow and heat transfer characteristics in microchannels. These effects were investigated for a helium gas at three different regimes: continuum, slip, and transition. The variation of *Kn* was achieved by reducing the fluid density while keeping the channel size constant. A benchmark case was performed to validate the DSMC code and the results were obtained for the slip and transition regimes. The pressure distribution, velocity profile, slip velocity and temperature, friction coefficient and Nusselt number were obtained and compared with the analytical solution. The results showed that the flow characteristics were changed significantly as *Kn* was increased. The friction coefficient was decreased from 24.0, for the continuum regime, to 14.0 for the transition regime and the slip velocity was increased for the entire length of the channel. The Nusselt number was decreased from 7.54, for the continuum regime, to 2.19 for the transition regime.

Nomenclature

C_w	cell width
cps	sum of the collision pair separation
c_r	relative speed
d	diameter of molecule
dt	time step

dx	distance in x direction
dy	distance in y direction
F	specular reflection coefficient
fu'	distribution function
H	characteristic length, channel height
h	heat transfer coefficient
K	Boltzmann constant
k	thermal conductivity
Kn	Knudsen number
L	length of the channel
n	number density
n_c	number of cells
NP	number of pairs
N_{smc}	number of simulated molecules in a cell
P	position of molecule
P	pressure
Pr	Prandtl number
Rf	random number
$R_{r,s}$	ratio of real molecules to simulated molecules
T	temperature
U	average stream velocity
u	velocity
u, v, w	molecular velocity in Cartesian coordinate system
V	volume
x, y	Cartesian coordinate
XC	location of collision

Subscripts

1	begin
2	end
avg	average
b	fluid bulk
c	cell
f	fraction
h, q	label name for a molecule
m	mean value
mp	most probable speed
New	updated
r	real molecule, relative
ref	reference
s	simulated molecule
slip	slip at the wall
SB	surface boundary
smc	simulated molecule per cell
x,y	Cartesian coordinate

$'$	thermal
wall	at the channel wall

Greek symbols

γ	specific heat ratio
α	thermal accommodation factor
θ	$2\pi Rf$
β	reciprocal of the most probable speed
σ_T	collision cross-section
Ψ	probability function
τ	shear stress
λ	mean free path
μ	dynamic viscosity

References

[1] Bird, G.A., Molecular Gas Dynamics, *Oxford Engineering Science*, Oxford University Press: Oxford, 1976.

[2] Nanbu, K., Theoretical basis of the direct simulation Monte Carlo Method. *Proceedings of the 15th International Symposium on Rarefield Gas Dynamics*, **1**, pp. 369–383, 1986.

[3] Nanbu, K. & Igarashi, S., Three-dimensional low-density flows in the spiral grooves of a turbo-molecular pump. *Computers Fluids*, **21(2)**, pp. 221–228, 1992.

[4] Riechelman, D. & Nanbu, K., Three-dimensional simulation of wavy Taylor vortex flow by direct simulation Monte Carlo method. *Physics Fluids*, **9**, pp. 811–813, 1997.

[5] Moss, J.N., Mitcheltree, R.A., Dogra, V.K. & Wilmoth, R.G., Direct simulation Monte Carlo and Navier–Stokes simulations of blunt body wake flow. *AIAA Journal*, **32**, pp. 1399–1406, 1994.

[6] Bird, G.A., *The DS2G Program User's Guide, Version 3.1*, G.A.B. Consulting Pty Ltd.: Killara, Australia, 1998.

[7] Oh, C.K., Oran, E.S. & Sinkovits, R.S., Computations of high-speed, high Knudsen number microchannel flows. *Journal of Thermophysics and Heat Transfer,* **11(4)**, pp. 497–505, 1997.

[8] Mavriplis, C., Ahn, J.C. & Goulard, R., Heat transfer and flow fields in short microchannels using direct simulation Monte Carlo. *Journal of Thermophysics and Heat Transfer,* **11(4)**, pp. 489–496, 1997.

[9] Piekos, E.S. & Breuer, K.S., Numerical modelling of micromechanical devices using the direct simulation Monte Carlo method. *ASME Journal of Fluids Engineering*, **118**, pp. 465–469, 1996.

[10] Kavehpour, H.P., Faghri, M. & Asako, Y., Effects of compressibility and rarefaction on gaseous flows in microchannels. *Numerical Heat Transfer*, Part A, **32**, pp. 677–696, 1997.

[11] Sayegh, R., Faghri, M., Asako, Y. & Sunden, B., Direct simulation Monte Carlo of gaseous flow in micro-channel. *ASME Proceeding of the National Heat Transfer Conference*, HTD99-256, 1999.
[12] Sayegh, R., Faghri, M., Asako, Y. & Sunden, B., Direct simulation Monte Carlo of thermal transport in micro-channel. *ASME Proceeding of the National Heat Transfer Conference*, NHTC2000-12105, 2000.
[13] Liou, W.W. & Fang, Y., Heat transfer in microchannel devices using DSMC. *Journal of Microelectromechanical Systems*, **10(2)**, pp. 274–279, 2002.
[14] Fang, Y. & Liou, W., Computations of the flow and heat transfer in microdevices using DSMC with implicit boundary conditions. *Journal of Heat Transfer*, **124**, pp. 338–345, 2002.
[15] Piekos, E. & Breuer, K.S., DSMC modelling of micromechanical devices. *ASME Journal of Fluids Engineering*, **118**, pp. 464–469, 1996.
[16] Beskok, A. & Karniadakis, G., A model for flows in channels, pipes and ducts at micro and nano scales. *Journal of Microscale Thermophysical Engineering*, **3**, pp. 43–77, 1999.
[17] Hadjiconstantinou, N.G. & Simek, O., Constant-wall-temperature Nusselt number in micro and nano-channels. *ASME Journal of Heat Transfer*, **124**, pp. 356–364, 2002.
[18] Bird, G.A., *Molecular Gas Dynamics and the Direct Simulation of Gas Flows*, Oxford University Press: Oxford, 1994.

CHAPTER 8

DSMC modeling of near-interface transport in liquid–vapor phase-change processes with multiple microscale effects

V.P. Carey
Mechanical Engineering Department
University of California, Berkeley, CA 94720, USA

Abstract

This chapter examines strategies for using direct simulation Monte Carlo (DSMC) methods for modeling near-interface transport during vaporization and condensation processes in microsystems. At the small scales of interest in these applications, changes in mechanisms give rise to departures from classical, macroscopic continuum models that are applicable to more conventional applications. The chapter begins with a summary examination of various mechanisms that come into play during vaporization and condensation in microscale systems. It then explores how treatments of these mechanisms can be incorporated into DSMC based models. Emphasis here is on approaches that properly account for multiple microscale effects that act in tandem and interact with one another to modify the transport during liquid–vapor phase change processes in microscale and nanoscale systems. Recent successful applications of models of this type are described, and strategies for extending this methodology to other liquid–vapor phase change processes are discussed.

1 Introduction

Conventional macroscale analysis of transport near a liquid–vapor interface usually invokes a continuum viewpoint with the interface modeled as a two-dimensional surface at which properties change discontinuously. Such treatments also generally include the assumption of local thermodynamic equilibrium at the interface. In many applications, transport in the immediate neighborhood of the interface is critical to the process being considered. When microscale features of near-interface transport are of interest, macroscopic treatments of the type described above are often inadequate, and alternative modeling methods must be employed. The direct simulation Monte Carlo (DSMC) method can be useful as an analysis tool if it can be modified to include appropriate treatment of microscale thermophysical mechanisms that affect the transport near the interface. This chapter summarizes available information regarding methods for modeling microscale and nanoscale mechanisms that affect near-interface transport in liquid–vapor phase change processes. Several physical effects that are minor in macroscale processes play a central role in microscale transport in vaporization and condensation processes. In some instances, a single microscale mechanism is responsible for the departure of the transport from classical continuum behavior. In many cases, however, multiple microscale effects come into play, and their combined effect dictates the nature of the resulting transport. This latter circumstance, which is the focus of this paper, is often a very challenging one to model because the individual mechanisms, as well as their interaction, must be treated accurately. This paper begins with a review of microscale effects that arise in vaporization and condensation processes and the analysis tools that can be applied to them. It then continues with an examination of strategies to construct DSMC based models for transport during liquid–vapor phase change in microscale systems for circumstances where microscale mechanisms strongly interact. An example model of this type will be considered in detail, and extension of the approach to other microscale systems will be discussed.

2 Phase equilibrium in microscale multiphase systems

2.1 Ultra-small bubbles and droplets

In modeling near-interface transport in microscale systems, it is often necessary to account for several well-known effects of system size on the equilibrium system vapor pressure and temperature. This arises, for example, in pure systems involving microdroplets, microbubbles, ultra-thin films and interfaces in microchambers. For very small droplets, the surface tension shifts the equilibrium vapor pressure from the flat interface value for a macroscopic system. Classical thermodynamics dictates that this shift can be computed from the following implicit relation (see [1])

$$P_{ve} = P_{sat}(T_d)\exp\left\{\frac{[P_{ve} - P_{sat}(T_d)]}{\hat{\rho}_l R T_d} + \frac{r_\sigma}{r_d}\right\} \tag{1}$$

where

$$r_\sigma = \frac{2\sigma_{lv}}{\hat{\rho}_l R T_d} \tag{2}$$

In the above relations, T_d is the droplet and surrounding vapor temperature ($T_d = T_v$), r_d is the droplet radius and $P_{sat}(T_d)$ is the flat-interface saturation pressure for the liquid at T_d. When the radius of the droplet is large compared to r_σ, the equilibrium vapor pressure at the droplet surface is that for a flat interface between bulk phases, assuming that local thermodynamic equilibrium exists there. As seen in Figure 1, r_d typically must be smaller than $10 r_\sigma$ before the departure from the flat-interface vapor pressure is significant.

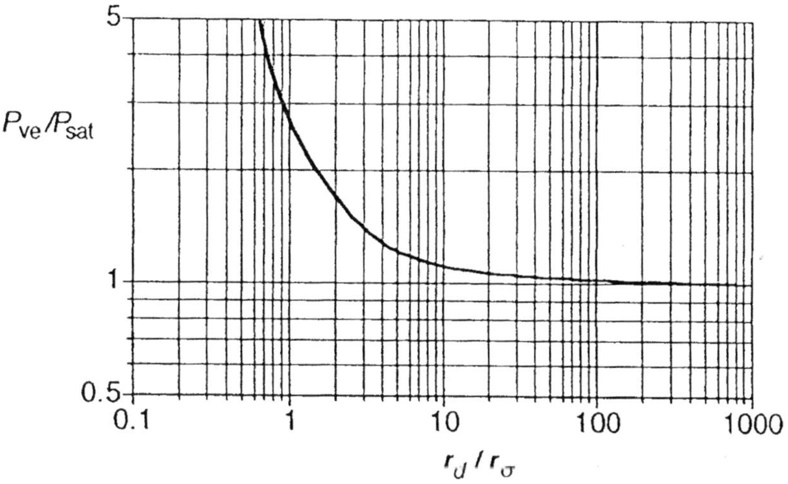

Figure 1: Variation of equilibrium vapor pressure with droplet radius.

Inverting eqn. (1) indicates that for a specified vapor pressure P_v and temperature T_v, the equilibrium droplet radius r_{de} is given by

$$r_{de} = \frac{2\sigma_{lv}}{(\hat{\rho}_l R T_v)\ln\{P_v / P_{sat}(T_v)\} - P_v + P_{sat}(T_v)} \tag{3}$$

Thus, only droplets having a radius equal to that given by the above relation will be in equilibrium with the surrounding vapor at T_v and P_v. Thermodynamic analysis also indicates that the equilibrium condition represented by eqn. (3) is unstable, with the consequence that droplets spontaneously grow if their radius is larger that r_{de} and they spontaneously shrink if their radius is less than r_{de}. Furthermore, for a droplet of finite radius, equilibrium can be achieved only if the vapor is supersaturated and the liquid is subcooled relative to the normal saturation state for a flat interface.

For small vapor bubbles in pure liquid, thermodynamic analysis yields the following implicit relation for the equilibrium vapor pressure inside the bubble:

$$P_{ve} = P_{sat}(T_l)\exp\left\{\frac{[P_{ve} - P_{sat}(T_l)]}{\hat{\rho}_l R T_l} + \frac{r_\sigma}{r_b}\right\} \tag{4}$$

where again

$$r_\sigma = \frac{2\sigma_{lv}}{\hat{\rho}_l R T_d} \tag{2}$$

and r_b is the bubble radius. Comparison of the above relation with eqn. (1) reveals that this relation for the equilibrium vapor pressure inside a small bubble is similar in form to that for the pressure surrounding a small droplet in equilibrium with surrounding vapor. Inverting eqn. (4) yields the following relation for the equilibrium radius r_{be} for specified ambient liquid conditions:

$$r_{be} = \frac{2\sigma_{lv}}{P_{sat}(T_l)\exp\{[P_l - P_{sat}(T_l)]/\hat{\rho}_l R T_l\} - P_l} \tag{5}$$

Thus only bubbles having a radius equal to that given by the above relation will be in equilibrium with the surrounding superheated liquid at T_l and P_l.

2.2 Ultra-thin liquid films

Another circumstance in which system size affects equilibrium conditions is the case of a very thin liquid film on a solid surface. If a solid surface is initially in contact with a gas or vapor phase and a small quantity of liquid is brought into contact with the surface, the liquid may bead-up into a droplet if it poorly wets the surface, or it may spread over the surface to form a thin liquid film. The latter behavior generally results when there is a strong attractive force interaction between the molecules in the liquid and the molecules of the solid near the interface. To analyze more fully the thin film that occurs for such circumstances, we consider the system shown in Figure 2 in which liquid is in contact with and fully wets the horizontal upward-facing solid surface.

When the film is very thin, these attractive forces act to pull liquid into the layer as if the pressure in the layer were reduced below the ambient pressure P_l by an amount P_d, which is known as the *disjoining pressure*. By convention, if the affinity of the liquid for the solid draws liquid into the film, P_d is taken to be negative.

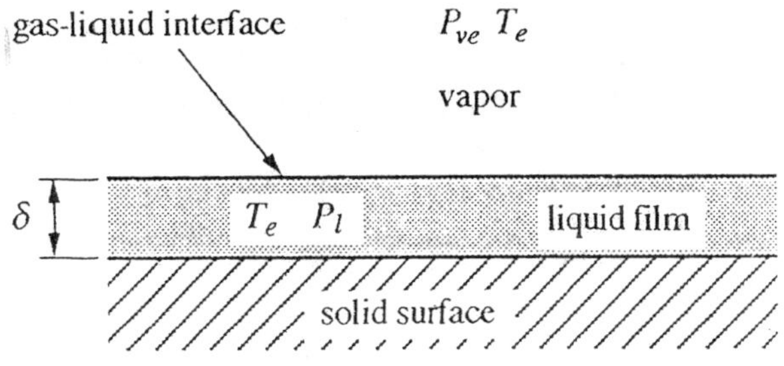

Figure 2: Equilibrium for a thin film on a solid surface.

For the circumstances shown in Figure 2, the local pressure P_l and the disjoining pressure effect (due to solid–liquid attraction) act in tandem to thicken the film. To maintain a thin liquid film, the gas pressure force acting on the film must balance both effects, which implies that at equilibrium

$$P_g = P_l + |P_d| = P_l - P_d \tag{6}$$

Note in eqn. (6) that because P_d is negative for a liquid that is attracted to the solid surface, $-P_d$ is positive for this system. Because the attractive forces between liquid molecules and those of the solid are expected to be stronger for liquid molecules that are closer to the interface, the disjoining pressure difference $-P_d$ required to counteract them is expected to increase as δ gets smaller. This is reflected in the variation of disjoining pressure with δ indicated in Figure 3 for carbon tetrachloride on glass.

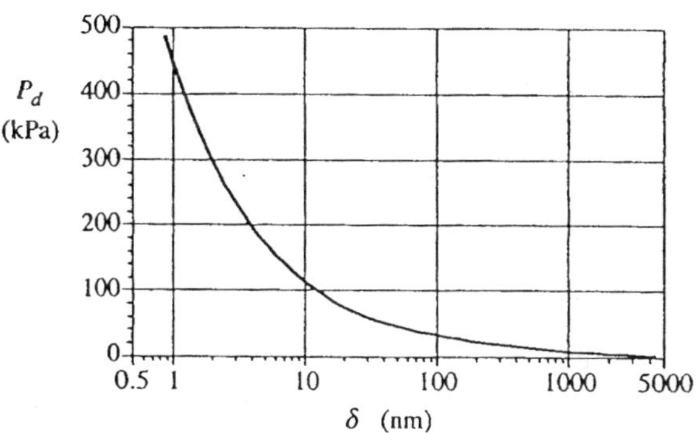

Figure 3: Variation of disjoining pressure with film thickness for carbon tetrachloride on glass at 77°C (from ref. [2]).

A detailed analysis of molecular forces and the reversible work required to establish a liquid microlayer can be used to develop a relation for the disjoining pressure (see Carey [1]). Doing so, a relation of the following type may be obtained:

$$-P_d(\delta) = \frac{2A_H}{\pi(n-2)(n-3)}\delta^{3-n} \tag{7}$$

In the above relation, n is the exponent in the long-range interaction potential assumed to characterize the force interaction between the liquid and solid molecules

$$\phi(r) = -\frac{C_\phi}{r^n} \tag{8}$$

Note that if we take this to be equivalent to the outer portion of the Lennard-Jones 12-6 potential

$$\phi(r) = 4\varepsilon_0\left[\left(\frac{\sigma_{LJ}}{r}\right)^{12} - \left(\frac{\sigma_{LJ}}{r}\right)^{6}\right] \tag{9}$$

then

$$C_\phi = 4\varepsilon_0\sigma_{LJ}^6,\ n = 6 \tag{10}$$

The constant A_H in eqn. (7) is the Hamaker constant defined as

$$A_H = \pi^2\rho_{N,s}\rho_{N,l}C_\phi \tag{11}$$

Where $\rho_{N,s}$ and $\rho_{N,l}$ are the number density of molecules in the solid and liquid phases, respectively. If the long-range attractive force potential is equivalent to the long-range portion of the Lennard-Jones 12-6 potential, then $n = 6$ and eqn. (7) reduces to

$$-P_d(\delta) = \frac{A_H}{6\pi}\delta^{-3} \tag{12}$$

The equilibrium conditions for a very thin film in contact with a solid surface and its vapor can be examined by considering the system shown in Figure 2. At a specified vapor ambient pressure P_{ve}, the system is at equilibrium at temperature T_e. Two necessary and sufficient conditions for equilibrium are that the temperature and chemical potential in the liquid and vapor phases must be equal. Invoking these conditions in a thermodynamic analysis of the system yields the relation

$$P_{ve} = P_{sat}(T_e)\exp\left\{\frac{\hat{v}_l\left[P_l - P_{sat}(T_e)\right]}{RT_e}\right\} \tag{13}$$

The analysis described above links the disjoining pressure to the film thickness δ by a relation of the form

$$P_d = -A\delta^{-B} \tag{14}$$

(See also the discussion of disjoining pressure in references [2] and [3].) It follows from the above relation that for the system in Figure 2

$$P_l = P_{ve} - A\delta^{-B} \tag{15}$$

Substituting this expression into eqn. (13) yields

$$P_{ve} = P_{sat}(T_e)\exp\left\{\frac{\hat{v}_l\left[P_{ve} - P_{sat}(T_e) - A\delta^{-B}\right]}{RT_e}\right\} \tag{16}$$

For a specified film thickness, this relation can be used to determine the equilibrium vapor pressure for a given temperature. Alternatively, for a specified system pressure and film thickness, the equilibrium temperature can be determined. Rearranging this relation, an expression for the equilibrium film thickness for a specified set of vapor pressure and temperature conditions can also be obtained:

$$\delta = \left[\frac{A}{P_{sat}(T_e)}\right]^{1/B}\left\{\frac{P_{ve}}{P_{sat}(T_e)} - 1 - \frac{RT_e}{\hat{v}_l P_{sat}(T_e)}\ln\left\{\frac{P_{ve}}{P_{sat}(T_e)}\right\}\right\}^{-1/B} \tag{17}$$

At a given vapor pressure, this equation indicates that if the film is thin enough, it can be superheated ($T_e > T_{sat}(P_v)$) and be in equilibrium with the vapor. This means that the liquid is in a superheated metastable state and the vapor is also superheated. It also implies that heating the solid surface, and hence the liquid film, to a temperature slightly above its normal saturation temperature will not necessarily evaporate the liquid film completely. Instead an equilibrium thin film may remain on the surface with a thickness that satisfies eqn. (17).

Eqn. (17) can be interpreted as indicating that a thin film can exist on a solid surface even when its vapor pressure in the surrounding gas is below the normal saturation pressure for the system temperature. This is consistent with the observed adsorption of thin liquid films onto surfaces such as metal, even when the vapor pressure of the adsorbed species is below the normal saturation value at the system temperature.

The liquid film thickness generally must be exceedingly small for the equilibrium saturation conditions to differ significantly from normal flat-interface saturation conditions. For carbon tetrachloride on glass, Potash and Wayner [2] recommend the values A = 1.782 Pa m^B and B = 0.6. Using these values with thermodynamic data for CCl_4, eqn. (16) was used to predict the fractional change in the equilibrium saturation pressure with film thickness at a system temperature of 77 °C. The resulting variation is shown in Figure 4. For a liquid film thickness of 1 mm (1000 μm) the fractional change in the vapor pressure is less than 10^{-4}, or less than 0.01%. A one percent change in the vapor pressure would require a film thickness of about 100 nm. For films thicker than 100 nm, the deviation from normal equilibrium conditions is insignificant.

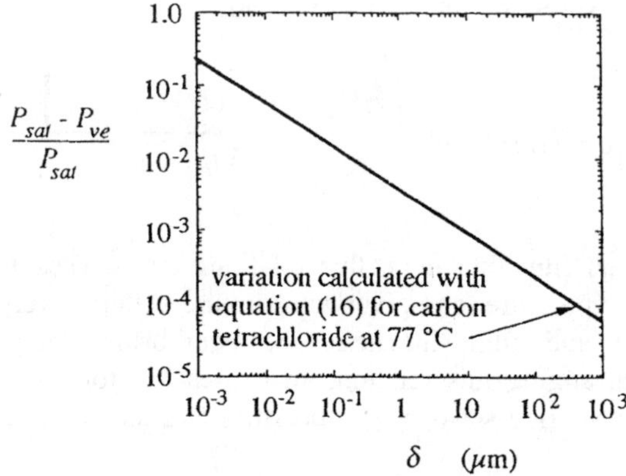

Figure 4: Fractional change in vapour pressure above a liquid film of carbon tetrachloride on glass.

For highly wetting liquids, the above considerations also affect the thermophysics at the location where a pool of liquid comes into contact with the wall of a container. Such circumstances are shown in Figure 5. A highly wetting fluid establishes an extended meniscus at the wall which includes an intrinsic meniscus region due to interfacial tension effects as well as a thin film which results from the liquid's tendency to spontaneously spread over the surface. If the wall temperature is above the normal saturation temperature, heat may be conducted across the liquid film in this region, resulting in evaporation at the interface. However, because the interface is curved in the intrinsic meniscus region, the wall superheat will have to exceed that required for equilibrium with the interface curvature present. In addition, because of the disjoining pressure effects described above, beyond the intrinsic meniscus a thin film can exist on the wall at equilibrium, even when the wall is superheated above the normal saturation temperature for the local vapor pressure. This thin film can exist without evaporating if its thickness is equal to that specified by eqn. (17).

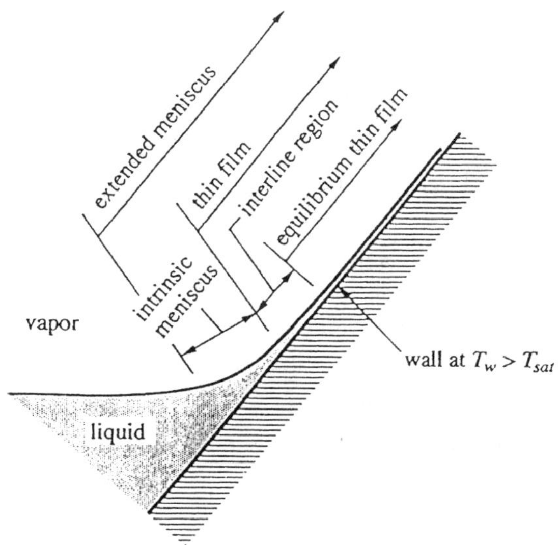

Figure 5: Equilibrium thin film on a solid surface

3 Molecular transport at interfaces

The mass and energy transfer during vaporization and condensation processes are important in a variety of technological applications. To fully understand the characteristics of such processes at a molecular level, it is useful to consider some of the results from kinetic theory that relate to the motion of vapor molecules near a liquid–vapor interface. We consider a plane surface in the vapor phase but immediately adjacent to the interface as shown in Figure 6. Even if no net vaporization or condensation occurs at the interface, a dynamic equilibrium is established in which the molecules from the vapor phase that enter the interfacial region and become part of the liquid phase are balanced, on the average, by an equal number of molecules that escape the liquid into the vapor region. When (net) condensation occurs, the flux of vapor molecules joining the liquid must exceed the flux of liquid molecules escaping into the vapor phase. When (net) vaporization occurs, the opposite must be true.

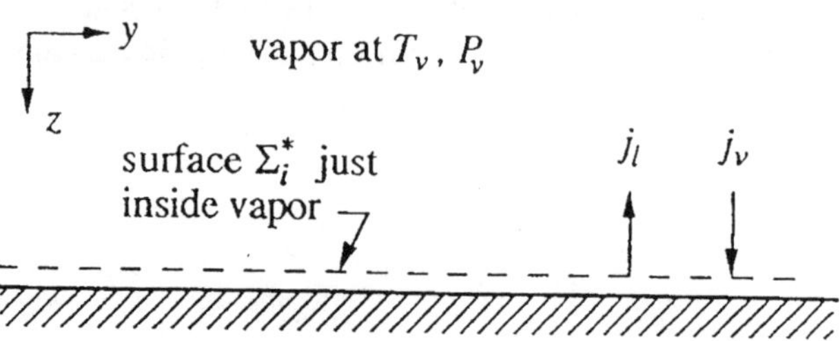

Figure 6: Molecular fluxes at the interface.

At least initially, we admit the possibility that the respective values of pressure and temperature in the two bulk phases shown in Figure 6 may be different. For the system shown in this figure, the net number flux of molecules across surface Σ_i^* immediately adjacent to the interface must be equal to the difference between the number fluxes j_l and j_v passing through the surface Σ_i^* in opposite directions just inside the vapor region:

$$j_{net} = j_v - j_l \tag{18}$$

To facilitate the evaluation of j_v and j_l in eqn. (18), we will assume that here the flux of molecules j_l is characterized by T_l and P_l whereas j_v is characterized by T_v and P_v. With this assumption, it would appear that j_v and j_l can be obtained by using the relation from kinetic theory for the flux of molecules across a plane surface in a ideal gas

$$j = \left(\frac{\overline{M}}{2\pi N_A k_B T} \right)^{1/2} \frac{P}{m} \tag{19}$$

In this relation $\overline{M}$ is the molecular weight of the substance and m is the mass per molecule. However, use of eqn. (19) to evaluate j_v and j_l is inadequate in two respects. First, eqn. (19) predicts the flux of molecules in a stationary gas, whereas in the system in Figure 6, the vapor must have a net bulk velocity $w = w_0$ in the z direction as a result of the phase change at the interface. The bulk velocity is toward the interface ($w_0 > 0$) for condensation and away from the interface ($w_0 < 0$) for vaporization.

By extending basic kinetic theory analysis, it can be shown (see Schrage [4]) that when the gas moves normal to a planar surface with a speed w_0, the flux of molecules through the plane in the direction of bulk motion is

$$j_{w+} = \Gamma_a \left(\frac{\overline{M}}{2\pi N_A k_B T} \right)^{1/2} \frac{P}{m} \tag{20}$$

and the flux of molecules in the direction opposite to that of the bulk motion is

$$j_{w-} = \Gamma_{-a} \left(\frac{\overline{M}}{2\pi N_A k_B T} \right)^{1/2} \frac{P}{m} \tag{21}$$

where

$$a = \frac{w_0}{\left(2N_A k_B T / \overline{M}\right)^{1/2}} \tag{22}$$

The factors Γ_a and Γ_{-a}, which correct for the effects of bulk gas motion, are given by the following relations

$$\Gamma_a = \exp\left(a^2\right) + a\pi^{1/2}\left[1 + \mathrm{erf}(a)\right] \qquad \text{(for } a > 0\text{)} \tag{23}$$

$$\Gamma_{-a} = \exp\left(a^2\right) - a\pi^{1/2}\left[1 - \mathrm{erf}(a)\right] \qquad \text{(for } a < 0\text{)} \tag{24}$$

The variations of Γ_a and Γ_{-a} with a are shown in Figure 7.

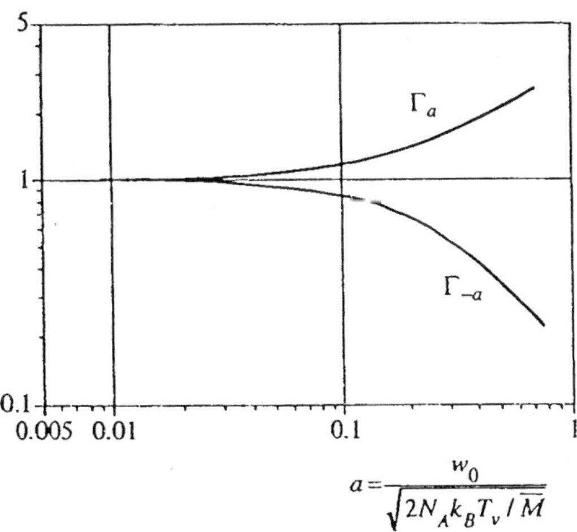

Figure 7: Variation of Γ_a and Γ_{-a} with a.

Because the vapor in Figure 6 is expected to be in bulk motion relative to the surface Σ_i^*, it is more appropriate to compute j_v using eqn. (20) or (21). As suggested by Schrage [4], the surface Σ_i^* is assumed to be an infinitesimally small distance from the interface so that there is no bulk motion effect on molecules emerging from the liquid and passing through the surface Σ_i^*. The net flux of molecules from the liquid j_l is therefore calculated using eqn. (19).

The second inadequacy in using eqn. (19) alone to compute fluxes at the interface is that doing so implicitly assumes that all molecules crossing the surface Σ_i^* in the negative z direction are actually from the liquid phase. Actually only a fraction $\hat{\sigma}_e$ is due to vaporization from the liquid. The remaining fraction $1-\hat{\sigma}_e$ is due to "reflection" of vapor molecules that strike the interface but do not condense. Note that in this context "vaporization" is used to describe molecules escaping the liquid phase and 'condensation' is used to described molecules from the vapor phase being absorbed into the liquid phase. The fraction of molecules crossing the surface Σ_i^* in the positive z direction that condense and are not reflected is designated as $\hat{\sigma}_c$. If no phase change occurs at the interface, equilibrium requires that

$$\hat{\sigma}_e = \hat{\sigma}_c = \hat{\sigma} \tag{25}$$

Usually $\hat{\sigma}_e$ and $\hat{\sigma}_c$ are assumed to be equal even for the dynamic case when (net) phase change occurs at the interface, although the validity of this assumption is suspect. As a result, $\hat{\sigma}$ is sometimes referred to as an *evaporation coefficient*, *condensation coefficient* or *accommodation coefficient*. The term accommodation coefficient has been perhaps most widely used for $\hat{\sigma}$ in recent years and will be used here.

Allowing for the bulk motion of the vapor and the role of the accommodation coefficient as described above, the portion of the incident molecular flux that actually enters the liquid phase upon striking the liquid, j_{vc}, is given by

$$j_{vc} = \begin{cases} \hat{\sigma}\, j_{w+} & \text{for condensation} \\ \hat{\sigma}\, j_{w-} & \text{for evaporation} \end{cases} \tag{26}$$

and the portion of the molecular flux crossing surface Σ_i^* that is due to molecules that actually emerged from the liquid phase, j_{le}, is given by

$$j_{le} = \hat{\sigma}\, j \tag{27}$$

The net molecule flux to or from the interface as a result of the phase change j_i is just equal to the difference between j_{vc} and j_{le}.

$$j_i = j_{vc} - j_{le} \tag{28}$$

Substituting eqns. (26) and (27) into (28), evaluating j_{w+} or j_{w-} at P_v and T_v and j at P_l and T_l and rearranging yields

$$j_i = \frac{\hat{\sigma}}{m}\left(\frac{\overline{M}}{2\pi N_A k_B}\right)^{1/2}\left[\frac{\Gamma P_v}{T_v^{1/2}} - \frac{P_l}{T_l^{1/2}}\right] \tag{29}$$

Macroscopically, the heat flux to the interface $J_{Q,i}$ must equal the net mole flux $j_i m / \overline{M}$ multiplied by the latent heat of vaporization $\Delta\hat{h}_{lv}$. Using eqn. (29) to evaluate j_i, the resulting relation for the heat flux is

$$J_{Q,i} = \frac{\hat{\sigma}\Delta\hat{h}_{lv}}{\left(2\pi N_A k_B\right)^{1/2}}\left[\frac{\Gamma P_v}{T_v^{1/2}} - \frac{P_l}{T_l^{1/2}}\right] \tag{30}$$

For constant ambient pressure, the surrounding vapor must move to replace condensing vapor or it must move away to make room for vapor generated by evaporation. This implies that the bulk velocity of the vapor is related to the net molecular flux at the interface as

$$w_0 = \frac{j_i \hat{v}_v}{N_A} = \frac{j_i k_B T_v}{P_v} \tag{31}$$

which implies that

$$a = \frac{j_i k_B T_v}{P_v}\left(\frac{\overline{M}}{2 N_A k_B T_v}\right)^{1/2} \tag{32}$$

Since $J_{Q,i} = j_i m \Delta\hat{h}_{lv} / \overline{M}$, it also follows that

$$a = \frac{J_{Q,i} \overline{M} k_B T_v}{m \Delta\hat{h}_{lv} P_v}\left(\frac{\overline{M}}{2 N_A k_B T_v}\right)^{1/2} \tag{33}$$

Because Γ is a function of a, eqn. (29) is actually an implicit relation for j_i, and eqn. (30) is actually an implicit relation for heat flux. If the liquid and vapor phases are in equilibrium (and the interface is planar), the pressure in each phase is expected to equal the saturation pressure at the temperature in that phase. If we assume that this is also true when net condensation or vaporization occurs, eqns. (29) and (30) can be written

$$j_i = \frac{\hat{\sigma}}{m}\left(\frac{\overline{M}}{2\pi N_A k_B}\right)^{1/2}\left[\frac{\Gamma P_{sat}(T_v)}{T_v^{1/2}} - \frac{P_{sat}(T_l)}{T_l^{1/2}}\right] \tag{34}$$

$$J_{Q,i} = \frac{\hat{\sigma}\Delta\hat{h}_{lv}}{\left(2\pi\overline{M}N_A k_B\right)^{1/2}}\left[\frac{\Gamma P_{sat}(T_v)}{T_v^{1/2}} - \frac{P_{sat}(T_l)}{T_l^{1/2}}\right] \tag{35}$$

With this assumption we have linked the molecular flux and the heat flux at the interface to the temperatures in the bulk phases T_v and T_l.

For vaporization and condensation processes at high temperatures, a is often small. For example, condensing or evaporating saturated pure water at 100 °C at a heat flux of 100 kW/m^2 corresponds to an a value of 1.3×10^{-4}. On the other hand, the values of a for vaporization or condensation processes at cryogenic temperatures (below 100 K) may be much higher. In the limit of small a, Γ_a and Γ_{-a} in eqns. (23) and (24) are well approximated by

$$\Gamma = 1 + a\pi^{1/2} \tag{36}$$

Substituting this relation into eqns. (34) and (35) and using eqns. (32) and (33), the following relations are obtained for the molecular flux and heat flux:

$$j_i = \frac{2\hat{\sigma}}{m(2-\hat{\sigma})}\left(\frac{\overline{M}}{2\pi N_A k_B}\right)^{1/2}\left[\frac{P_{sat}(T_v)}{T_v^{1/2}} - \frac{P_{sat}(T_l)}{T_l^{1/2}}\right] \tag{37}$$

$$J_{Q,i} = \left[\frac{2\hat{\sigma}}{(2-\hat{\sigma})}\right]\frac{\Delta\hat{h}_{lv}}{\left(2\pi\overline{M}N_A k_B\right)^{1/2}}\left[\frac{P_{sat}(T_v)}{T_v^{1/2}} - \frac{P_{sat}(T_l)}{T_l^{1/2}}\right] \tag{38}$$

An alternate form of the heat flux relation can be obtained if the relations

$$\Delta P_{sat,vl} = P_{sat}(T_v) - P_{sat}(T_l) \tag{39}$$

$$\Delta T_{vl} = T_v - T_l \tag{40}$$

are substituted into eqn. (38) to eliminate P_l and T_l.

$$J_{Q,i} = \left[\frac{2\hat{\sigma}}{(2-\hat{\sigma})}\right]\frac{\Delta\hat{h}_{lv}}{\left(2\pi\overline{M}N_A k_B\right)^{1/2}}\left[\frac{P_{sat}(T_v)}{T_v^{1/2}} - \frac{P_{sat}(T_v) - \Delta P_{sat,vl}}{\left(T_v - \Delta T_{vl}\right)^{1/2}}\right] \tag{41}$$

We next expand the second term in the square brackets in terms of $\Delta T_{vl} / T_v$ using the assumption that $\Delta P_{vl} / P_v << 1$ and $\Delta T_{vl} / T_v << 1$ so that terms of order $(\Delta P_{vl} / P_v)^2$ may be neglected. This leads to the relation

$$J_{Q,i} = \left[\frac{2\hat{\sigma}}{(2-\hat{\sigma})}\right] \frac{\Delta \hat{h}_{lv}}{\left(2\pi \overline{M} N_A k_B T_v\right)^{1/2}} \left[\frac{\Delta P_{sat,vl}}{\Delta T_{vl}} - \frac{P_{sat}(T_v)}{2T_v}\right] \Delta T_{vl} \tag{42}$$

Using the Clapeyron equation from classical thermodynamics to evaluate $\Delta P_{sat,vl} / \Delta T_{vl}$ as

$$\frac{\Delta P_{sat,vl}}{\Delta T_{vl}} \cong \left(\frac{dP}{dT}\right)_{sat} = \frac{\Delta \hat{h}_{lv}}{T_v \Delta \hat{v}_{lv}}$$

we can further simplify eqn. (42) to

$$J_{Q,i} = \left[\frac{2\hat{\sigma}}{(2-\hat{\sigma})}\right] \frac{\left(\Delta \hat{h}_{lv}\right)^2}{\left(2\pi \overline{M} N_A k_B T_v\right)^{1/2} T_v \Delta \hat{v}_{lv}} \left[1 - \frac{P_{sat}(T_v) \Delta \hat{v}_{lv}}{2\Delta \hat{h}_{lv}}\right] \Delta T_{vl} \tag{43}$$

Eqn. (43) is an explicit relation between the heat flux associated with a condensation or vaporization process and the temperature difference across the interface that is required to drive the process. Note that this result applies for small a.

The above analysis of interfacial transport assumes that the fluxes of condensing and vaporizing molecules can be derived from kinetic theory for each flux separately and the results superimposed to obtain the net heat flux. The analysis does not consider nonequilibrium interactions between the molecules leaving the interface and those approaching the interface, which may have different mean energy levels because of the difference between the vapor and liquid temperatures. Inclusion of these effects makes analysis of the system considerably more difficult (see Wilhelm [5]).

Despite its approximate nature, the results of the above analysis provide a useful framework for assessing the effects of interfacial transport on vaporization and condensation processes. It should be noted that the equation relating $J_{Q,i}$ and ΔT_{vl} derived above applies equally well to vaporization and condensation with the convention that $J_{Q,i}$ is positive for condensation and negative for vaporization. In engineering analysis of vaporization and condensation processes, it is usually assumed that the vapor and liquid temperatures are both equal to the saturation temperature at the local pressure. The above results indicate that there is some deviation from this assumption of local thermodynamic equilibrium. Fortunately for most cases of practical interest, this deviation is very small.

Inspection of the relations derived above clearly indicates that the relation between heat flux and driving temperature difference depends directly on the value of the accommodation coefficient $\hat{\sigma}$. Quoted values of $\hat{\sigma}$ in the literature vary widely. The tabulation assembled by Paul [6] lists values ranging from 0.02 to 0.04 for some liquids to values very near 1.0 for others. Mills [7] has suggested that molecular accommodation should be imperfect only when the interface is impure. On the other hand, recent molecular dynamic simulation studies by Yasuoka *et al.* [8] imply that the accommodation coefficient for a pure water interface should be near 0.4. Paul [6] noted that in almost every case where $\hat{\sigma}$ was found to be appreciably less than one, the average condition of the vapor molecules was different from that for the liquid phase because of association, disassociation or polymerization. This suggests that if a fluid is virtually pure and that association, disassociation and polymerization do not occur to any significant degree, the accommodation coefficient should be close to unity. Because extreme purity is unlikely in most engineering systems, a value of $\hat{\sigma}$ less than one is expected in common applications. Values reported in the literature should therefore be regarded as typical values of $\hat{\sigma}$ rather than constants applicable to all systems, since system contamination may alter values for a particular system.

Eqn. (43) indicates that a finite temperature difference is required to drive heat flow across an interface in the presence of vaporization or condensation. This is commonly interpreted as an additional resistance to heat flow. In many cases, this resistance is low and may justifiably be neglected in engineering analysis of such systems. However, eqn. (43) indicates that this resistance may be large at high heat flux levels and/or at low accommodation coefficient values.

4 High Knudsen number and nonequilibrium effects

In some important microscale evaporation or condensation processes, the system characteristic length scale may be comparable to the mean free path of the vapor molecules near the interface. When this is true, the heat and mass transfer in the vapor may be in the free molecular regime or in the transition regime between free molecular and continuum transport. An example is the condensation growth of water microdroplets in the atmosphere immediately following nucleation which is an important element of the microphysics of cloud formation in the atmosphere. Fog formation in air-conditioning evaporators and precipitation resulting from supersaturated conditions in industrial process flows may also fall into this category. For the initial stage of droplet growth in such circumstances, the size of the droplets is often comparable to the mean free path of the surrounding gas molecules. As the droplet grows, the near-interface heat and mass transfer associated with the process traverses the transition regime between free molecular and continuum transport. As indicated in Figure 8, the Knudsen layer region immediately adjacent to the droplet interface exhibits noncontinuum transport in these circumstances. Since neither free molecular theory nor continuum theory is accurate in the transition regime, prediction of the condensation growth of microdroplets in this size range is particularly challenging. Previous efforts to model this type of transport have typically used an analytical method based on kinetic theory, or a DSMC method.

High Knudsen number noncontinuum effects may also arise in evaporation and condensation in vapor bubbles in microporous structures and in two-phase systems in microchannels of ultra-miniature condensers and evaporators at low pressures. As in the case of transport near droplets, the best options for these circumstances are likely to be kinetic theory based models or DSMC methods.

Noncontinuum transport can also result when process conditions preclude a high enough collision rate to ensure local thermodynamic equilibrium. This can happen near a liquid–vapor interface if the region adjacent to the interface is a Knudsen layer like that shown in Figure 8. Molecules emitted from the liquid surface may not experience enough collisions to establish the usual Boltzmann distributions of kinetic and rotational energy. The net result can be significant departure from local thermodynamic equilibrium and a failure of continuum transport models that are based on local thermodynamic equilibrium. Carey and Hawk [9] demonstrated that this effect is approximately modeled by the DSMC method for evaporating microdroplets.

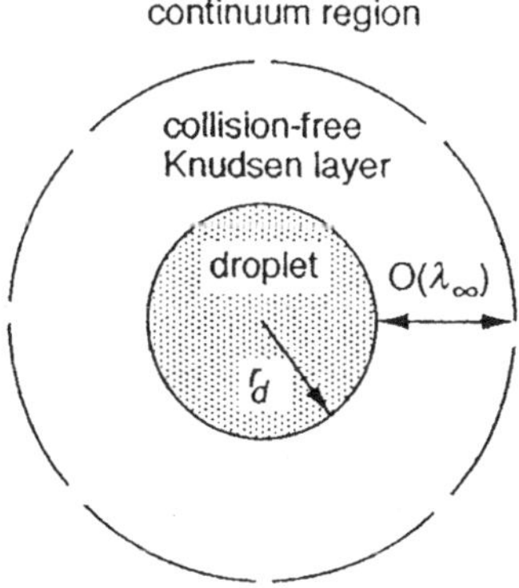

Figure 8: Transport domains near a droplet at moderate to large Knudsen number.

The primary options for modeling the high Knudsen number and nonequilibirum effects described in this section are extensions of kinetic theory and DSMC methods. Of these two, the DSMC method is arguably the more flexible and easier to implement. It can more easily accommodate complex system geometries, and with rapidly increasing computing power, the range of liquid–vapor phase change processes that can be handled by DSMC based methods is rapidly increasing.

5 Variation of interfacial tension with interface curvature

Molecular theories of capillarity, in combination with thermodynamic analysis, predict that interfacial tension will vary with radius of curvature when the radius of curvature becomes comparable to the thickness of the interfacial region (see Tolman [10] and Rowlinson and Widom [11]). From this line of analysis it can be argued that for small droplets or bubbles, the surface tension varies with radius of curvature according to a relation of the form

$$\sigma_{lv} = \sigma_{lv,\infty}\left(1 - \frac{2\delta_T}{r_i}\right) \tag{44}$$

where $\sigma_{lv,\infty}$ is the interfacial tension for a flat interface ($r_i \to \infty$) and δ_T is the so-called Tolman length scale, below which the radius of curvature affects the magnitude of the interfacial tension. This effect can be important, for example, during post nucleation growth of embryonic bubbles or droplets that are just larger than critical size.

A number of recent investigations have used molecular dynamic simulations to examine the effect of interfacial size and curvature on interfacial tension. The information thus provided is somewhat inconclusive at this point. Nijmeijer *et al.* [12] found that their simulations for a Lennard-Jones 12-6 fluid indicated that the magnitude of the Tolman Length scale was in the range $\pm 0.7\,\sigma_{LJ}$, where σ_{LJ} is the length scale in the Lennard-Jones potential

$$\phi(r) = 4\varepsilon_{LJ}\left[\left(\frac{\sigma_{LJ}}{r}\right)^{12} - \left(\frac{\sigma_{LJ}}{r}\right)^{6}\right] \tag{45}$$

Haye and Brune [13] also developed molecular dynamic simulations for a Lennard-Jones 12-6 fluid, but by using a relation for δ_T proposed by Blokhuis and Bedeaux [14], their results indicate that δ_T / σ_{LJ} = 0.20 ±0.05. The analytical model developed by Hadjiagapiou [15] using density functional theory predicts a small negative value for δ_T which itself varies with radius. In yet another recent molecular simulation study for a Lennard-Jones fluid, Chen [16] concluded that the interfacial tension also varies with interfacial area when the characteristic size of the interface becomes on the order of 10–20 molecular diameters. Below that size, the surface tension was found to increase above that for an interface of infinite size.

6 Liquid phase and interfacial region effects

Recent molecular dynamic simulation studies have provided insight into liquid phenomena that may affect near interface transport in microscale systems. A molecular dynamics model system like that shown in Figure 9 was also used by

Dang and Tsun-Mei [17] to study the liquid–vapor interface and water clusters using many-body potentials.

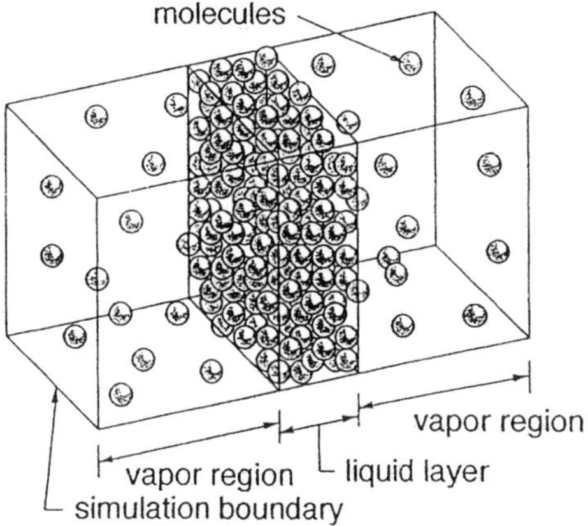

Figure 9: System model used in molecular dynamic simulation studies of liquid–vapor interfacial phenomena.

This type of model has also been used by Matsumoto *et al.* [18] to study interfacial tension in water–methanol mixtures and by Tarek *et al.* [19] to study interfacial tension in ethanol–water solutions. In the Tarek *et al.* [19] simulations, the liquid solution phase was surrounded by air. An example of one of the number density profiles obtained by Tarek *et al.* [19] for a bulk ethanol mole fraction of 0.1 is shown in Figure 10.

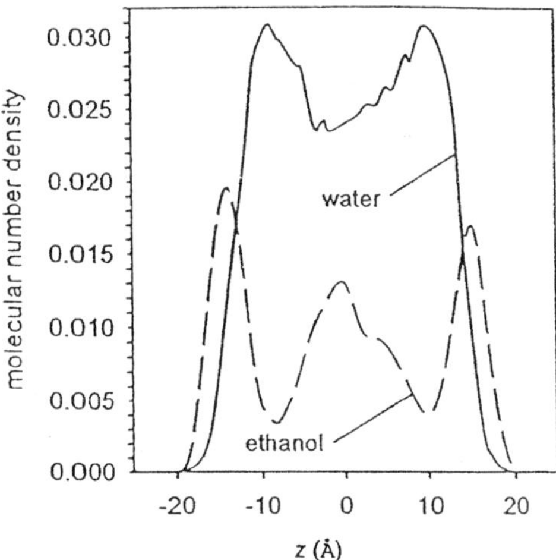

Figure 10: Number density profile obtained by Tarek *et al.* [19] for a water–ethanol mixture with a mean ethanol mole fraction of 0.1.

The number density variations of water and ethanol are seen to reflect the fact that ethanol concentrates at the interface locations. Another interesting result of the simulations of Tarek *et al.* [19] is the time variation of the interfacial tension (interfacial free energy per unit area) determined in their simulation. An approximate fit to the time variation of the interfacial free energy during one of the simulations of Tarek *et al.* [19] is shown in Figure 11. The simulation indicates that over a time scale of a few picoseconds and surface areas with length scales of a few molecular diameters, the interfacial free energy per unit area may fluctuate substantially. If transport times are extremely short, this time variation may affect the results to a degree that it must be properly taken into account in any modeling scheme.

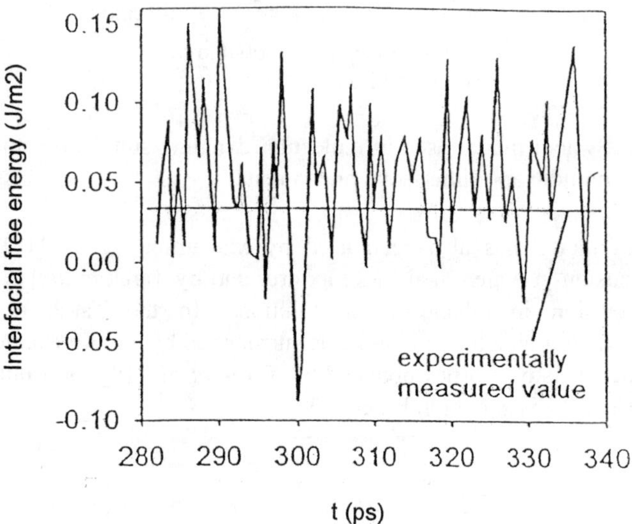

Figure 11: Variation of the interfacial free energy per unit area with time in the simulation of Tarek *et al.* [19] for a water–ethanol mixture with a mean ethanol mole fraction of 0.1.

7 DSMC modeling of combined effects during vaporization and condensation

The discussion above considered several different physical effects that may arise in microscale systems during vaporization and condensation processes. Theoretical analysis and experimental exploration of these has usually considered them as separate independent effects. In many microscale systems of interest in applications, these effects occur in tandem, and often the interaction among these mechanisms can play a central role in the transport near the interface during the phase change process. In this section an example of how these combined mechanisms affect transport will be examined in detail. This section will specifically examine DSMC based schemes for modeling combined effects. These schemes typically used the DSMC method to model the vapor phase transport near the liquid–vapor interface, with the interaction of vapor molecules with the interface treated as a boundary condition imposed on the DSMC scheme. In contrast, molecular dynamic simulation models of near-interface transport typically treat liquid and vapor molecules all in the same computational domain. A model of the molecular interaction force potential is required for molecular simulation models, whereas for DSMC schemes, probability distributions for the outcomes of molecular interactions are needed. In this respect, DSMC schemes are less fundamental, incorporating a more statistically averaged treatment. Their advantage is that they can be used to model larger mesoscale systems with moderate computational resources.

As noted above, this section examines the use of a DSMC based method to explore how the combined mechanism discussed in the previous section affects transport. Extension of the general methodology to other cases will also be discussed.

7.1 Post nucleation growth of microdroplets

Post nucleation growth of liquid microdroplets in a supersaturated surrounding vapor or gas mixture is dependent on at least three microscale mechanisms that strongly affect transport at the liquid–vapor interface. In general, the motion of the interface during growth of the droplet is sufficiently slow that a quasi-equilibrium treatment of the transport in the vapor near the interface is appropriate. Two conditions must be satisfied for a quasi-equilibrium model to accurately predict transport at any instant during the droplet growth process. First, the time scale associated with internal conduction inside the droplet must be small compared to the characteristic time associated with transport in the gas. Since the liquid thermal conductivity is large compared to that of the vapor, this condition is satisfied for the systems of interest here. This line of reasoning implies that the temperature is essentially uniform within the droplet but may vary slowly with time as the droplet grows. In addition, the quasi-equilibrium model treats the radius as fixed during the simulation, even though the droplet is actually growing during the condensation process. This is a good approximation if the characteristic time of the droplet growth process $(r_d/(dr_d/dt))$ is large compared with the characteristic time for transport in the gas. This requirement is met for many systems of practical interest. The droplet size changes significantly over milliseconds while times on the order of microseconds are required to establish steady transport in the gas.

The microphysics of condensation in the presence of a noncondensable gas is significantly more complex than condensation of the pure vapor phase of the liquid in the droplet. Condensation removal of vapor at the droplet interface and mass diffusion result in a water vapor concentration at the interface which can be significantly different from that in the far ambient. For a quasi-equilibrium process, the interface temperature must equal the saturation temperature of the liquid species at the partial pressure of the vapor in the gas at the interface. Thus, the droplet surface temperature cannot be specified a priori. Any transport analysis scheme for this type of droplet condensation process must therefore include a means of determining the interface temperature as an integral part of the analysis. Also, to handle an arbitrary mixture, treatment of molecules that store energy by rotation as well as translation must be incorporated into the model.

In the very early stages of post-nucleation growth of liquid microdroplets in a supersaturated surrounding vapor or gas mixture, three microscale factors associated with the size of the droplet affect its growth. The first of these is that the droplet size is comparable to or smaller than the mean free path of the molecules in the surrounding gas. For such circumstances, the Knudsen layer region within about one mean free path of the droplet surface is essentially a collision-free zone (see Figure 8). Immediately after nucleation, the droplet size may be so small that the extent of the Knudsen layer is large compared to the droplet and the transport is in the free-molecular regime. As the droplet grows, the size of the droplet becomes comparable to and then larger than the thickness of the Knudsen layer with the associated changes from free molecular transport to the transition regime and then to continuum transport.

The second factor is that for very small droplets, the surface tension shifts the equilibrium vapor pressure from the flat interface value. Classical thermodynamics dictates that this shift is given by eqn. (1) discussed above. Note that the vapor pressure is a key factor in predicting the molecular flux at the interface using eqns. (19), (29), (34) or (37). Eqn. (1) also implies that for a specified vapor pressure, the equilibrium saturation temperature will be shifted at small droplet radii. This shift will also directly affect the driving potential for heat transfer at the interface.

The third factor affecting droplet growth is that for extremely small droplets, the interfacial tension may be a function of the droplet radius. As indicated by eqn. (44), molecular thermodynamic analysis suggests that the interfacial tension may decrease below the flat interface value at smaller droplet radius values. Molecular dynamic simulations, such as those by Nijmeijer *et al.* [12] and Haye and Brune [13] described above, suggest that other variations of surface tension with drop size may also be possible. Because the surface tension appears as a factor in the relation for the equilibrium vapor pressure (1), its variation affects the equilibrium conditions and the net molecular flux at the surface.

There have been a number of investigations of noncontinuum effects on condensation growth of droplets. Investigations by Loyalka [20], Lang [21], Chernayak and Margilevskiy [22], Young [23, 24], Peters and Paikert [25] and Peters and Meyer [26] have explored analytical extensions of kinetic theory as a means of predicting heat and mass transfer between a condensing droplet and a surrounding gas at arbitrary Knudsen numbers. Widder and Titulaer [27] formulated an analysis of transport to a condensing droplet in a supersaturated mixture of a light noncondensable gas and a heavy vapor species. Their analysis was constructed in terms of the Navier-Stokes equations and the Klien–Kramers equation for Brownian motion of the vapor molecules. More recently, a molecular simulation model has been used by Carey *et al.* [28, 29] to model condensation growth of water microdroplets in supersaturated mixtures of water vapor and a noncondensable gas.

In some of the investigations mentioned above, the droplets were sufficiently large that the effects of interface tension on equilibrium vapor pressure were not considered. None considered the variation of interfacial tension with droplet radius at the extremely small droplet sizes immediately following nucleation. The earlier investigations by Carey *et al.* [28, 29] and a very recent study by the same group [30] provide insight into the individual and combined roles of microscale mechanisms associated with post-nucleation growth of water droplets in a supersaturated argon-water gas mixture. The conditions considered in these studies were those in a series of experiments by Peters and Paikert [25]. These conditions were chosen to facilitate comparison of the model predictions with droplet growth rates inferred from the experiments.

In all three investigations [28–30], a DSMC particle simulation method was used to model transport in the gas mixture surrounding the droplet. The particle simulation method used was a derivative of the direct simulation Monte Carlo technique pioneered by Bird [31]. Some of the refinements developed by McDonald [32] and Baganoff and McDonald [33] were also incorporated into the scheme. The details of the particle simulation method used are described in these three references. Some of the features relevant to modeling microscale effects will be summarized here. To minimize memory requirements and computational time, the simulation domain was limited to one octant around the droplet, as shown in Figure 12. In addition, to avoid the computational costs of extending the simulation out to large radial distances, in these studies the simulation domain was limited to a radial distance of four droplet radii. For free molecular conditions, the simulation boundary condition was set to the ambient conditions at $r = 4r_d$. For transition regime conditions, the simulation boundary condition was matched to the continuum transport solution beyond $r = 4r_d$.

For transition regime conditions, the temperature and water vapor concentration gradients decrease rapidly with distance from the droplet, approximately proportional to r^2. Consequently, while the temperature may drop and the water concentration may rise significantly over a few mean free paths near the sphere, beyond $r = 4r_d$ it will take many free paths for the temperature to change significantly. The local Knudsen number $\lambda(dT/dr)/(T_d - T_\infty)$ decreases rapidly with distance from the sphere, causing the temperature and concentration profiles to approach those for continuum transport at large r. This line of reasoning, which divides the region near the sphere into a noncontinuum region close to the body and a continuum transport region farther away, is the basis for the matching procedure used at $r = 4r_d$.

The octant simulation region shown in Figure 12 was bounded by the surface of the sphere at $r = r_d$, the outer boundary at $r = 4r_d$ and specularly reflecting planar surfaces at $x = 0$, $y = 0$ and $z = 0$. The specular surfaces which lie along planes of symmetry have the effect of giving the octant the same statistical behavior as a complete sphere surrounded by particles. Each particle in the simulation represents N_{mpp} molecules. The region bounded by these surfaces was filled with cubic cells which were used to collect collision candidate pairs and define regions over which particle properties were averaged to obtain macroscopic thermodynamic characteristics of the gas adjacent to the droplet.

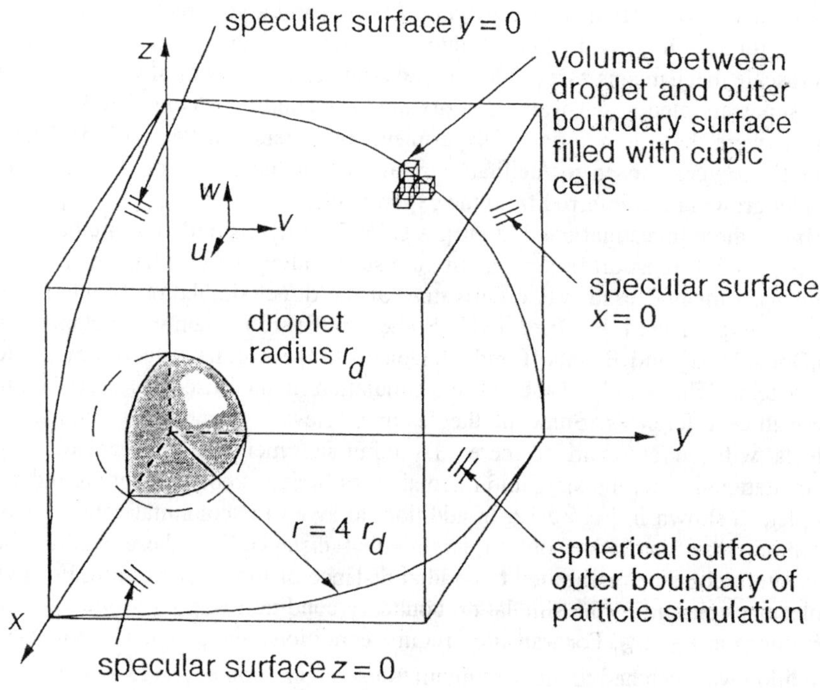

Figure 12: Computational domain for simulation used by Carey *et al.* [28–30].

Particles that move beyond the outer boundary when the particles are advanced along their trajectory were simply removed from the simulation. During each time step, particles of each species are added to the simulation just inside the $r = 4r_d$ boundary. The rate of water particle addition $\dot{n}_w$ and the rate of nitrogen particle addition $\dot{n}_n$ are computed as

$$\dot{n}_w = N_{mpp}(\pi/2)(4r_d)^2 j_w \tag{46}$$

$$\dot{n}_n = N_{mpp}(\pi/2)(4r_d)^2 j_n \tag{47}$$

where j_w and j_n are the fluxes of water and nitrogen molecules across a surface predicted by kinetic theory for a mixture at a pressure P_∞, temperature T_4 and water mole fraction of $X_{w,4}$:

$$j_w = \left(\frac{k_B T_4}{2\pi m_w}\right)^{1/2} \left(\frac{X_{w,4} P_\infty}{k_B T_4}\right) \tag{48}$$

$$j_n = \left(\frac{k_B T_4}{2\pi m_n}\right)^{1/2} \left(\frac{(1 - X_{w,4}) P_\infty}{k_B T_4}\right) \tag{49}$$

This treatment allows the simulation domain to establish whatever mean particle density is necessary to balance the inward and outward fluxes at the outer boundary.

Based on the observation that the liquid thermal conductivity is large compared to that of the vapor, conduction inside the droplet was idealized as being very fast. The droplet Biot number, which essentially equals the ratio of the gas conductivity to the liquid conductivity, is approximately 0.27. For the conditions of interest in this study, temperature variation inside the droplet was estimated to be less than 0.6 K. The temperature was therefore taken to be uniform within the droplet and transient conduction effects in the liquid droplet were not considered in the energy balance at the droplet surface.

As noted in an earlier section, when molecules from a surrounding gas enter the interfacial region near a solid or liquid phase, experimental evidence and molecular simulation studies indicate that only a fraction of the incident molecules thermally interact with the molecules of the condensed phase. In the DSMC models of Carey *et al.* [28–30], this behavior is stochastically represented by specifying an accommodation coefficient σ_t which is taken to equal the fraction of incident molecules that thermally interact with the droplet surface.

If a particle strikes the droplet surface during a given time step, a random number $\Re$ uniformly distributed on [0,1] is compared to the specified accommodation coefficient. If $\Re > \sigma_t$, the particle is diffusely scattered from the surface. Its speed and rotational energy are left unchanged, but its direction is randomly set by sampling a distribution that, on the average, isotropically scatters particles from the surface. If $\Re \le \sigma_t$, the particle interacts thermally with the surface. Each thermally interacting particle was treated as if it merges into the liquid droplet, delivering its energy, including an energy of vaporization, to the droplet. Computationally, particles that merge into the droplet are removed from the simulation. The number of particles absorbed in this manner is counted, and the energy delivered to the droplet by these particles is summed.

In a subsequent portion of the time step calculations, particles are added to the simulation domain just beyond $r = r_d$. Before executing the emission of particles from the droplet surface, the temperature of the droplet surface is computed. Kinetic theory predicts that for an interface at equilibrium, the flux of emitted water vapor molecules is given by

$$j_w = \sigma_t \frac{P_{ve}(T_d)}{\left(2\pi m_w k_B T_d\right)^{1/2}} \tag{50}$$

For the spherical microdroplets considered here, the equilibrium vapor pressure P_{ve} in the above relation is a function of the droplet temperature T_d and the droplet radius r_d as given by eqn. (1).

In the limit of long times, the steady-state energy exchange at the interface must satisfy conservation of energy

$$\sum_{i=1}^{N_{steps}} (E_{gain})_i = 3k_B T_d N_{mpp} \sum_{i=1}^{N_{steps}} (N_{non})_i + \frac{\pi}{2} r_d^2 j_w \Delta t \left(\hat{u}_{lv} + \frac{1}{2} k_B T_d \right) \tag{51}$$

In the above equation, $(N_{non})_i$ is the number of noncondensable molecules which thermally interact with the surface and then are re-emitted into the gas and $(E_{gain})_i$ is the energy delivered by molecules colliding with the droplet surface in time step i. The left side of the above equation is the total energy delivered by molecules colliding with the surface through N_{steps} time steps of the simulation. The right side is the energy carried away by molecules leaving the droplet surface. The first term on the right represents energy carried away by noncondensable gas molecules re-emitted to the gas, and the second represents energy removed by water molecules. Combining eqns. (1) and (50), using the fact that $P_{ve} - P_{sat}(T_d) << 2\sigma_{lv} / r_{de}$ and substituting into eqn. (51), yields

$$\sum_{i=1}^{N_{steps}} (E_{gain})_i = 3k_B T_d N_{mpp} \sum_{i=1}^{N_{steps}} (N_{non})_i$$

$$+ \frac{\pi r_d^2 \sigma_t P_{sat}(T_d)}{2\left(2\pi m_w k_B T_d\right)^{1/2}} \Delta t \left(\hat{u}_{lv} + \frac{1}{2} k_B T_d \right) \exp\left(\frac{2\sigma_{lv}}{\rho_l r_d R T_d} \right) \tag{52}$$

Use of eqns. (1), (50) and (51) in the model requires a means of evaluating the interfacial tension that accounts for possible dependency on droplet radius. For water, Pruppacher and Klett [34] derived a relation for surface tension dependency on droplet radius that can be cast in the form

$$\sigma_{lv} = \sigma_{lv,\infty} \left(1 + \frac{2\delta_T}{r_d} \right)^{-1} \tag{53}$$

where the recommended value of δ_T is 0.157 nm. This relation was used in this investigation, with the flat-interface interfacial tension $\sigma_{lv,\infty}$ determined from

$$\sigma_{lv} = 0.07583 - 0.0001477(T_d - 273.2) \tag{54}$$

where T_d is in K and σ_{lv} is in N/m. This relation is an established surface tension relation for a water liquid–vapor interface at equilibrium (see Carey [35]) which is extrapolated here to nonequilibrium conditions.

It should be noted that combining eqns. (1) and (53), using the fact that $P_{ve} - P_{sat}(T_d) << 2\sigma_{lv} / r_{de}$ and solving for the equilibrium radius yields

$$r_d = \frac{2\sigma_{lv,\infty} / \rho_l R T_d}{\ln\left[P_{ve} / P_{sat}(T_d)\right]} - 2\delta_T \tag{55}$$

This relation indicates that the equilibrium radius of a droplet, which is the critical radius for a nucleating embryo, is decreased by $2\,\delta_T$ due to the variation of surface tension with radius. As discussed in the next section, this has an important effect on the early growth after nucleation. For the smallest droplets considered in this study, the interfacial tension is more properly interpreted as the mean interfacial free energy per unit area of the droplet. Its use in this model analysis is consistent with this interpretation.

In eqn. (52) P_{sat} and $\hat{u}_{lv}$ are functions of temperature. For the purposes of the simulation calculations P_{sat} was computed using the following relation which is a curve-fit to data for water between 10 °C and 70 °C based on the Clapeyron equation.

$$P_{sat}(T_d) = e^{(A - B / T_d)} \tag{56}$$

In eqn. (56), $A = 18.71$ and $B = 5240$ for T_d in K and P_{sat} in kPa. Values of $\hat{u}_{lv}$ were determined using the following linear fit to data for water between 10 °C and 70 °C:

$$\hat{u}_{lv} = 7.016 \times 10^{-20}\left[1 - (T_d - 283.2)/825.4\right] \tag{57}$$

where $\hat{u}_{lv}$ is in J/molecule and T_d is in K. In the simulation, the summations of incident energy and number of incident noncondensable molecules were updated at each time step. The droplet temperature was computed at each time step by iteratively solving eqns. (52), (53), (54), (56) and (57) simultaneously. A guess of the surface temperature was provided to initiate the simulation. As the number of time steps increases, the surface temperature computed in this manner converges to a value which is interpreted as being the actual physical temperature of the droplet for the quasi-steady process being modeled in the simulation.

Once the surface temperature of the droplet was determined for a given time step, particles were computationally emitted from the droplet surface. First, N_{non} noncondensable particles were emitted, so that there is no net transfer of the noncondensable species to the droplet during the time step. Then, particles of the condensable species were emitted from the droplet surface until an energy balance was achieved.

Each time a particle was emitted from the droplet surface, the energy carried away from the droplet by the particle was added to a cumulative total for the time step. When this total just exceeded the amount delivered to the sphere during the time step, the particle addition process was stopped. The amount by which the energy for the added particles exceeded that for the incoming particles was subtracted from the energy of the incoming particles in the next time step. This imposed a zero net energy flux condition on the droplet interface.

For both condensable and noncondensable particles, the location at which a particle is added to the simulation was determined by randomly sampling a distribution function which produces a uniform flux of particles over the spherical surface of the droplet. The velocity components and rotational energy were randomly sampled from appropriate Boltzmann distributions for molecules crossing a fixed surface in a gas at the specified sphere temperature T_d. (See Carey *et al.* [28] for more details.)

The energy of saturated water vapor molecules, on average, exceeds that of saturated liquid molecules by $\hat{u}_{lv}$. At temperature T_d, the mean energy of a water vapor molecule is $3k_B T_d$. The energy of an arbitrary water molecule in the gas with translational energy ε_{tr} and rotational energy ε_{rot} thus exceeds the mean energy of the liquid water molecules in the droplet by $\varepsilon_{tr} + \varepsilon_{rot} - 3k_B T_d + \hat{u}_{lv}$. In the energy balance calculations at the droplet interface, the mean energy of the saturated liquid molecules was taken as the level of zero energy for water molecules. The energy delivered to the droplet by one interacting water particle from the surrounding gas is therefore given by

$$\Delta E_{w,in} = N_{mpp}\left(\varepsilon_{tr} + \varepsilon_{rot} - 3k_B T_d + \hat{u}_{lv}\right) \tag{58}$$

where N_{mpp} is the number of molecules per particle in the simulation. This relation was used to compute the energy delivered to the droplet by a water particle striking and interacting with the droplet. The energy delivered by one nitrogen particle is given by $N_{mpp}\left(\varepsilon_{tr} + \varepsilon_{rot}\right)$. Based on similar reasoning, the following relation was used to compute the energy removed from the droplet by an emitted water particle

$$\Delta E_{w,em} = N_{mpp}\left(\varepsilon_{tr} + \varepsilon_{rot} - 3k_B T_d + \hat{u}_{lv}\right) \tag{59}$$

where ε_{tr} and ε_{rot} were obtained by sampling the appropriate Boltzmann distributions at T_d.

In the simulation calculations in these studies, the simulation domain shown in Figure 12 was filled with cubic cells which formed a uniform Cartesian mesh. The radius of the droplet in the simulations ranged from 0.24 nm to 1000 nm. The effects of cell and time step sizes on the results of the simulation were explored and consequently the cell size was chosen to be on the order of one mean free path. The mean free path for the system considered here is on the order of 90 nm. For the droplet sizes considered, this corresponds to a range of Knudsen number Kn_D from 0.044 to 186. The time step was chosen so a particle with average speed traveled approximately half a mean free path in one time step. Generally, the results were found to be insensitive to the choice of these parameters as long as they were close to the values indicated by these guidelines. The side of the cell varied from 0.24 nm for the smallest droplets to 0.05 μm for the largest droplets considered in this study. The number of cells used was a compromise between accuracy and computational effort. The ratio of the number of molecules per particle N_{mpp} was chosen in each simulation to compromise between computational accuracy and storage limitations on the computer. The values used in the calculations reported here ranged from 6.5×10^{-6} to 1784 molecules per particle. Simulations were typically run for 16,000 to 800,000 time steps, with a time step typically being 20–200 picoseconds. For the smallest droplets, time steps were on the order of 1 picosecond. Altering the time step by ±50% had a negligible effect (< 0.1%) on transport parameters predicted by the simulation.

After running the simulation for a sufficiently long time, the simulation predicts the equilibrium temperature and concentration fields adjacent to the surface of the droplet for arbitrarily specified values of the ambient pressure and the temperature and water vapor concentration at $r = 4r_d$. For transition regime conditions, the matching procedure mentioned above was then be used to determine the corresponding values of the temperature and water vapor concentration in the far ambient. Note that for free molecular transport ($Kn_D >$ 2), calculations to determine the water vapor concentration and temperature in the far field were unnecessary since the ambient conditions are those specified at the edge of the simulation ($r = 4r_d$). The net heat flux from the droplet surface (carrying away the latent heat of the net condensation effect) and the net mass flux of water vapor at the droplet surface m''_d were computed from the simulation results using the following relations

$$q''_d = \frac{\Delta N_{net} \hat{u}_{lv} N_{mpp}}{(4\pi r_d^2 / 8) N_{steps} \Delta t} \tag{60}$$

$$m_d'' = \frac{\Delta N_{net} N_{mpp}}{(4\pi r_d^2 / 8) N_{steps} \Delta t} \tag{61}$$

where N_{steps} is the number of time steps in the simulation and ΔN_{net} is the net number of water particles transferred to the droplet due to particle interactions during the simulation to that time step. The growth rate of the droplet was then computed as

$$\frac{dr_d}{dt} = \frac{m_d''}{\rho_l} \tag{62}$$

where m_d'' is computed using eqn. (61).

The DSMC model analysis described above was used in all three of the studies by Carey and co-workers [28–30]. The earliest of these studies [28] focused on larger droplets and assessed the accuracy of the simulation method by comparing its predictions with previously obtained data for droplet condensation in supersaturated argon–water gas mixtures. For an accommodation coefficient value of 1.0, the simulation predictions were found to agree well with droplet growth rates inferred from measurements in a supersaturated mixture of argon and water vapor for droplet radii greater than 0.4 μm. In that study no efforts were made to explore the transport or droplet growth rate trends at smaller radius values which characterize droplets immediately after nucleation. The subsequent study [29] considered smaller droplet radii where surface tension alteration of equilibrium conditions was important, but did not consider the possible variation of surface tension with droplet radius. The most recent study [30] considered droplet radius values at or below the critical radius and included the model described above to account for the variation of surface tension with radius of curvature.

Note that the most recent study by Carey [30] incorporates several microscale effects. The DSMC model directly handles noncontinuum (transition or free molecular) transport in the gas phase adjacent to the interface. The boundary condition treatment at the interface uses the molecular flux relation, with a specified accommodation coefficient value in the energy and mass balances used to iteratively determine the droplet temperature in the quasi-equilibrium model. Surface tension effects on the equilibrium conditions at the interface are included in the energy balance as well. In addition, eqn. (53) with a recommended value of 0.157 nm for δ_T was used to predict the variation of surface tension with droplet radius at very small droplet sizes.

Carey *et al.* [28] found that simulation calculations for $\sigma_t = 1.0$ best fit the droplet growth rate data from the experiment of Peters and Paikert [25]. Two values of the accommodation coefficient, 1.0 and 0.6, were therefore used in the simulation calculations in subsequent studies. The simulation calculations in these studies were performed for one of the argon–water vapor mixtures tested experimentally by Peters and Paikert [25]. These calculations span the range of droplet radius from values near the critical radius r_C where nucleation occurs, to 1.0 μm where transition conditions merge into the continuum transport regime. Figure 13 shows the variation of the droplet temperature with time during the simulation calculation for a droplet radius of 0.4 μm, based on samples at 1000 step intervals. It can be seen that the iteration of the droplet temperature achieved by solving eqn. (52) at successive time steps does converge quickly to a value that remains steady with time. This steady value is interpreted as the droplet temperature that corresponds to the droplet size in the simulation during the growth process.

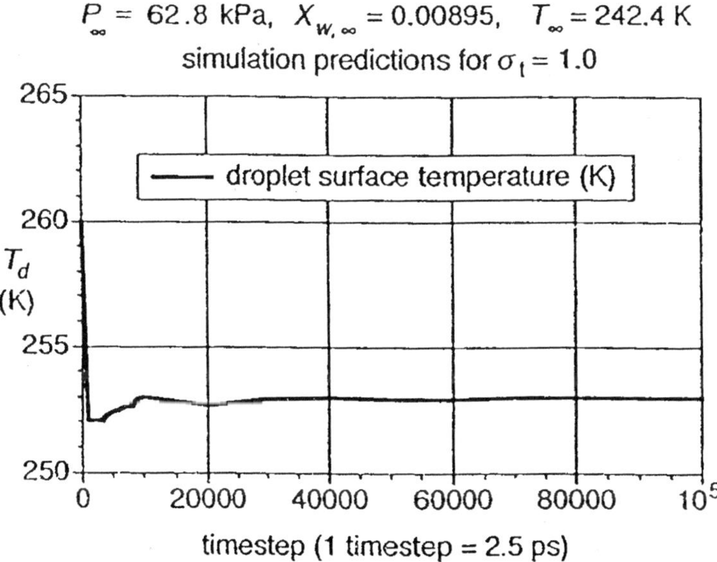

Figure 13: Variation of droplet temperature at 1000 step intervals during the simulation of Carey [29] for a water droplet with a radius of 40 nm in an argon–water gas mixture.

These DMSC simulations provided detailed information about the temperature and concentration fields near the embryo droplets. An example of the predicted fields for a droplet with a radius of 20 nm is shown in Figure 14. For this case the Knudsen number is high, corresponding to free molecular transport, and the temperature field is flat, with noticeable temperature slip near the interface.

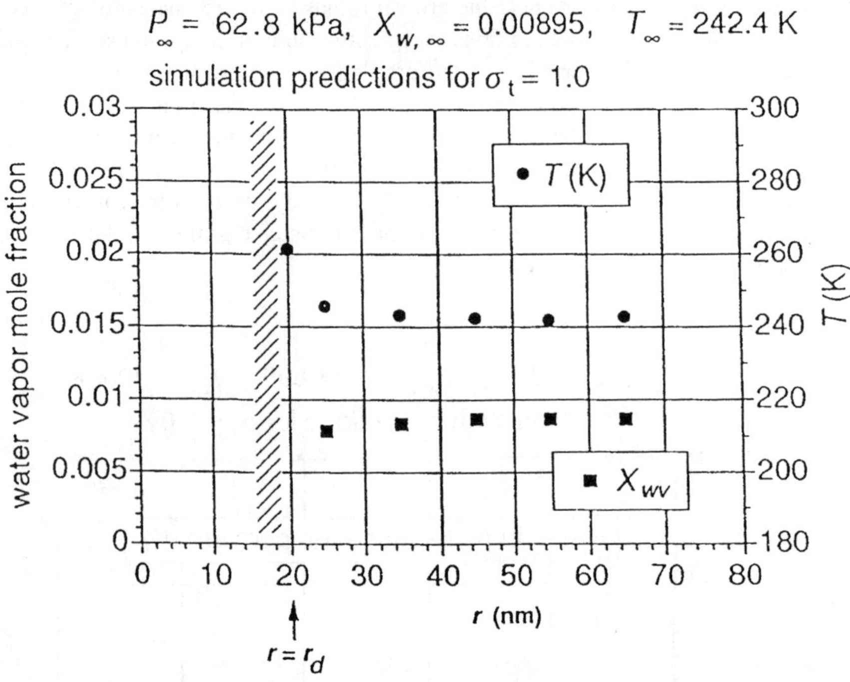

Figure 14: Temperature and concentration fields predicted by the DSMC model of Carey [29] for a water droplet with a radius of 20 nm.

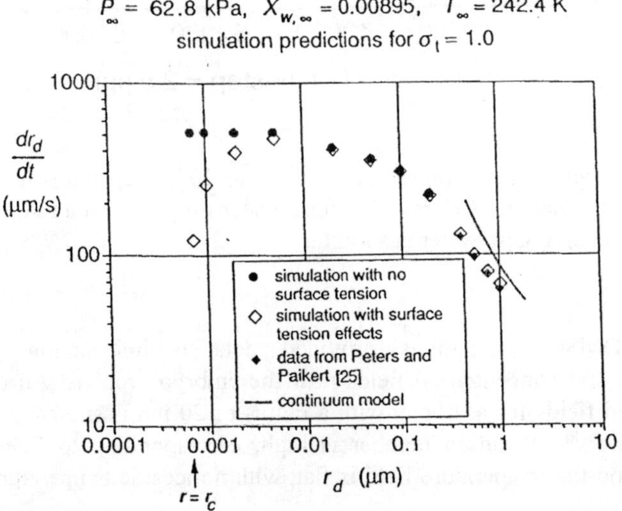

Figure 15: Comparison of droplet growth rates for conditions in the experiments of Peters and Paikert [25].

The simulation prediction of the variation of droplet growth rate with droplet size for one of the argon–water vapor mixtures studied by Peters and Paikert [25] is plotted in Figure 15. Results are shown for droplet radii ranging from just above the critical radius (for homogeneous nucleation, see Carey [35]) to droplets with a radius of 1 μm. Simulation results are shown with the effect of surface tension included and with it artificially removed (σ_{lv} set to zero). For both sets of predictions in this figure, the Tolman length was set to zero so no effect of radius on interfacial tension is included. It can be seen that these predictions of the growth rate variation with radius are the same except at very low radii. The surface tension effect on equilibrium vapor pressure results in a strong reduction in the droplet growth rate as the critical radius is approached. Also shown in Figure 15 is the prediction of standard continuum theory using Fourier's law and Fick's law to predict the simultaneous heat and mass transfer associated with quasi-steady condensation at the indicated conditions. The governing equations and boundary conditions and the method used to solve them are described by Carey [29] and are not repeated here.

The simulation predictions of droplet growth rate are seen to agree well with the experimentally determined growth rates and to approach the continuum predictions at larger droplet sizes. Of particular interest is that the simulation model predicts that the droplet growth rate achieves a maximum value in the transition regime which is attributable to the interaction of surface tension effects and transport variation with droplet size. The extremum in the growth rate is a consequence of the variations of the heat transfer coefficient and equilibrium vapor pressure with droplet size. For the small droplet sizes characteristic of conditions immediately after onset of nucleation, the heat transfer coefficient is large, but the driving temperature difference is small because of the surface-tension-induced vapor pressure shift at small droplet sizes. As the droplet grows in size, the vapor pressure shift diminishes and the driving temperature difference increases faster than the heat transfer coefficient decreases. This initially produces an increase in the growth rate as the droplet grows. Eventually the reduction in the heat transfer coefficient dominates, causing the droplet growth rate to peak and then decline with increasing droplet size. It is noteworthy that these results are consistent with the predictions of the kinetic theory model of Peters and Meyer [26] which also predicts a maximum in the droplet growth rate with increasing size for water droplets growing in supersaturated pure steam.

Simulation predictions of the droplet growth rate for accommodation coefficients of 1.0 and 0.6 with the Tolman length set to 0.15 nm are shown in Figure 16. Also shown for comparison are the predictions for $\delta_T = 0$ and $\sigma_t = 1.0$. It can be seen that for $\sigma_t = 1.0$ and $\sigma_t = 0.6$, the full model predicts growth rates that vary from zero at the critical radius to values that approach the continuum behavior at large droplet radius. Both variations pass through maximum growth rates at radii between 5 and 10 nm.

The results for $\sigma_t = 1.0$ indicate that, as expected, the effect of surface tension variation with radius is limited to droplet sizes below 5 nm. The effect of this variation is to increase the growth rate for a given droplet size. Note that the critical embryo size (equilibrium size) is smaller with the radius effect included.

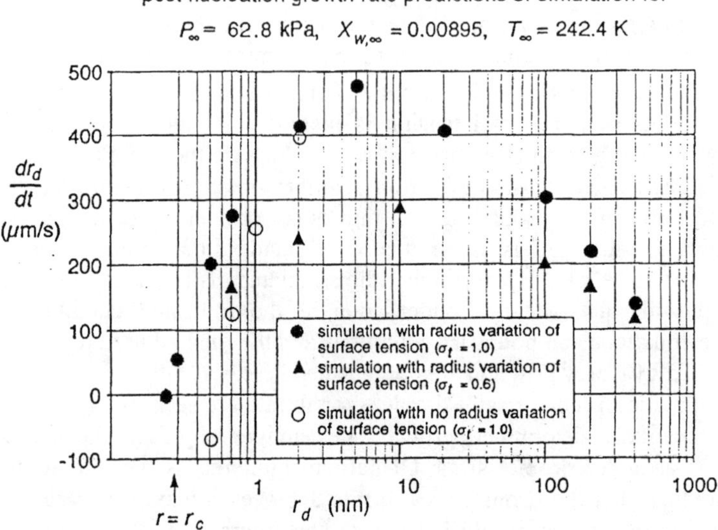

Figure 16: Comparison of droplet growth rates predicted by the DSMC models of Carey [30].

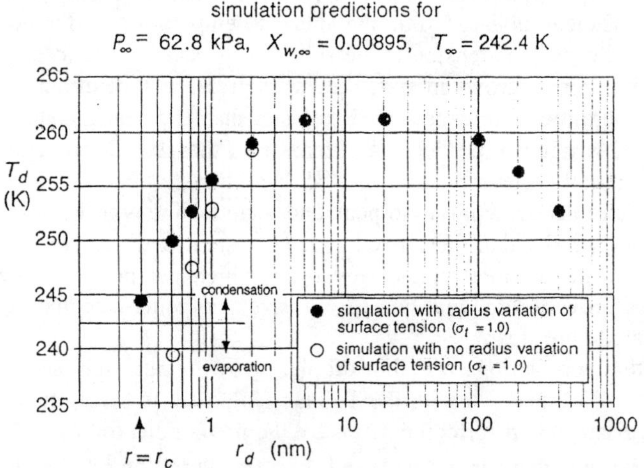

Figure 17: Comparison of droplet temperatures predicted by the DSMC models of Carey [30].

Simulation predictions of the droplet temperature at different droplet radii for σ_t = 1.0 are shown in Figure 17. Predictions are shown for simulations with the effect of surface tension variation included and with it artificially removed (δ_T set to zero). Surface tension effects are seen to strongly affect the droplet temperature at the smallest radius values immediately following nucleation. A peak in the droplet temperature occurs in the transition regime.

The model for this specific system demonstrates how the several microscale mechanisms that affect this system interact to produce the variation in droplet growth rate indicated in Figure 16. As seen in Figure 15, the continuum model predicts a progressively increasing droplet growth rate as the droplet radius decreases. When noncontinuum (high Kn_D) effects are included, the growth rate asymptotically approaches a constant as the radius approaches zero as the transport across the Knudsen layer near the interface approaches the ballistic limit.

At extremely small droplet radius values, the interfacial tension alters the thermodynamic equilibrium conditions in a manner that reduces the droplet temperature established during the quasi-equilibrium growth process. This reduces the driving potential for heat transfer from the droplet, thereby reducing the growth rate. As the critical radius is approached, the droplet growth rate tends towards zero. If a droplet radius less than the critical radius is specified for the simulation, the predicted growth rate is negative. These predictions are consistent with the thermodynamic prediction that droplets of less than critical size will spontaneously shrink whereas those of greater than critical size will spontaneously grow. Inclusion of the model for the dependence of surface tension on droplet radius slightly shifts the interface equilibrium condition at radii near the critical radius. Consistent with eqn. (55) and the results shown in Figures 16 and 17, the variation of the interfacial tension with radius in this model shifts the critical radius for the onset of nucleation to a slightly lower radius. Accurate prediction of the near-interface transport during this type of droplet growth clearly requires a model that incorporates treatments of the multiple microscale effects involved, and correctly treats interactions among those mechanisms.

7.2 Other processes involving multiple microscale effects

In addition to the post-nucleation droplet condensation process, DSMC based models have recently been developed for several other vaporization and condensation processes that involve multiple microscale effects.

Carey and Hawks [9] used a DSMC simulation to model transport near an evaporating microdroplet of pure liquid nitrogen in superheated nitrogen gas. In this study, the droplets were sufficiently large that surface tension effects on the equilibrium vapor pressure were negligible and the interface temperature was taken to be the saturation temperature for the ambient pressure. The model was used to predict the droplet evaporation rate for superheated nitrogen vapor at a pressure of about 0.2 atm for droplet diameters from 0.2 μm to 1.5 μm at ambient temperatures ranging from 120 K to 400 K. This range of conditions corresponds to Knudsen number and Jakob number ranges of $0.09 < Kn_D < 1.2$

and $0.2 < Ja < 1.7$. Based on the observations regarding accommodation coefficients discussed above, Carey and Hawks [9] selected $\sigma_t = 0.9$ for use in the simulation calculations, with the implicit assumptions that the liquid nitrogen is virtually pure and that association, disassociation and polymerization do not occur to any significant degree.

The Carey and Hawks [9] model accounted for three important microscale effects. First, the DSMC method directly accounts for the noncontinuum (low Kn_D) transport near the droplet interface. Second, the microscale molecular fluxes were used to determine the net evaporation rate at the droplet interface. The third microscale feature is the departure from thermodynamic equilibrium in the vapor close to the droplet interface. This is a consequence of the Knudsen layer near the droplet surface and the fact that a nitrogen molecule requires a number of collisions before the rotational energy storage approaches thermodynamic equilibrium for the local conditions. A large number of molecules are emitted from the liquid having translational and rotational energies consistent with a Boltzmann distributions at the droplet temperature. Because few collisions occur in the Knudsen layer near the droplet, molecules in this region, on the average, have radial velocity components and rotational energies that are higher than the values associated with Boltzmann distributions at the local temperature. Note that this third microscale feature is actually a consequence of the interaction of the first two.

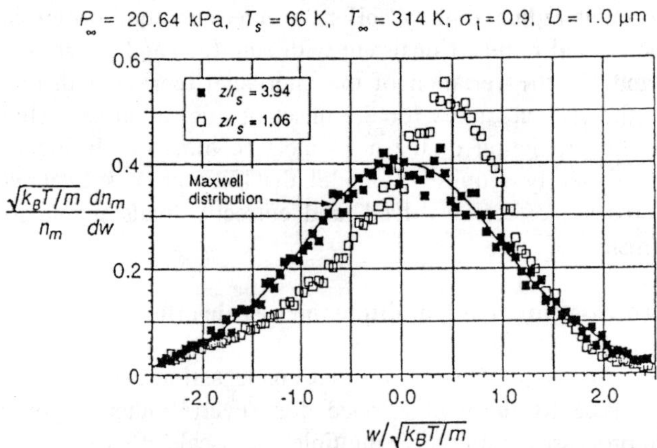

Figure 18: Distribution of the w component of velocity obtained from the simulation of Carey and Hawks [9] at cells along the z axis.

Figure 18 shows the w velocity distribution at two cell locations along the z axis predicted by the Carey and Hawks [9] model for conditions corresponding to $Ja = 1.22$ and $Kn_D = 0.33$. One location is adjacent to the droplet surface ($r = r_d$) and the other is near the outer edge of the simulation at $r = 4r_d$. Note that along the z axis, w is the velocity component in the radial direction. The points in Figure 18 are based on a bin size Δw equal to $0.05\sqrt{k_B T / m}$, where T is the local temperature in the cell. The w velocity exhibits a shift to positive values near the droplet surface as a consequence of the net radial velocity associated with evaporation at the surface of the droplet.

At the outer edge of the simulation, the mean radial velocity has dropped virtually to zero and the w distribution agrees well with the well-known Boltzmann distribution. At lower *Ja* values, the surface-normal velocity component is shifted to a lesser degree. At both locations indicated in Figure 18 the u and v velocity distributions agreed well with the Maxwell distribution for the local temperature.

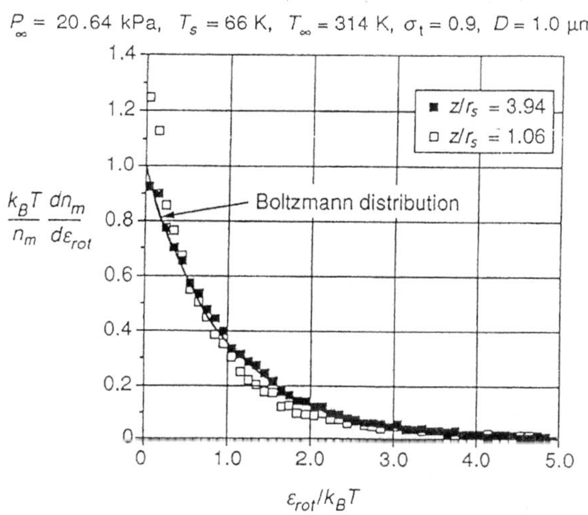

Figure 19: Distribution of the rotational energy obtained from the simulation of Carey and Hawks [9] at cells along the z axis.

The rotational energy distributions for the simulation at the same two cell locations are shown in Figure 19. In gathering these statistics, the bin size $\Delta\varepsilon_{rot}$ was taken to be $0.1 k_B T$. Far from the droplet, the rotational energy distribution agrees well with the Boltzmann distribution for the local temperature. Near the droplet, the distribution deviates from the corresponding Boltzmann distribution by as much as 20%. It appears that this deviation is a result of the fact that only a fraction of the collisions are inelastic (involving an exchange of rotational energy). Since the mean free path is on the order of the cell size in these simulations, molecules emitted from the droplet surface may travel many cells away from the surface before they undergo enough collisions to equilibrate the rotational energy to the local conditions. Thus, the system departs significantly from local thermodynamic equilibrium in the region close to the droplet interface.

The simulation computations of Carey and Hawks [9] indicate that noncontinuum effects and interface vapor generation at the droplet surface are important mechanisms in the transport which controls the evaporation rate. Both effects tend to reduce the rate of heat transfer to the droplet compared to what would occur if continuum transport occurred with no interface vapor generation. Increased generation of vapor at the interface also tends to reduce the temperature slip at the droplet surface. The interaction of the noncontinuum effect and the interface vapor generation resulted in the departure from local thermodynamic equilibrium observed in vapor adjacent to the interface.

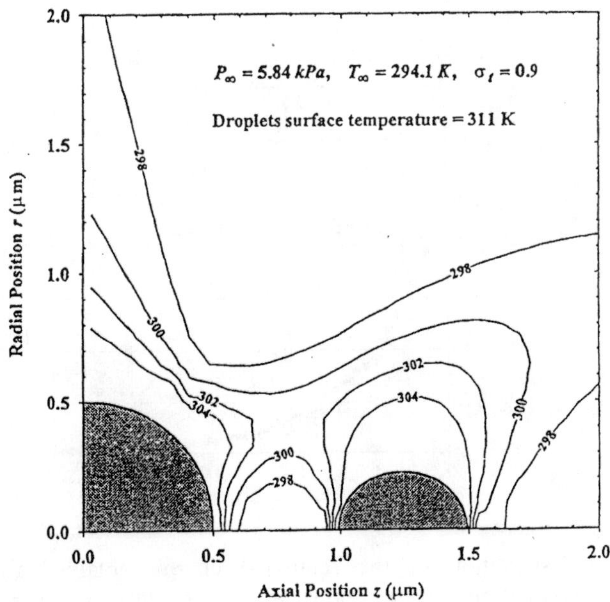

Figure 20: Temperature field predicted by the simulation of Carey and Oyumi [36] for condensation growth of three droplets in supersaturated steam.

Carey and Oyumi [36] used a similar DSMC model to explore interactions among multiple microdroplets undergoing condensation growth in supersaturated steam. Simulation predictions indicate that the heat flux from the droplet surface associated with the condensation process is strongly altered when two droplets are in close proximity. Figure 20 illustrates the temperature field near the droplets predicted by the simulation when two smaller droplets are close to a third larger droplet for one of the circumstances considered in this study. (The temperature field is symmetric about both axes, so only a quarter is shown in Figure 20.) The departure from the spherical symmetry for a single droplet is very distinct. This model includes the microscale elements present in the single droplet case, plus the interaction of the transport fields when the separation distance between droplets is comparable to the mean free path in the vapor.

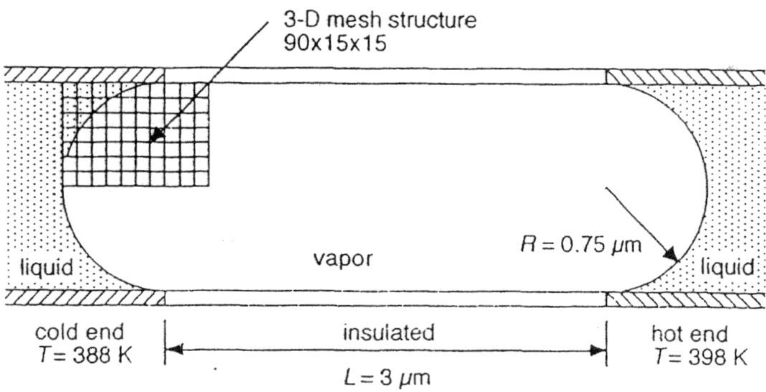

Figure 21: Schematic of DSMC computational domain for the microbubble heat pipe modeled by Jiang and Carey [37].

A final noteworthy example of DSMC based models with multiple microscale effects is the recent work of Jiang and Carey [37]. This investigation used a DSMC simulation to model transport within the water microbubble heat pipe shown schematically in Figure 21. The microbubble heat pipe has two spherical cap ends, one held at a hotter temperature and the other at a cooler temperature. The cylindrical center section is adiabatic, with a radius of 750 nm. The dimensions of the microbubble are comparable to the mean free path of the vapor molecules and the molecular exchange at the interfaces is modeled by kinetic theory relations and the relations for interfacial tension effects on equilibrium vapor pressure. Figure 22 shows the predicted variations of the temperature and axial velocity along the centerline of the heat pipe. The flow within the heat pipe is compressible and the temperature and velocity of the vapor varies substantially with the bubble. For this system, the interaction of the microscale mechanisms gives rise to a fairly complex transport process. The simulation predicts an overall effective thermal conductivity of 25.3 W/mK for this system.

The DSMC methodology used in the studies described above can be extended to other microscale systems. One possible example is extension of this type of model to vaporization or condensation in micropassages of microevaporators or microcondensers. Micro heat exchangers of this type have been considered for microscale thermal control systems and as components in microrefrigeration devices. These micro heat exchangers may contain round or rectangular passages like those indicated schematically in Figure 23. This figure indicates the expected morphology of the system for an annular flow vaporization or condensation process. At low pressures in very small passages, the mean free path in the vapor may be comparable to the passage dimensions.

The DSMC method could be used to account for noncontinuum transport effects in the vapor phase. If the passage dimensions are small, the effects of the interface radius of curvature and surface tension can be accounted for by the analysis models described in previous sections of this chapter.

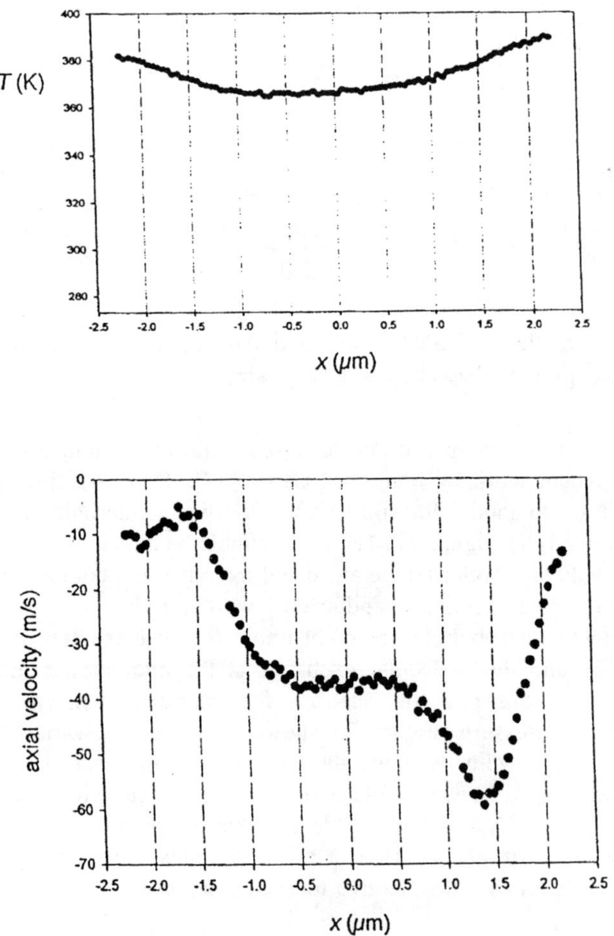

Figure 22: Centerline velocity and temperature profiles predicted by the DSMC microbubble heat pipe model developed by Jiang and Carey [37].

As indicated in Figure 23, in rectangular passages the interface may be curved in corner regions and a thin film with a flat interface may exist on the walls between the corners. If the film is very thin, the analysis described earlier for disjoining pressure effects can be used to account for their effects on interface equilibrium temperature, vapor pressure and molecular flux. For these circumstances the resulting model may be a DSMC based simulation with noncontinuum high Knudsen number transport in the vapor, curvature and surface tension effects in the passage corners and thin film disjoining pressure effects on flat portions of passage limits. This would be another potential application of a DSMC based model with multiple micoscale effects.

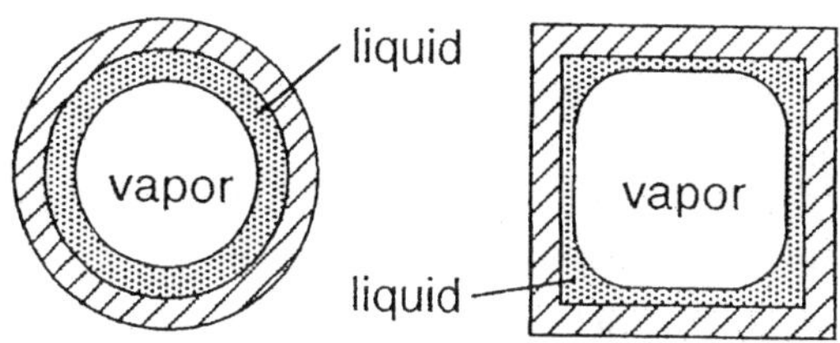

Figure 23: Schematic of annular flow evaporation or condensation in micropassages.

8 Concluding remarks

Effective modeling of liquid–vapor phase change processes in microscale systems requires a methodology that can account for the individual effects of multiple microscale mechanisms as well as their interaction. As noted in the earlier sections of this paper, this can pose a formidable challenge for many systems of practical interest. The most useful tools for analyzing such systems include molecular extensions of kinetic theory, molecular dynamic simulation and the direct simulation Monte Carlo methods that have been the focus of this chapter. The previous studies described in this chapter illustrate that DSMC methods, when combined with appropriate models of microscale mechanisms, can effectively model transport in vaporization or condensation processes in microscale systems. The results of these studies indicate that the interaction of the microscale mechanisms can be of central importance to the transport in such systems. Continued development of models that can accurately account for these interactions will enhance our ability to understand and predict heat and mass transfer in microscale vaporization and condensation processes.

Nomenclature

$c_{p,v}$	vapor specific heat at constant pressure
D	droplet diameter, $2r_d$
D_v	mass diffusion coefficient in vapor
h_{lv}	latent heat of vaporization

j	molecular number flux
$J_{Q,i}$	energy flux at interface
Ja	Jakob number, $c_{p,v}(T_\infty - T_d)/h_{lv}$
k_v	gas thermal conductivity
k_B	Boltzmann constant
Kn_D	Knudsen number, $\lambda/2r_d$
m	mass of a molecule
$\overline{M}$	molecular mass
n_m	number density of molecules
P	pressure
P_{ve}	equilibrium vapor pressure at the droplet interface
P_∞	pressure in the far ambient vapor
r_c	critical droplet radius
r_d	radius of droplet
r_s	radius of droplet, bubble or sphere
$\Re$	a random number uniformly distributed on [0,1]
T	absolute temperature
T_∞	temperature in the far ambient vapor
$\hat{u}_{lv}$	latent energy per water molecule merged into droplet
X_w	mole fraction of water vapor in gas mixture

Greek

δ_T	Tolman length
Δt	time interval for one time step of simulation
ε	molecular energy component
ε_{LJ}	Lennard-Jones energy parameter
λ_∞	molecular mean free path in far ambient
ρ_l	liquid density
σ_{lv}	liquid–vapor interfacial tension
σ_{LJ}	Lennard-Jones length parameter
σ_t	energy accommodation coefficient

Subscripts

d	corresponding to the liquid droplet surface
em	corresponding to particles emitted by the droplet
i	corresponding to the liquid–vapor interface
in	corresponding to particles incident on the droplet
l	liquid
rot	rotational
tr	translation
v	vapor

∞	corresponding to far ambient conditions
0	corresponding to conditions in the gas at the droplet surface
4	corresponding to $r/r_d = 4$

References

[1] Carey, V.P., *Statistical Thermodynamics and Microscale Thermophysics*, Cambridge University Press: New York, 1999.

[2] Potash Jr., M. & Wayner Jr., P.C., Evaporation from a two-dimensional extended meniscus. *Int. Journal of Heat and Mass Transfer*, **15**, p. 1851, 1972.

[3] Deryagin, B.V. & Zorin, A.M., Optical study of the adsorption and surface condensation of vapors in the vicinity of saturation on a smooth surface. *Proc. 2nd Int. Congress on Surface Activity* (London), **2**, p. 145, 1957.

[4] Schrage, R.W., *A Theoretical Study of Interphase Mass Transfer*, Columbia University Press: New York, 1953.

[5] Wilhelm, D.J., *Condensation of Metal Vapors and the Kinetic Theory of Condensation*, Argonne National Laboratory Report ANL-6948, 1964.

[6] Paul, B., Compilation of evaporation coefficients. *ARS Journal*, **32**, p. 1321, 1962.

[7] Mills, A.F., *The Condensation of Steam at Low Pressures*, technical report NSF GP-2520, series no. 6, issue no. 39, Space Sciences Laboratory, University of California at Berkeley, 1965.

[8] Yasuoka, K., Matsumoto, M. & Kataoka, Y., Molecular simulation of evaporation and condensation I. Self condensation and molecular exchange. *Proc. ASME/JSME Thermal Engineering Joint Conf.*, **2**, p. 459, 1995.

[9] Carey, V.P. & Hawks, N.E., Stochastic modeling of molecular transport to an evaporating microdroplet in a superheated gas. ASME *J. Heat Transfer*, **117**, pp. 432–439, 1995.

[10] Tolman, R.C., Effect of droplet size on surface tension. *J. Chem. Phys.*, **17**, p. 333–343, 1949.

[11] Rowlinson, J.S. & Widom, B., *Molecular Theory of Capillarity*, Oxford University Press: New York, 1989.

[12] Nijmeijer, M.J.P., Bruin, C., van Woerkom, A.B., Bakker. A.F. & van Leeuwen, J.M.J., Molecular dynamics of the surface tension of a drop. *J. Chem. Phys.*, **96**, pp. 565–576, 1992.

[13] Haye, M.J. & Bruin, C., Molecular dynamics study of the curvature correction to the surface tension. *J. Chem. Physics*, **100**, pp. 556–559, 1989.

[14] Blokhuis. E.M. & Bedeaux, D., Derivation of microscopic expressions for the rigidity constants of a simple liquid–vapor interface. *Physica A*, **184**, pp. 42–70, 1992.

[15] Hadjiagapiou, I., Density functional theory for spherical drops. *J. Phys.: Condensed Matter*, **6**, pp. 5303–5322, 1994.

[16] Chen, L.-J., Area dependence of the surface tension of a Lennard-Jones fluid from molecular dynamics simulations. *J. Chem. Physics*, **103**, pp. 10214–10216, 1992.

[17] Dang, L.X. & Tsun-Mei, C., Molecular dynamics study of water clusters, liquid and liquid–vapor interface of water with many-body potentials. *J. Chem. Physics*, **106**, pp. 8149–8159, 1997.

[18] Matsumoto, M., Takaoka, Y. & Kataoka, Y., Liquid–vapor interface of water-methanol mixture. *J. Chem. Physics*, **98**, pp. 1464–1472, 1993.

[19] Tarek, M., Tobias, D.J. & Klein, M.L., Molecular dynamics of an ethanol-water solution. *Physica A*, **231**, pp. 117–122, 1996.

[20] Loyalka, S.K., Condensation on a spherical droplet, II. *J. Colloid Interf. Sci.*, **87**, pp. 216–224, 1982.

[21] Lang, H., Heat and mass exchange of a droplet in a polyatomic gas. *Physics Fluids*, **26**, pp. 2109–2114, 1983.

[22] Chernyak, V.G. & Margilevskiy, A.Y., The kinetic theory of heat and mass transfer from a spherical particle in a rarefied gas. *Int. J. Heat Mass Transfer*, **32**, pp. 2127–2134, 1989.

[23] Young, J.B., The condensation and evaporation of liquid droplets in a pure vapour at arbitrary Knudsen number. *Int. J. Heat Mass Transfer*, **34**, pp. 1649–1661, 1991.

[24] Young, J.B., The condensation and evaporation of liquid droplets at arbitrary Knudsen number in the presence of an inert gas. *Int. J. Heat Mass Transfer*, **36**, pp. 2941–2956, 1993.

[25] Peters, F. & Paikert, B., Measurement and interpretation of growth and evaporation of monodispersed droplets in a shock tube. *Int. J. Heat Mass Transfer*, **37**, pp. 293–302, 1994.

[26] Peters, F. & Meyer, K.A.J., Measurement and interpretation of growth and evaporation of monodispersed water droplets suspended in pure vapor, *Int. J. Heat Mass Transfer*, **38**, pp. 3285–3293, 1995.

[27] Widder, M.E. & Titulaer, U.M., Kinetic boundary layers in gas mixtures: systems described by nonlinearly coupled kinetic and hydrodynamic equations and applications to droplet condensation and evaporation. *J. Stat. Physics*, **70**, pp. 1255–1279, 1993.

[28] Carey, V.P., Oyumi, S.M. & Ahmed, S., Post nucleation growth of water microdroplets in supersaturated gas mixtures: a molecular simulation study. *Int. J. Heat Mass Transfer*, **40**, pp. 2393–2406, 1997.

[29] Carey, V.P., Surface tension effects on post-nucleation growth of water microdroplets in supersaturated gas mixtures. ASME *J. Heat Transfer*, **122**, pp. 294–302, 2000.

[30] Carey, V.P., DSMC modeling of interface curvature effects on near-interface transport. *Microscale Thermophysical Engineering*, **6**, pp. 75–83, 2002.

[31] Bird, G.A. *Molecular Gas Dynamics and the Direct Simulation of Gas Flows*, Oxford University Press: Oxford, 1994.

[32] McDonald, J.D., *A Computationally Efficient Particle Simulation Method Suited to Vector Computer Architectures*, Report SUDAAR 589, Department of Aeronautics and Astronautics, Stanford University, 1989.
[33] Baganoff, D. & McDonald, J.D., A collision-selection rule for a particle simulation method suited to vector computers. *Phys. Fluids A*, **2**, pp. 1248–1259, 1990.
[34] Pruppacher, H.R. & Klett, J.D. *Microphysics of Clouds and Precipitation*, D. Reidel Publishing Company: Dordrecht, 1980.
[35] Carey, V.P., *Liquid–Vapor Phase Change Phenomena*, Taylor and Francis: New York, 1992.
[36] Carey, V.P. & Oyumi, S.M., Condensation growth of single and multiple water microdroplets in supersaturated steam: Molecular simulation predictions. *Microscale Thermophysical Engineering*, **1**, 31–38, 1997.
[37] Jiang, P. & Carey, V.P., Parallelized DSMC modeling of transport near liquid–vapor interfaces in a micro bubble heat pipe. *Proceedings of the 2001 ASME Congress and Exposition,* ASME, New York, paper IMECE2001/HTD-24131, 2001.

CHAPTER 9

Molecular dynamics simulation of nanoscale heat and fluid flow

T. Ohara
Institute of Fluid Science, Tohoku University, Japan

Abstract

Molecular dynamics (MD) simulation is a powerful tool to analyze molecular-scale characteristics of heat and fluid flow, especially those determined by dynamics of molecular motion. A number of studies are performed in the field of heat transfer and fluid flow, which include thermophysical properties such as viscosity and thermal conductivity, interfacial phenomena such as condensation and evaporation, surface tension, and thermal energy and momentum transfer through the interfaces. This chapter describes the basic theory and procedure of MD simulation, some tricks and techniques needed in the simulation, and some applications to micro/nanoscale problems.

1 Introduction

Molecular-scale analysis of thermal and fluid phenomena is becoming important in the vast movement of science and engineering toward micro/nanoscales nowadays. As background, people are now interested in the fundamental mechanism of thermal and fluid phenomena such as energy and momentum transfer and interfacial phenomena. Also various molecular-scale phenomena are now really engineering problems, which owes much to the development of micromachining technique and microscale measurement.

Molecular dynamics (MD) simulation is a major computation tool to analyze the molecular-scale characteristics and mechanism of heat transfer and fluid flow phenomena, because it gives a time-series data of position and velocity of all the molecules in the system. MD simulation is sometimes the only choice, especially for the problems governed by dynamic characteristics of molecular motion. The computation of molecular motion itself is not a difficult task; the fundamental theory has been well established in physics and its application technique has matured in physical chemistry and other fields. However, it is not only proven simulation systems that we use when we solve a problem in thermal and fluid engineering; we often face difficulties when we apply MD simulation in a new manner. First of all, we should define the problem to be solved by MD simulation. The scale that is treated by MD simulation is practically limited to be only of the order of 10 nm in space and 10 ns in time even with up-to-date computers, while our object to be engineered is much larger in most cases. In such a case where we cannot simulate the whole phenomenon, we should decompose the problem into research elements that can be treated by MD simulation. After that, we need a simulation system. If we follow an existing research procedure that is proven in physics or physical chemistry, we can adopt its simulation system without difficulty. However, sometimes we should establish a new research procedure of our own. In such a case, we should construct a simulation system with boundary conditions, which are valid from both the viewpoints of molecular physics and thermal and fluid engineering. We do the simulation with an appropriate potential function model for molecules, and then we sometimes face the biggest problem; what we get by MD simulation is only the transient position and velocity of all the molecules, which is not what we really want to know. Although there is a well-established way to get macroscopic values such as temperature and pressure from such molecular data, we should think about how to extract data which is meaningful for our engineering problem.

In this chapter, the basic theory and procedure of MD simulation, and some tricks and techniques needed in the simulation, are described. Some applications to micro/nanoscale problems are also mentioned.

2 Basic equations and finite difference scheme

2.1 Basic equation for translational motion of molecules

In MD simulation, the motion of molecules is solved by classical mechanics. The motion of individual molecules is governed by Newton's equation of motion. i.e.,

$$\frac{d^2\mathbf{r}_i}{dt^2} = \frac{\mathbf{F}_i}{m_i}. \tag{1}$$

Here, m_i is the mass of molecule i. $\mathbf{r}_i$ and $\mathbf{F}_i$ are the position vector and force vector acting on the molecule.

Generally, $\mathbf{F}_i$ is given by the gradient of the potential of the system, ϕ, as follows:

$$\mathbf{F}_i = -\frac{d\phi}{d\mathbf{r}_i}. \tag{2}$$

In what follows, we think about a system under no macroscopic body force. For the potential of the system, ϕ, the pair potential approximation is used in most cases, where ϕ is assumed to be the sum of the pair potential ϕ_{ij} between every pair of molecules in the system. i.e.,

$$\phi = \sum_i \sum_{j>i} \phi_{ij} , \tag{3}$$

where ϕ_{ij} is an effective pair potential which includes not only the real pair potential between the two molecules but also multi-body effect so that ϕ of eqn. (3) has a correct value for the potential of the system. Eqns. (2) and (3) result in the assumption

$$\mathbf{F}_i = \sum_j \mathbf{F}_{ij}, \quad \mathbf{F}_{ij} = -\frac{\partial \phi_{ij}}{\partial r_{ij}} \frac{\mathbf{r}_{ij}}{r_{ij}} , \tag{4}$$

which expresses that the force acting on a molecule, $\mathbf{F}_i$, is the sum of the intermolecular force between every pair of molecules, $\mathbf{F}_{ij}$. Here, $\mathbf{r}_{ij} = \mathbf{r}_i - \mathbf{r}_j$ and $r_{ij} = |\mathbf{r}_{ij}|$. This assumption minimizes the computational load to obtain $\mathbf{F}_i$ for all the molecules and gives a clear definition for some macroscopic quantities such as energy and momentum flux, which will be discussed later.

Equations (1) and (2) construct an NVE ensemble with constant number of molecules, volume, and energy of the system. These equations are also utilized to simulate systems with heat and/or momentum fluxes with special boundary conditions, which is called nonequilibrium molecular dynamics (NEMD) in its broad sense. On the other hand, some other ensembles, such as those with constant temperature and/or pressure, are constructed by the basic equations with some special terms added thereto [1]. Although handling of these ensembles and its physical meaning are well established in statistical mechanics, care should be taken in case of using these ensembles in a problem of thermal and fluid engineering. Also the basic equations with special terms of fictitious external forces are utilized to generate fluxes of mass, momentum and energy in a bulk material [2, 3], which is NEMD in its strict sense. This NEMD is the most efficient method to measure thermophysical properties of the molecular system, such as the diffusion coefficient, viscosity and thermal conductivity. However, it should be noted that only the limit of zero fictitious external force corresponds to real phenomena except for the sheared liquid case. We cannot observe the state of the system under a certain flux by this method. For such a purpose, we should adopt the above NEMD by using equations (1) and (2) without any additional terms and with special boundary conditions.

2.2 Finite difference scheme

In the simulation, eqn. (1) is integrated by the finite difference method. The force acting on every molecule is determined by eqn. (4), with which eqn. (1) is integrated to give the new velocity and position of the molecule. Various general time-marching schemes are applied, such as the predictor–corrector method and the Runge–Kutta method. The Verlet algorithm [4, 5] shown below is often used because of its simplicity and stability especially in simulations where the structure of the material, i.e., the position of the molecules, is the most important.

$$\mathbf{r}_i(t+\Delta t) = 2\mathbf{r}_i(t) - \mathbf{r}_i(t-\Delta t) + \Delta t^2 \frac{\mathbf{F}_i(t)}{m_i},$$
$$\mathbf{v}_i(t) = \frac{\mathbf{r}_i(t+\Delta t) - \mathbf{r}_i(t-\Delta t)}{2\Delta t}. \tag{5}$$

Here, Δt and $\mathbf{v}_i$ are the time step and translational velocity vector of molecule i, respectively. The time step should be sufficiently small compared with the characteristic time of the vibrational molecular motion. The typical time step for the translational motion of molecules is of the order of 1 fs (femtosecond = 10^{-15} s). It should be noted that the characteristic time of rotational motion of polyatomic molecules and its intramolecular vibration is much smaller than the translational motion of the molecules. For example, the time step to solve rotational motion of small polyatomic molecules such as water and carbon dioxide is of the order of 0.1 fs.

The velocity of molecules does not appear explicitly in the equation of motion of the Verlet algorithm. On the other hand, the following leap-frog algorithm [4, 5], a modified form of the Verlet algorithm, involves the velocity being determined by solving the equation of motion and hence it is often applied in the simulation of thermal phenomena where the kinetic energy of molecules is important:

$$\mathbf{v}_i\left(t+\frac{\Delta t}{2}\right) = \mathbf{v}_i\left(t-\frac{\Delta t}{2}\right) + \Delta t \frac{\mathbf{F}_i(t)}{m_i},$$
$$\mathbf{r}_i(t+\Delta t) = \mathbf{r}_i(t) + \Delta t\, \mathbf{v}_i\left(t+\frac{\Delta t}{2}\right). \tag{6}$$

2.3 Rotational motion of polyatomic molecules

The orientation of molecules is also important in the case of molecules having nonspherical potential surfaces, so that the rotational motion of molecules should be obtained. In the case that intramolecular vibration is negligible, which is appropriate at normal temperatures, molecular motion is divided into translational motion of the center of mass of the molecule and rotation about the center of mass [6]. The intermolecular potential for polyatomic molecules gives a torque acting on the molecules and the molecular rotation is solved based on the torque. One of the schemes to obtain the transient orientation of molecules that is defined by the Euler angle ϕ, θ, and ψ is as follows. The angular momentum of a molecule $\mathbf{L}$ is solved by the following equation with torque $\boldsymbol{\tau}$ acting on the molecule.

$$\frac{d\mathbf{L}}{dt} = \boldsymbol{\tau}\,. \tag{7}$$

Here, $\mathbf{L}$ and $\boldsymbol{\tau}$ are defined in space-fixed coordinates. In the above equation, the subscript i is omitted for simplicity but we are still thinking about individual molecules. On the other hand, we have other coordinates here, which are fixed in the molecular frame and coincide with the principal axis of inertia of the molecule, which are denoted by (x', y', z'). The angular momentum of the molecule in (x', y', z') is given by a transformation of coordinates using the matrix

$$\mathbf{A} = \begin{pmatrix} \cos\phi\cos\psi - \sin\phi\cos\theta\sin\psi & \sin\phi\cos\psi + \cos\phi\cos\theta\sin\psi & \sin\theta\sin\psi \\ -\cos\phi\sin\psi - \sin\phi\cos\theta\cos\psi & -\sin\phi\sin\psi + \cos\phi\cos\theta\cos\psi & \sin\theta\cos\psi \\ \sin\phi\sin\theta & -\cos\phi\sin\theta & \cos\theta \end{pmatrix}, \tag{8}$$

as

$$\mathbf{L}' = \mathbf{A}\mathbf{L}\,. \tag{9}$$

Angular velocities $(\omega_{x'}, \omega_{y'}, \omega_{z'})$ are derived from the components of $\mathbf{L}'$, $(L_{x'}, L_{y'}, L_{z'})$, and the moment of inertia about the axis of molecule-fixed coordinates $(I_{x'}, I_{y'}, I_{z'})$ as

$$\omega_\alpha = L_\alpha / I_\alpha, \quad \alpha = x', y', z'\,. \tag{10}$$

Now the rate of change of the Euler angle of the molecule is given by

$$\begin{aligned} \dot{\phi} &= \frac{d\phi}{dt} = (\omega_{x'}\sin\psi + \omega_{y'}\cos\psi)/\sin\theta\,, \\ \dot{\theta} &= \frac{d\theta}{dt} = \omega_{x'}\cos\psi - \omega_{y'}\sin\psi\,, \\ \dot{\psi} &= \frac{d\psi}{dt} = \omega_{z'} - (\omega_{x'}\sin\psi + \omega_{y'}\cos\psi)\cos\theta/\sin\theta\,. \end{aligned} \tag{11}$$

Use of eqn. (11) in the computation may suffer from inaccuracy around $\sin\theta = 0$. To avoid this, a set of quaternions (a_1, a_2, a_3, a_4) is introduced by transforming the Euler angle as follows.

$$a_1 = \sin\frac{\theta}{2}\sin\frac{\psi - \phi}{2}, \quad a_2 = \sin\frac{\theta}{2}\cos\frac{\psi - \phi}{2},$$
$$a_3 = \cos\frac{\theta}{2}\sin\frac{\psi + \phi}{2}, \quad a_4 = \cos\frac{\theta}{2}\cos\frac{\psi + \phi}{2}. \tag{12}$$

Then, the rotational motion of molecules is solved by eqn. (13) instead of eqn. (11):

$$\begin{pmatrix} \dot{a}_1 \\ \dot{a}_2 \\ \dot{a}_3 \\ \dot{a}_4 \end{pmatrix} = \frac{1}{2}\begin{pmatrix} -a_3 & -a_4 & a_2 & a_1 \\ a_4 & -a_3 & -a_1 & a_2 \\ a_1 & a_2 & a_4 & a_3 \\ -a_2 & a_1 & -a_3 & a_4 \end{pmatrix}\begin{pmatrix} \omega_{x'} \\ \omega_{y'} \\ \omega_{z'} \\ 0 \end{pmatrix}. \tag{13}$$

Here, the transformation matrix that is needed in eqn. (9) is given by

$$\mathbf{A} = \begin{pmatrix} -a_1^2 + a_2^2 - a_3^2 + a_4^2 & 2(a_3a_4 - a_1a_2) & 2(a_2a_3 + a_1a_4) \\ -2(a_1a_2 + a_3a_4) & a_1^2 - a_2^2 - a_3^2 + a_4^2 & 2(a_2a_4 - a_1a_3) \\ 2(a_2a_3 - a_1a_4) & -2(a_1a_3 + a_2a_4) & -a_1^2 - a_2^2 + a_3^2 + a_4^2 \end{pmatrix}. \tag{14}$$

2.4 Deformation of molecules and intramolecular vibration

For most polyatomic molecules, intramolecular vibration according to the deformation of molecules is quantized to a significant degree, so that it cannot be treated correctly by classical mechanics. However, it is useful if the change of the time-averaged molecular shape according to phase and state changes is incorporated in the simulation system. Also calculation of shape change is essential for simulation of complex polymers. For these purposes, two methods can be employed in the simulation: one is the use of an interatomic potential function for all atoms included in the molecules, where the motion of all the atoms is solved by eqns. (1) and (2) just as in the classical mechanics of particles. The other choice is the constraint dynamics, which treat only selected degrees of freedom. For example, most polyatomic molecules are rigid for stretching of interatomic bonds while they are relatively flexible for bending motion. In such a case, bending motion of molecules is calculated under the constraint for distance between atoms in a molecule. SHAKE [7] and RATTLE [8] are utilized as algorithms of constraint dynamics, which is summarized in ref. [9].

3 Intermolecular potential model

With the pair potential approximation, the overall potential field made by the whole configuration of molecules is given by summing up the intermolecular potential for all pairs in the system. Therefore, the intermolecular potential model does not give the correct potential energy between two molecules when these molecules are isolated in space. What is required for the intermolecular potential model is to give the correct overall potential of the system when it is summed up as eqn. (3) for all pairs of molecules in a high density region. This is the concept of the effective pair potential model. It is difficult to satisfy the above requirement for wide ranges of states such as temperature and density, which is the reason that any intermolecular potential models are excellent in some cases and not in other cases. In fact, potential models are generally tuned to fit target values with some exceptions of *ab initio* (nonempirical) models. In this section, the Lennard-Jones potential model, which is the most basic one for spherical molecules and utilized in most simulations, is described first. Most real engineering materials need more complex intermolecular potential models and we often have to select an appropriate model from among many proposed in the literature. We will see the circumstances about water as an example.

3.1 Lennard-Jones potential model for spherical molecules

The Lennard-Jones (12-6) intermolecular potential model (LJ potential) described by eqn. (15) is widely utilized for simple liquids with spherical molecules.

$$\phi_{ij}(r_{ij}) = 4\varepsilon\left\{\left(\frac{\sigma}{r_{ij}}\right)^{12} - \left(\frac{\sigma}{r_{ij}}\right)^{6}\right\}, \tag{15}$$

where r_{ij} represents the distance between the two molecules i and j. The parameters ε and σ have dimensions of energy and length, respectively. The values are $\sigma = 0.341$ nm and $\varepsilon = 1.654 \times 10^{-21}$ J [10, 11] for argon for example. The values are determined corresponding to species of molecules in an empirical way, mostly to fit viscosity or the second virial coefficient obtained for the potential model by a known statistical mechanical theory to measured values for the real material. Therefore, several different values are proposed for the same molecule corresponding to the fitting targets. The values of ε and σ are reported for many species including simple polyatomic molecules. It should be noted that properties of materials other than the fitting target may not be reproduced well by the potential model, especially in the case of polyatomic molecules.

For the interaction between different species, the first guess of the parameters is made by the Lorentz–Berthelot mixing rule [11, 12]

$$\sigma_{ij} = \frac{1}{2}(\sigma_{ii} + \sigma_{jj}), \quad \varepsilon_{ij} = (\varepsilon_{ii}\,\varepsilon_{jj})^{1/2}. \tag{16}$$

where i and j indicate different species.

Figure 1 shows the LJ potential as a function of intermolecular distance. The LJ potential with its simple form represents well both the strong repulsive forces when two molecules are close to each other and the stable distance where the repulsive force and attractive force are balanced. Therefore, it is the first choice if one intends to reproduce thermal and fluid phenomena that are common to all fluids, such as phase change and interfacial phenomena. The potential decays rapidly at long distances which allows the use of a cut-off distance such as 3–4σ, beyond which contributions of the potential are neglected. See refs. [13, 14] for handling of the slight residual potential and force at the cut-off distance.

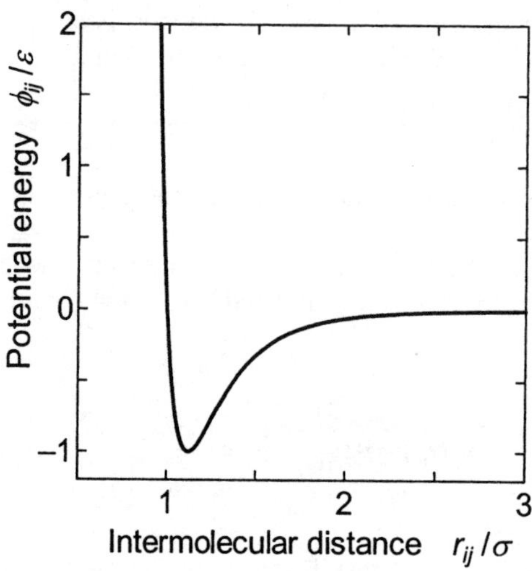

Figure 1: The Lennard-Jones (12-6) potential model (eqn. 15).

The equations of motion with the LJ potential model and simulation conditions such as temperature and density are often expressed in their nondimensional form, which is reduced by length σ, energy ε, mass m, time $(\varepsilon / m\sigma^2)^{1/2}$, force ε/σ, and temperature ε / k_B, where k_B is the Boltzmann constant. The equation of state (EOS) and saturation curve of an LJ fluid are reported in refs. [15–19] for two-dimensional fluid, and refs. [20–22] for three-dimensional fluid. The critical point is $T^* = k_B T/\varepsilon = 0.52 - 0.54$, $\rho^* = \rho\sigma^2/m = 0.31 - 0.37$ for a two-dimensional fluid and $T^* = 1.32 - 1.35$, $\rho^* = \rho\sigma^3/m = 0.30 - 0.35$ for a three-dimensional fluid. Generally, one cannot expect that the EOS for a model fluid by the LJ potential with parameters for a specific molecular species agrees with that of the real fluid. Therefore, it is difficult to determine simulation conditions without knowledge of the EOS for the model fluid, which cannot be replaced by experimental values. For example, one cannot do a simulation of a saturated liquid with only a knowledge of the saturation curve of the real fluid if the saturation density of the model fluid is not known.

3.2 Intermolecular potential model for a complex molecule: water

In the case of simulation for a specific material, we should select an appropriate intermolecular potential model among those proposed in the literature. Here, we will see the circumstances about water as an example. No one doubts the importance of water in chemistry, life science and engineering. Also water is a challenging target to model the interaction between molecules, including the electrostatic force, and to reproduce the anomalous tetrahedral structure of liquid water due to the hydrogen bond. With this background, a number of intermolecular potential models have been proposed in the literature. The majority involves a rigid molecule model in which the deformation of the molecular shape is ignored. In the beginning, a model molecule having an oxygen at the center of a tetrahedron and two hydrogen and lone pair electrons at four vertices was proposed (ST2 potential [23]), which is a straightforward extension of the well-known tetrahedral structure of solid and liquid water. All of the later models have adopted molecular configurations with everything on a plane; i.e., a molecule with electric charges placed on the three nuclei (SPC [24], SPC/E [25], etc.) or one with two positive charges on hydrogen and a negative charge on a bisector of the angle HOH on the same plane with HOH (MCY [26], RWK [27], TIP4P [28], CC [29], etc.). Figure 2 shows these configurations. The angle HOH is 109.47° (ST2, SPC, SPC/E, etc.), which is the tetrahedral angle, or 104.52° (MCY, RWK, TIP4P, CC, etc.), which is a measured value for an isolated water molecule. The OH bond length is 0.09572 nm (MCY, RWK, TIP4P, CC, etc.), which is a measured value, or 0.1 nm (ST2, SPC, SPC/E, etc.). Most of the above models are empirical ones except MCY and CC which are *ab initio* models determined by quantum chemical calculation. The empirical models adopt the LJ potential between oxygen nuclei and the Coulomb potential between electric charges, while the *ab initio* models consist of many exponential functions between oxygen–oxygen, hydrogen–hydrogen and oxygen–hydrogen nuclei and the Coulomb potential between charges.

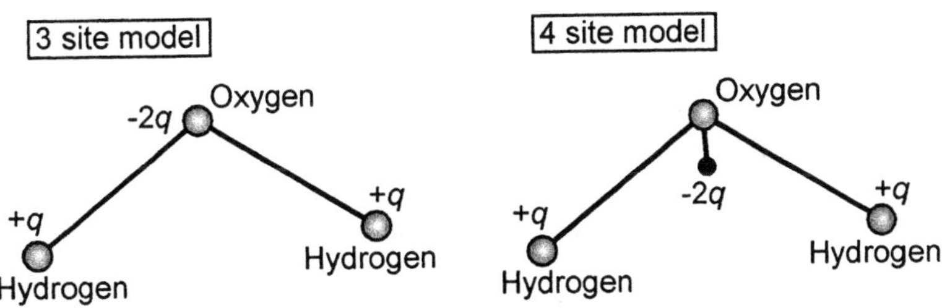

Figure 2: Configuration of water molecules.

A number of water models with a flexible molecule have also been proposed. Although intramolecular vibration of molecules is mostly quantized and cannot be treated properly by classical mechanics, the flexible molecular model is effective to reproduce the change of the average shape of molecules due to temperature and density. Most of the flexible models are based on the above rigid molecule model and an intramolecular part that responds to deformation of the molecule is added. In the case of water, stretching of O-H bonds is much stiffer than bending of HOH, so that many models consider only the bending. A few exceptions to the above rigid molecule-based flexible model are those with spherical potentials between oxygens, hydrogens, and oxygen–hydrogen [30–32], with which the calculation of molecular motion is reduced to that of atomic motion with the simple system of eqns. (1)–(3).

Another potential model having additional function is the polarizable water model [32–34], for which the dipole moment of a molecule is determined as a function of the electric field. This type of model requires high computational load to solve the induced dipole and electric field simultaneously; these are related to each other, so that there are few reports in which polarizable models are used.

At present, there is no potential model that reproduces all the characteristics of real water precisely. A great majority of the proposed potential models are effective potential models that have been tuned to fit target values. A few exceptions are nonempirical *ab initio* models such as MCY and CC but satisfactory results have not been reported for these models. Therefore, it is true that for all materials we should select a potential model carefully according to the purpose of the simulation. We can find a lot of validation studies in the literature reporting thermophysical properties and the EOS resulting from the models, which should be referred to when we select a potential model. It is important whether the EOS and the saturation curve for the model have been reported or not, because they are indispensable to determine simulation condition as was mentioned above.

With respect to water, the most successful potential model is the SPC/E (simple point charge/extended) potential [25] in spite of its simple configuration of molecule and mathematical expression. Satisfactory results have been reported about saturation liquid density and gas-liquid surface tension [35], and EOS and dielectric constant [36]. The EOS and saturation curve are reported also for CC [37] and SPC [38] water.

It should be noted here that the Coulomb potential between electric charges does not fall off fast enough so that we cannot use a cut-off distance. Such long-range forces should be counted for every pair of charges, i.e., all the combinations of charges, in the system. Therefore, MD simulation of molecules with electric charges, such as water, requires much more computational capacity. Handling of the Coulomb potential under periodic boundary conditions will be discussed later.

4 Macroscopic quantities

What is obtained by MD computation is only the position and velocity of the molecules in the system. On the other hand, we need macroscopic quantities in most cases to understand and interpret the phenomena; they include state quantities such as temperature, internal energy and pressure, and transport quantities such as heat and momentum fluxes. These quantities are obtained by averaging the MD simulation data. It should be noted that it is sometimes critical to define the macroscopic quantities in a microscopic MD simulation system. Although it is well established for the ordinary steady-state thermally equilibrium system, it still has problems under unsteady nonequilibrium conditions, such as pressure distribution during propagation of a shock wave. The problem is related to the time- and space-scales in which the concept of the macroscopic quantities is valid.

4.1 Quantity of state

4.1.1 Energy and temperature

Temperature is related to the kinetic energy of molecules. Temperature defined according to each degree of freedom of molecular motion is often utilized to measure how the total energy is partitioned to each degree of freedom. In most cases, the temperature of each individual molecule is not useful even if it can be defined. The temperature is determined based on the average kinetic energy of the molecules over a region and over a period of time. For example, the "translational temperature", denoted by T_α^{tr} $(\alpha = x, y, z)$, which is a temperature based on the kinetic energy of translational motion of molecules, E_α^{tr}, is defined by the following equation.

$$E_\alpha^{\mathrm{tr}} = \sum_{i=1}^{N} \frac{1}{2} m_i v_{\alpha,i}'^{\,2} = \frac{1}{2} N k_{\mathrm{B}} T_\alpha^{\mathrm{tr}}, \qquad (17)$$

where N is the number of molecules in the domain for which the average value of T_α^{tr} is defined.

In eqn. (17), $v'_{\alpha,i}$ is a velocity related to thermal energy. If the velocity of the center of mass of the system

$$\bar{v}_\alpha = \sum_{i=1}^{N} m_i v_{\alpha,i} \Big/ \sum_{i=1}^{N} m_i \qquad (18)$$

is not zero, the system is moving in space. In the case of fluids, $\bar{v}_\alpha$ is a macroscopic flow velocity. To fit the concept of the macroscopic continuum equations used in thermal and fluid engineering, velocities of molecules are divided into that of the center of mass of the system, and those of molecules relative to the center of mass; they are handled in a different way. The former is a flow velocity that appears in the Navier–Stokes equation, while the latter ones are related to the thermal energy that is treated by the energy conservation equation. In order to obtain the macroscopic temperature that appears in the energy conservation equation, the $v'_{\alpha,i}$ in eqn. (17) should be $v_{\alpha,i} - \bar{v}_{\alpha}$.

In the case of polyatomic molecules, the rotational temperature, T_{β}^{rot} $(\beta = x', y', z')$, is also defined in the same manner:

$$E_{\beta}^{\mathrm{rot}} = \sum_{i=1}^{N} \frac{1}{2} I_{\beta,i} {\omega_{\beta,i}}^{2} = \frac{1}{2} N k_{\mathrm{B}} T_{\beta}^{\mathrm{rot}} \tag{19}$$

The degree of freedom for the rotational motion is not always three; it is two for diatomic molecules and other linear molecules such as nitrogen and carbon dioxide. In the thermal equilibrium state, the kinetic energy is shared equally by each degree of freedom so that all T_{α}^{tr} and T_{β}^{rot} are equal; this is the macroscopic temperature. In the simulation of an equilibrium state, it should be ensured that averages of T_{α}^{tr} and T_{β}^{rot} over a sufficiently long period of time are all equal within statistical error, and then, the average of them is the system temperature in a macroscopic sense. i.e.,

$$\frac{\nu^{\mathrm{tr}} + \nu^{\mathrm{rot}}}{2} N k_{\mathrm{B}} T = \sum_{\alpha=1}^{\nu^{\mathrm{tr}}} E_{\alpha}^{\mathrm{tr}} + \sum_{\beta=1}^{\nu^{\mathrm{rot}}} E_{\beta}^{\mathrm{rot}} , \tag{20}$$

where T expresses the macroscopic equilibrium temperature and ν denotes the degree of freedom for the molecule. The statistical error depends also on the average scale or number of molecules, N, so that care should be taken concerning the scale of the average domain especially when the local temperature is defined under a nonuniform temperature distribution. See ref. [39] for a basic discussion of estimation of error.

The internal energy of the system is given by the sum of the kinetic energy of molecules and the potential energy of the system:

$$E^{\mathrm{total}} = \sum_{\alpha} E_{\alpha}^{\mathrm{tr}} + \sum_{\beta} E_{\beta}^{\mathrm{rot}} + \phi . \tag{21}$$

The intramolecular vibrational energy should also be considered in eqn. (21) if the flexible molecular model is applied. However, it should be noted that the intramolecular vibration is quantized in most cases of real molecules and cannot be properly treated by classical mechanics with the flexible molecular model.

The internal energy is conserved during an MD simulation with constant NVE. Although E_{α}^{tr}, E_{β}^{rot}, and ϕ fluctuate with time, the fluctuation should be cancelled and E^{total} is maintained at a constant value within the computational error.

4.1.2 Pressure

The pressure P is given by the following equation based on the Virial theorem [40] using the positions of the molecules in the system and forces acting on them:

$$P = \frac{Nk_B T}{V} + \frac{1}{3V}\sum_{i=1}^{N} \mathbf{r}_i \cdot \mathbf{F}_i \,. \tag{22}$$

Here V is the system volume. Under the pair potential approximation (eqn. (3)), the following equation is used.

$$P = \frac{Nk_B T}{V} - \frac{1}{3V}\sum_{i=1}^{N-1}\sum_{j>i}^{N} r_{ij}\frac{d\phi_{ij}}{dr_{ij}} \,. \tag{23}$$

4.2 Transport properties

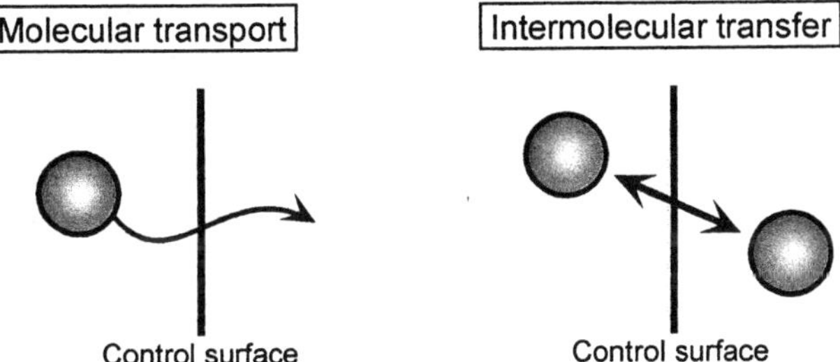

Figure 3: Mechanism of momentum transport. The contribution of molecular transport (left), which is expressed by the first term on the right side of eqn. (24), is caused by the motion of molecules that go across the control surface, while the intermolecular momentum transfer (right) expressed by the second term on the right side of eqn (24) is caused by the interaction between molecules which changes the momentum of the molecules.

4.2.1 Momentum flux

The momentum flux that occurs in sheared fluids is obtained by summing up all the molecular momentum across a control surface. With respect to the momentum in the direction of z, its flux in the direction of x, J_{zx}^m, is obtained at a certain area of control surface S_{yz} by the following equation [41]:

$$J_{zx}^m S_{yz} = \sum_i (mv_{z,i}/1)\frac{v_{x,i}}{|v_{x,i}|} + \sum_i\sum_{j>i} F_{z,ij}\frac{x_{ij}}{|x_{ij}|} \,. \tag{24}$$

The first term on the right side of eqn. (24) represents the momentum transport which occurs by the motion of the molecule itself, i.e., the momentum of each molecule is carried via molecular motion across the control surface. The summation is over the momentum of all the molecules which pass through the control surface in a unit period of time. The term (/1) represents the dimension of time^{-1}. The second term represents the rate of momentum transfer through the control surface due to the interaction between molecules; i.e., the change of molecular momentum due to intermolecular forces acting on the molecule. The double summation is the sum of the interaction between all pairs of molecules which interact across the control surface. See Figure 3 for the outline of these two contributions. The contribution of the second term is dominant in the case of liquids, while momentum transport in gases is contributed only by the first term.

In most practical cases in which momentum transfer due to fluid shear is examined by MD simulation, J_{zx}^{m} is averaged over a certain length in the direction of the flux, x. Thus, the following equation gives J_{zx}^{m} averaged over a control volume V [41]:

$$J_{zx}^{m}V = \sum_{i} m v_{z,i} v_{x,i} + \sum_{i}\sum_{j>i} F_{z,ij}\, x_{ij}^{*} \tag{25}$$

The first summation on the right side of eqn. (25) is over all molecules that are contained in the control volume V. The double summation in the second term is over all pairs of molecules with the condition that the line connecting the two molecules i and j penetrates any yz plane included in the control volume. In cases where only part of the connecting line is contained in the control volume, then x_{ij}^{*} is the x-component of the portion of the connecting vector contained in the control volume.

4.2.2 Heat flux

The heat flux J_{x}^{e} that occurs under a temperature gradient is obtained by the following equations in the same way as eqns. (24) and (25):

$$J_{x}^{e} S_{yz} = \sum_{i} (E_i/1)\frac{v_{x,i}}{|v_{x,i}|} + \frac{1}{2}\sum_{i}\sum_{j} (\mathbf{v}_i \cdot \mathbf{F}_{ij} + \boldsymbol{\omega}_i \cdot \boldsymbol{\tau}_{ij})\frac{x_{ij}}{|x_{ij}|}, \tag{26}$$

$$J_{x}^{e} V = \sum_{i} E_i v_{x,i} + \frac{1}{2}\sum_{i}\sum_{j} (\mathbf{v}_i \cdot \mathbf{F}_{ij} + \boldsymbol{\omega}_i \cdot \boldsymbol{\tau}_{ij})\, x_{ij}^{*}. \tag{27}$$

Here, E_i represents energy held by a molecule i, which is given by the sum of the kinetic energy and potential energy. In the second term on the right side of eqns. (26) and (27), which represent transport of molecular energy by the work due to intermolecular forces, a term related to rotational motion of molecules appears in contrast to the momentum flux in eqns. (24) and (25). The rotation terms are ignored in the case of monatomic molecules. Another difference between heat flux and momentum flux is the asymmetry concerning i and j in the double summation of the second term on the right side. The physical meaning of this term, including the factor, 1/2, has been discussed in refs. [42, 43].

4.2.3 Some methods to measure transport properties

The methods to measure transport properties fall roughly into three groups. The first one is the NEMD with the fictitious external force terms in the equation of motion for molecules. This is the most efficient method with a normal simulation system with three-dimensional periodic boundary conditions (see next section). However, the results are valid only when the fictitious external force is zero, so that the valid data at zero external force is obtained by extrapolation of the several simulation results with a certain magnitude of the external force. The second method is an equilibrium MD simulation during which dissipation of the transported quantity is tracked. The Green–Kubo formula [44, 45] relates the autocorrelation function of the transported quantity to the transport coefficients. The third method is a kind of NEMD with special boundary conditions and the equation of motion without any additional term. Although the computational load of this method for measuring the transport properties is much heavier than the NEMD with the fictitious external force terms in the equation of motion, this is the only way to observe the phenomena under a certain magnitude of real flux.

5 Boundary conditions and simulation system

The standard MD simulation for the equilibrium condition, which employs a system with a cubic basic cell and periodic boundary conditions in three dimensions, has been well established in physics and physical chemistry. This section will start with some description about this type of system. In this case, the validity of the system and its application limit are well known.

Although a number of studies have employed this system and various significant results have been obtained, it is true that problems in the field of heat and fluid flow also require other systems for their simulation, especially in the studies of nonequilibrium phenomena. In such a case, researchers select an appropriate system, or sometimes they invent a system. In the latter half of this section, some nonstandard systems will be mentioned.

5.1 Initial condition

In most cases of MD simulation in the field of heat and fluid flow, especially those with equilibrium state, information about the position and velocity of individual molecules are not needed. Therefore, establishing an initial condition is generating a set of molecules that are uniformly distributed in space and have a desired macroscopic quantity such as temperature, density and pressure. In the case of constant-NVE simulation, which is used most frequently, the volume of the system and the number of molecules in it are given based on the density as a simulation condition. The position and velocity of individual molecules at the start of the MD calculation do not need to be given carefully because the equilibrium state with the Maxwell–Boltzmann velocity distribution will be achieved automatically by the repeated interaction among molecules. In many cases, the position of molecules is given by random numbers and the velocity by normal random numbers but this is not essential; any other position and velocity will also work.

When a simulation with constant NVE is started with an unrealistic molecular position and velocity, a relaxation process goes until the system reaches an equilibrium state. During the process, the total energy, $E^{\text{total}} = E^{\text{tr}} + E^{\text{rot}} + \phi$, is kept constant and the distribution to the kinetic and potential energies changes. The resulting equilibrium temperature is determined by the total energy introduced by the initial condition and generally it is different from the desired temperature. To establish an equilibrium state at a desired temperature, which is the initial condition of the MD simulation, the total energy should be adjusted. Although the equation system with special terms to construct the NVT ensemble also works for that purpose, the velocity scaling method is the simplest way. Based on the present temperature, velocities of all the molecules are scaled so as to change the temperature toward the target temperature. The quickest way is scaling the velocities by $\sqrt{T_{\text{t}}/T}$ where T_{t} and T are the target temperature and the present temperature, respectively. A certain quantity of energy is injected to or ejected from the system in this procedure and E^{total} approaches its value at the target temperature. Several tens of thousands of simulation time steps will be needed to stabilize E^{total}, and then the resulting set of the position and velocity of the molecules can be utilized as an initial condition of the NVE run.

The initial condition described above is obtained by a relaxation run of the MD simulation at a steady state. In contrast to that, the initial condition of an unsteady simulation, such as the superheated liquid state just before nucleation of a bubble, is hard to construct. Based on the understanding of the essence to be observed in the simulation, artificial manipulation that does not influence the essence should be selected carefully.

5.2 Periodic boundary condition

A periodic boundary condition helps a system with a limited number of molecules to express those with an infinite size virtually. It is used in (1) simulations of an equilibrium state, (2) NEMD simulation where momentum and/or heat flux are generated by a fictitious external force introduced in the equation of motion, and (3) NEMD with the equation of motion having no additional term where energy and/or momentum flux is generated by special boundary conditions. In the last case, the periodic boundary condition is applied only in the directions normal to the energy and momentum flux, since the special boundary condition is applied in another direction.

As shown in Figure 4, a cell is introduced first, which contains all the molecules of the system considered. Its replica cells are assumed to be adjacent to the basic cell. The replica cells appear continuously in all the directions. The center cell and the replica cells are called the basic cell and image cells, respectively. The positions of all the molecules in the image cells are the same as those in the basic cell; a molecule that quits the basic cell through a boundary enters the basic cell through the opposite boundary.

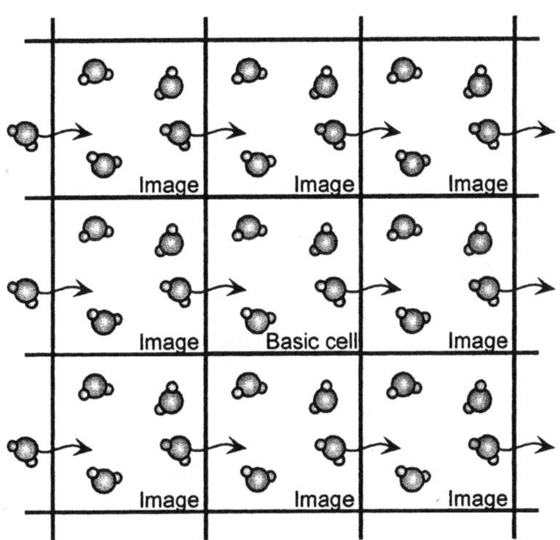

Figure 4: Periodic boundary condition. The cell in the center is the basic cell where the motion of all the molecules therein is solved.

Molecules in the basic cell interact with all the other molecules in the basic cell and all the molecules in all the image cells. Although there are an infinite number of molecules to be counted, this condition is lightened if the intermolecular force is short range, such as the LJ potential. In this case, the force acting on a molecule i by a molecule j is assumed to be contributed by only a single image of the molecule j, which is closest to the molecule i among all the images of the molecule j. This is the minimum image convention which is valid under the condition that the cell size is larger than twice the cut-off distance for the intermolecular potential.

In the case of a long-range interaction such as the Coulomb potential of water, contributions by the infinite number of molecules in the image cells should be considered. For that purpose, the Ewald method [46, 47], in which the contribution of the Coulomb potential due to electric charges that appear periodically in space is transformed to a function with much faster decay, is widely utilized.

Although the periodic boundary condition is a convenient one, it should be noted that the virtually infinite space produced by the periodic boundary condition is only a periodic appearance of the basic cell; it is accompanied by a tight restrictions that fluctuations or characteristics with a wavelength larger than the basic cell cannot be reproduced.

5.3 Nonequilibrium system with a velocity and/or temperature gradient

Although the heat flux itself can be observed by the NEMD using the fictitious external force term in the basic equation, it is valid only for bulk material and at the limit where the fictitious force is zero. Therefore, nonequilibrium systems with a "real" macroscopic temperature and/or velocity gradient are sometimes utilized to observe thermal and fluid phenomena under a certain magnitude of heat and/or momentum flux. Figure 5 shows a typical configuration of the basic cell in this case.

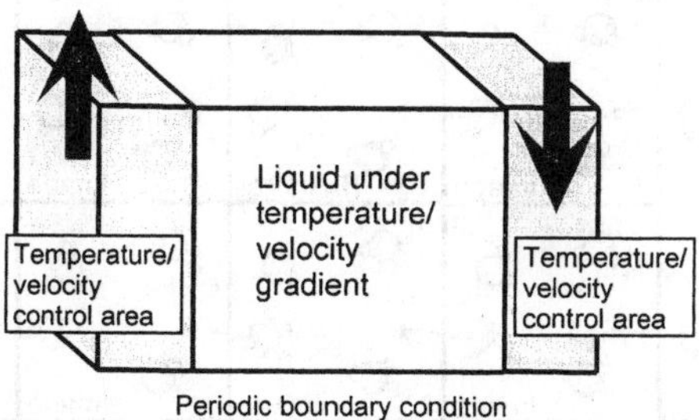

Figure 5: Configuration of NEMD simulation cell to give a real temperature and/or velocity gradient to the liquid. A temperature gradient can be given to a solid in a similar way.

The cell is a box that is long in one direction and the temperature and/or velocity gradient is given in the longitudinal direction. The periodic boundary condition is applied in the other two directions. There are two methods that are utilized to control the temperature and velocity at each end of the cell; one is installing a control area, and the velocity of molecules in this area is controlled to have a desired temperature and macroscopic flow velocity [42, 43]. The other is attaching a solid wall to each end of the cell, whose temperature and velocity is controlled [41]. (See below for the solid wall.) In case the former one is applied, one of the following boundary conditions is applied at each end of the cell to prevent molecules from quitting the cell: (1) mirror boundary condition to reflect molecules, (2) short-range potential such as the LJ potential applied to all the molecules in the system, which is a function of the distance between the molecule and the end of the cell [42, 43], (3) periodic boundary conditions with the system shown in Figure 6 [48]. In any case where a velocity gradient is given to the fluid system, viscous heating of the fluid occurs. The generated heat is removed from the system by applying a temperature control scheme to the fluid phase, or otherwise, no such temperature control is applied to the fluid and the generated heat sinks into the temperature control area or the temperature-controlled solid wall.

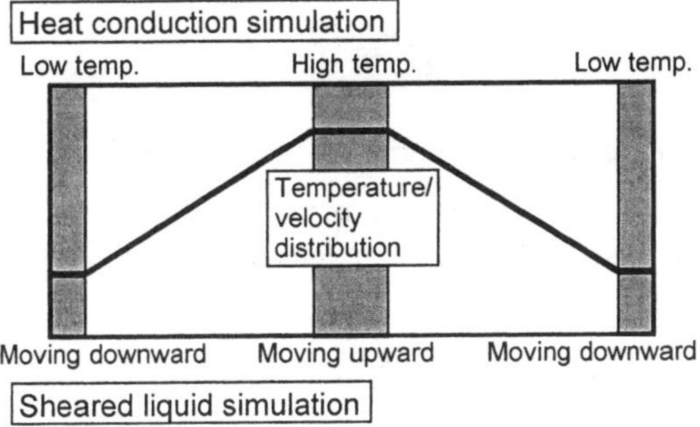

Figure 6: Configuration of NEMD simulation cell with three-dimensional periodic boundary condition.

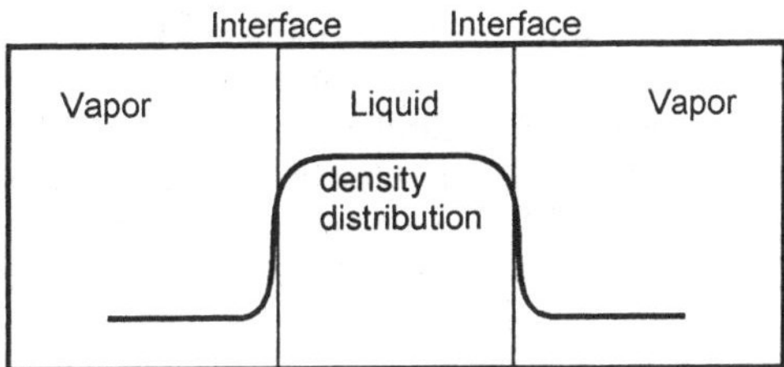

Figure 7: Standard cell configuration for simulation of liquid–vapor interface

5.4 Liquid–vapor coexistence system and determination of saturation curve

To observe phenomena at the liquid–vapor interface, including evaporation and condensation, density distribution, and surface tension, a liquid–vapor coexistence system shown in Figure 7 is widely utilized with a three-dimensional periodic boundary condition. The interface of various model liquids such as an LJ liquid [49] and water [35, 50] has been studied. In the standard method to examine the density distribution and the thickness of the interface, the simulation cell is divided into many slabs having a thickness of 0.01–0.1nm in the direction normal to the interface and the density obtained in each slab is curve-fitted by the following function,

$$\rho(z) = \frac{\rho_l + \rho_v}{2} - \frac{\rho_l - \rho_v}{2} \tanh\left[\frac{z - z_0}{d}\right], \tag{28}$$

where z is the coordinate in the direction normal to the interface. ρ_l and ρ_v are the densities of bulk liquid and vapor, respectively. d is utilized as a measure of the thickness of the interface. Surface tension is given by the following equation for the system of Figure 7:

$$\gamma = \frac{1}{2}\left[P_{zz} - \frac{1}{2}(P_{xx} + P_{yy}) \right] V / S_{xy}\,. \tag{29}$$

Here, V and S_{xy} express the volume of the cell and area of the interface, respectively. The 1/2 on the right side of eqn. (29) is due to the fact that two interfaces are in the system. The pressure tensor in eqn. (29) is given by

$$VP_{\alpha\alpha} = \sum_{i=1}^{N} m_i v_{\alpha,i}^{\ 2} + \sum_{i=1}^{N-1} \sum_{j>i}^{N} \alpha_{ij} f_{\alpha,ij}\,, \quad (\alpha = x, y, z)\,. \tag{30}$$

Otherwise, the distribution of the local surface tension [35, 51] is also utilized.

The above liquid–vapor coexistence system is convenient to determine the saturation density of the model fluid [35, 50], which is essential to study the saturated liquid and phase change as discussed above. For the same purpose, Monte Carlo simulation using a Gibbs ensemble [52] is also utilized. The most thorough method to examine the phase change of the model fluid is Kataoka's method [37], where a series of MD simulations is performed for bulk liquid and vapor at several hundreds of state points including the saturation curve and supercritical region, and the EOS is determined based on the resulting pressure and energy. The metastable state region is also determined by this method.

5.5 Solid wall and solid–liquid interface

Liquid in the vicinity of a solid surface changes its structure and transport properties. Thermal resistance has been recognized at the solid–liquid interface with high heat flux. Such solid–liquid interface phenomena have been investigated by MD simulation. The total influence of the solid wall on liquid molecules is expressed by a wall function in which all the interaction from the solid molecules is included. For example, the potential acting on a LJ liquid molecule by the FCC (1,0,0) surface of a LJ solid wall is given by [53]

$$\phi(z) = 2\pi\varepsilon\left(\frac{4}{5}\right)\left[\frac{2}{5}\left(\frac{\sigma}{z}\right)^{10} - \left(\frac{\sigma}{z}\right)^{4}\right] \tag{31}$$

where z is the distance between the liquid molecule and the solid wall.

Heat and momentum flux across the solid–liquid interface cannot be simulated by using the wall function. For that purpose, a wall that is structured with molecules is utilized and the interaction between solid molecules and liquid ones is given by the ordinary intermolecular potential model. The temperature of the solid wall is controlled by velocity scaling or using phantom molecules [54], where a system of phantom molecules expresses the integrated interference from a semi-infinite bulk solid at constant temperature and works as a heat bath when it is attached to a few layers of ordinary solid molecules.

6 MD application to heat and fluid flow

From the viewpoint of coding, MD simulation is not a tough computation. All parts of the MD computation is easily vectorized on vector computers and/or parallelized on shared-memory parallel computers. Memory usage is very small that the limited bandwidth for memory access is not a hard barrier. Due to these characteristics, 10–20 Gflop computations can be easily realized on up-to-date vector and/or parallel computers. However, the scale gap discussed in the introduction still remains. In most cases, we cannot simulate the whole thermal and fluid phenomena that we experience in our engineering device.

One possible way to fill this gap is pursuing a link to other computation methods with more macroscopic basic equations. The link to Monte Carlo simulation is achieved by models of molecular collision [55]. The link to continuum dynamics depends on thermophysical properties. Application of MD to transport of thermal energy in a solid, which has been studied extensively by phonon transport theory, was discussed [56] and attempts to link MD to phonon transport continue [57].

MD studies on heat and fluid flow problems have been increasing in these ten years. The majority of studies are related to interfacial phenomena. The molecular mechanism of condensation and evaporation at the liquid–vapor interface was studied with the system shown in Figure 7 [58, 59], where the molecular exchange phenomenon, i.e., a molecule condensed into the interface hits a liquid molecule and makes it evaporate instead of the incident molecule, has been found. Condensation of water molecules to aqueous electrolyte solution was also analyzed with a similar system [60]. A simulation of liquid droplets on a solid surface, including ambient vapor phase, has been performed [61] with an atomically structured wall and liquid molecules on it, some of which evaporate into the ambient space. The contact angle is well correlated with the molecular interactions. For heat transfer at the solid–liquid interface, thermal resistance at the interface has been found [62], and the mechanism of this thermal resistance was analyzed with a focus on the intermolecular energy transfer between solid and liquid molecules [63].

Other studies cover a wide range of thermal and fluid problems. Those include formation of fullerene [64], formation of CO_2 clathrate-hydrate [65], and momentum transfer and viscous heating in a lubrication thin liquid layer [66].

7 Future development

Molecular dynamics simulation in the field of heat and fluid flow is just like an experiment for a phenomenon that is reproduced very easily but with very difficult measurements. One performs MD simulation easily and gets an enormous amount of data on individual molecules, but faces a difficulty in deriving useful results for engineering purposes. It is required to accumulate proper research methods of our own, which are specific to thermal and fluid engineering.

Future studies will be made not only in the basic area to explore fundamental mechanisms, but also in the application field to make direct contributions to useful devices. From the viewpoint of methodology, a promising method is quantum molecular dynamics (QMD) [67, 68], which is essential to solve both quantized motion of molecules such as those at very low temperature and the electronic state of molecules to determine the interaction between molecules. The application of QMD in thermal and fluid engineering is still in its infancy. Although there is no standard method now which is generally applicable to thermal and fluid problems, various problems are to be solved, such as the sputtering process in semiconductor production, Joule heating in electronic devices, and noncombustion reactions such as fuel cells.

References

[1] Allen, M.P. & Tildesley, D.J., *Computer Simulation of Liquids*, Oxford University Press: Oxford, p. 227, 1987.

[2] Cummings, P.T. & Evans, D.J., Nonequilibrium molecular dynamics approaches to transport properties and non-Newtonian fluid rheology. *Ind. Eng. Chem. Res.*, **31(5)**, pp. 1237–1252, 1992.

[3] Ref. [1], p. 240.

[4] Ref. [1], p. 78.

[5] Berendsen, H.J.C. & van Gunsteren, W.F., Practical algorithms for dynamic simulations. *Molecular Dynamics Simulation of Statistical Mechanical Systems*, ed. G. Ciccotti & W.G. Hoover, North Holland: Amsterdam, pp. 43–65, 1986.

[6] Ref. [1], p. 84.

[7] Ryckaert, J.P., Ciccotti, G. & Berendsen, H.J.C., Numerical integration of the Cartesian equations of motion of a system with constraints: molecular dynamics of n-alkanes. *J. Comp. Phys.*, **23**, 327–341, 1977.

[8] Andersen, H.C., Rattle: a "velocity" version of the shake algorithm for molecular dynamics calculations. *J. Comp. Phys.*, **52**, pp. 24–34, 1983.

[9] Ciccotti, G. & Ryckaert, J.P., Molecular dynamics simulation of rigid molecules. *Comp. Phys. Rep.*, **4**, pp. 345–392, 1986.

[10] Hirschfelder, J.O., Curtiss, C.F. & Bird, R.B., *Molecular Theory of Gases and Liquids*, John Wiley: New York, p. 1110, 1964.

[11] Ref. [1], p. 21.

[12] Sprik, M., Effective pair potentials and beyond. *Computer Simulation in Chemical Physics*, ed. M.P. Allen & D.J. Tildesley, Kluwer: Dordrecht, pp. 211–259, 1993.

[13] Berendsen, H.J.C. & van Gunsteren, W.F., Practical algorithms for dynamic simulations. In: *Molecular-Dynamics Simulation of Statistical Mechanical Systems*, ed. G. Ciccotti, & W.G. Hoover, North Holland: Amsterdam, pp. 43–65, 1986.

[14] Ref. [1], p. 145.

[15] Tsien, F. & Valleau, J.P., A Monte Carlo study of the two-dimensional Lennard-Jones system. *Molecular Phys.*, **27(1)**, pp. 177–183, 1974.

[16] Barker, J.A., Henderson, D. & Abraham, F.F., Phase diagram of the two-dimensional Lennard-Jones system; evidence for first-order transitions. *Physica*, **106A**, pp. 226–238, 1981.

[17] Bruin, C., Bakker, A.F. & Marvin, B., Molecular dynamics study of the two-dimensional Lennard-Jones equation of state. *J. Chem. Phys.*, **80(11)**, pp. 5859–5860, 1984.

[18] Singh, R.R., Pitzer, K.S., de Pablo, J.J. & Prausnitz, J.M., Monte Carlo simulation of phase equilibria for the two-dimensional Lennard-Jones fluid in the Gibbs ensemble. *J. Chem. Phys.*, **92(9)**, pp. 5463–5466, 1990.

[19] Smit, B. & Frenkel, D., Vapor–liquid equilibria of the two-dimensional Lennard-Jones fluid(s). *J. Chem. Phys.*, **94(8)**, pp. 5663–5668, 1991.

[20] McDonald, I.R. & Singer, K., An equation of state for simple liquids. *Molecular Phys.*, **23(1)**, pp. 29–40, 1972.

[21] Nicolas, J.J., Gubbins, K.E., Streett, W.B. & Tildesley D.J., Equation of state for the Lennard-Jones fluid. *Molecular Phys.*, **37(5)**, pp. 1429–1454, 1979.

[22] Smit, B., Phase diagrams of Lennard-Jones fluids. *J. Chem. Phys.*, **96(11)**, pp. 8639–8640, 1992.

[23] Stillinger, F.H. & Rahman, A., Improved simulation of liquid water by molecular dynamics. *J. Chem. Phys.*, **60(4)**, pp. 1545–1557, 1974.

[24] Berendsen, H.J.C., Postma, J.P.M., van Gunsteren, W.F. & Hermans, J., *Intermolecular Forces*, Reidel: Dordrecht, p. 331, 1981.

[25] Berendsen, H.J.C., Grigera, J.R., & Straatsma, T.P., The missing term in effective pair potentials. *J. Phys. Chem.*, **91(24)**, pp. 6269–6271, 1987.

[26] Matsuoka, O., Clementi, E. & Yoshimine, M., CI study of the water dimer potential surface. *J. Chem. Phys.*, **64(4)**, pp. 1351–1361, 1976.

[27] Reimers, J.R., Watts, R.O. & Klein, M.L., Intermolecular potential functions and the properties of water. *Chem. Phys.*, **64(1)**, pp. 95–114, 1982.

[28] Jorgensen, W.L., Chandrasekhar, J. & Madura, J.D., Comparison of simple potential functions for simulating liquid water. *J. Chem. Phys.*, **79(2)**, pp. 926–935, 1983.

[29] Carravetta, V. & Clementi, E., Water–water interaction potential: an approximation of the electron correlation contribution by a functional of the SCF density matrix. *J. Chem. Phys.*, **81(6)**, pp. 2646–2651, 1984.

[30] Lemberg, H.L. & Stillinger, F.H., Central-force model for liquid water. *J. Chem. Phys.*, **62(5)**, pp. 1677–1690, 1975.

[31] Stillinger, F.H. & Rahman, A., Revised central force potentials for water. *J. Chem. Phys.*, **68(2)**, pp. 666–670, 1978.

[32] Halley, J.W., Rustad, J.R. & Rahman, A., A polarizable, dissociating molecular dynamics model for liquid water. *J. Chem. Phys.*, **98(5)**, pp. 4110–4119, 1993.

[33] Sprik, M. & Klein, M.L., A polarizable model for water using distributed charge sites. *J. Chem. Phys.*, **89(12)**, pp. 7556–7560, 1988.
[34] Kuwajima, S. & Warshel, A., Incorporating electric polarizabilities in water–water interaction potentials. *J. Phys. Chem.*, **94**, pp. 460–466, 1990.
[35] Alejandre, J., Tildesley, D.J. & Chapela, G.A., Molecular dynamics simulation of the orthobaric densities and surface tension of water. *J. Chem. Phys.*, **102(11)**, pp. 4574–4583, 1995.
[36] Guissani, Y. & Guillot, B., A computer simulation study of the liquid–vapor coexistence curve of water. *J. Chem. Phys.*, **98(10),** pp. 8221–8235, 1993.
[37] Kataoka, Y., Studies of liquid water by computer simulations. V. Equation of state of fluid water with Carravetta–Clementi potential. *J. Chem. Phys.*, **87(1)**, pp. 589–598, 1987.
[38] Pablo, J.J. & Prausnitz, J.M., Molecular simulation of water along the liquid–vapor coexistence curve from 25°C to the critical point. *J. Chem. Phys.*, **93(10),** pp. 7355–7359, 1990.
[39] Ref. [1], p. 191.
[40] Ref. [1], p. 46.
[41] Ohara, T. & Suzuki, D., Intermolecular momentum transfer in a simple liquid and its contribution to shear viscosity. *Microscale Thermophysical Engineering*, **5(2)**, pp. 117–130, 2001.
[42] Ohara, T., Intermolecular energy transfer in liquid water and its contribution to heat conduction: A molecular dynamics study. *J. Chem. Phys.*, **111(14)**, pp. 6492–6500, 1999.
[43] Ohara, T., Contribution of intermolecular energy transfer to heat conduction in a simple liquid. *J. Chem. Phys.*, **111(21)**, pp. 9667–9672, 1999.
[44] Ref. [1], p. 58.
[45] Allen, M.P., Back to basics. *Computer Simulation in Chemical Physics*, ed. M.P. Allen & D.J. Tildesley, Kluwer: Dordrecht, pp. 49–92, 1993.
[46] Ref. [1], p. 156.
[47] Hansen, J.-P., Molecular dynamics simulation of Coulomb systems in two and three dimensions. In: *Molecular-Dynamics Simulation of Statistical Mechanical Systems*, ed. G. Ciccotti, & W.G. Hoover, North Holland: Amsterdam, pp. 89–129, 1986.
[48] Ikeshoji, T. & Hafskjold, B., Non-equilibrium molecular dynamics calculation of heat conduction in liquid and through liquid-gas interface. *Mol. Phys.*, **81(2)**, pp. 251–261, 1994.
[49] Nijmeijer, M.J.P., Bakker, A.F., Bruin C. & Sikkenk, J.H., A molecular dynamics simulation of the Lennard-Jones liquid-vapor interface. *J. Chem. Phys.*, **89(6)**, pp. 3789–3792, 1988.
[50] Matsumoto, M. & Kataoka, Y., Study on liquid–vapor interface of water. I. Simulational results of thermodynamic properties and orientational structure. *J. Chem. Phys.*, **88(5)**, pp. 3233–3245, 1988.

[51] Berry, M.V., The molecular mechanism of surface tension. *Phys. Education*, **6**, pp. 79–84, 1971.
[52] Panagiotopoulos, A.Z., Direct determination of phase coexistence properties of fluids by Monte Carlo simulation in a new ensemble. *Molecular Phys.*, **61(4)**, pp. 813–826, 1987.
[53] Magda, J.J., Tirrell, M. & Davis, H.T., Molecular dynamics of narrow, liquid-filled pores. *J. Chem. Phys.*, **83(4)**, pp. 1888–1901, 1985.
[54] Maruyama, S., Molecular dynamics method for microscale heat transfer (Chapter 6). *Advances in Numerical Heat Transfer*, **2**, ed. W.J. Minkowycz & E.M. Sparrow, Taylor and Francis: London, pp. 189–226, 2000.
[55] Tokumasu, T. & Matsumoto, Y., Dynamic molecular collision (DMC) model for rarefied gas flow simulations by the DSMC method. *Phys. Fluids*, **11**, pp. 1907–1920, 1999.
[56] Tien, C.L., Lukes, J.R. & Chou, F.-C., Molecular dynamics simulation of thermal transport in solids. *Microscale Thermophysical Engineering*, **2(3)**, pp. 133–137, 1998.
[57] Some attempts to analyze MD simulation data from the viewpoint of phonon transport have been described in the US–Japan seminar on Nanotherm, Berkeley, 2002.
[58] Yasuoka, K., Matsumoto, M. & Kataoka, Y., Evaporation and condensation at a liquid surface I. Argon. *J. Chem. Phys.*, **101(9)**, pp. 7904–7911, 1994.
[59] Matsumoto, M., Yasuoka, K. & Kataoka, Y., Evaporation and condensation at a liquid surface II. Methanol. *J. Chem. Phys.*, **101(9)**, pp. 7911–7917, 1994.
[60] Daiguji, H. & Hihara, E., Molecular dynamics study of the water vapor absorption into aqueous electrolyte solution. *Microscale Thermophysical Engineering*, **3(2)**, pp. 151–165, 1999.
[61] Maruyama, S., Kurashige, T., Matsumoto, S., Yamaguchi, Y. & Kimura, T., Liquid droplet in contact with a solid surface. *Microscale Thermophysical Engineering*, **2(1)**, pp. 49–62, 1998.
[62] Maruyama, S. & Kimura, T., A study on thermal resistance over a solid–liquid interface by the molecular dynamics method. *Thermal Science and Engineering*, **7(1)**, pp. 63–68, 1999.
[63] Ohara, T. & Suzuki, D., Intermolecular energy transfer at a solid–liquid interface. *Microscale Thermophysical Engineering*, **4(3)**, pp. 189–196, 2000.
[64] Maruyama, S. & Yamaguchi, Y., A molecular dynamics demonstration of annealing to a perfect C_{60} structure. *Chem. Phys. Lett.*, **286(3–4)**, pp. 343–349, 1998.
[65] Hirai, S., Okazaki, K., Tabe, Y. & Kawamura, K., CO_2 clathrate-hydrate formation and its mechanism by molecular dynamics simulation. *Energy Management Conversion*, **38**, S301–306, 1997.

[66] Ohara, T. & Yatsunami, T., Energy and momentum transfer in an ultra-thin liquid film under shear between solid surfaces. *Microscale Thermophysical Engineering*, **7(1)**, pp. 1-13, 2003.
[67] Zolotoukhina, T.N. & Kotake, S., QMD energy transfer in the process of diatomic-surface collision. *Proc. Eurotherm Seminar No. 57, Microscale Heat Transfer*, ed. J.B. Saulnier & D. Lemonnier, pp. 105–112, 1999.
[68] Shibahara, M. & Kotake, S., Quantum molecular dynamics study of light-to-heat absorption mechanism in atomic systems. *Int. J. Heat Mass Transfer*, **41(6–7)**, pp. 839–849, 1998.

Heat Transfer in Gas Turbines

*Editors: **B. SUNDÉN**, Lund Institute of Technology, Sweden and **M. FAGHRI**, University of Rhode Island, USA*

Gas turbine engines and systems are designed to convert the energy of a fuel into some form of useful power, such as mechanical shaft power, electrical power, or the high-speed thrust of a jet. Research and development designed to perfect efficient cooling in gas turbines require fundamental and applied investigations of heat transfer.

Containing invited contributions from some of the most prominent specialists working in this field today, this unique title reflects current active research and covers a broad spectrum of heat transfer phenomena in gas turbines. All of the chapters follow a unified outline and presentation to aid accessibility and the book provides invaluable information for both graduate researchers and R&D engineers in industry and consultancy.

Contents: Heat Transfer Issues in Gas Turbine Systems; Combustion Chamber Wall Cooling - The Example of Multihole Devices; Conjugate Heat Transfer - An Advanced Computational Method for the Cooling Design of Modern Gas Turbine Blades and Vanes; Enhanced Internal Cooling of Gas Turbine Airfoils; Computations of Internal and Film Cooling; Heat Transfer Predictions of Stator/Rotor Blades; Recuperators and Regenerators in Gas Turbine Systems; Experimental Heat Transfer in Stationary Rib-Roughened Rectangular Channels; Experimental Heat Transfer in Roughened Leading- and Trailing-Edge and in Spanwise Rotating Channels.

Series: Developments in Heat Transfer, Vol 8

ISBN: 1-85312-666-7 2002 536pp
£159.00/US$247.00/€258.77

Condensation Heat Transfer Enhancement

***V.G. RIFERT**, Kiev, Ukraine and **H.F. SMIRNOV**, State Academy of Refrigeration, Ukraine*

In this monograph analysis and generalisation of the results of theoretical and experimental investigations of heat exchange during film condensation with variable methods of enhancement are described. Comparison of different calculation methods of enhanced condensation and effect of temperature difference, vapour shear, presence of non-condensable gases and heat exchange surface non-isothermness on average heat transfer coefficients is included.

Series: Developments in Heat Transfer, Vol 10

ISBN: 1-85312-538-5 2003 apx 300pp
apx £99.00/US$149.00/€161.12

WIT*Press*
Ashurst Lodge, Ashurst, Southampton, SO40 7AA, UK.
Tel: 44 (0) 238 029 3223
Fax: 44 (0) 238 029 2853
E-Mail: witpress@witpress.com

Find us at
http://www.witpress.com

Save 10% when you order from our encrypted ordering service on the web using your credit card.

Measurement of Heat Flux

Editors: ***P. OOSTHUIZEN****, Queen's University, Ontario, Canada and* ***D. NAYLOR****, Ryerson Polytechnic University, Canada*

Written by experts working at the forefront of this field, this unique book features a comprehensive and up-to-date survey and review of modern methods of measuring surface heat flux. The main topics covered are optical methods, periodic and transient methods, inverse heat conduction methods, various transient techniques, heat flux gauges, and methods of calibration.
An invaluable guide for researchers and engineers in industry and research institutes, the text also provides an excellent basis for graduate level courses on this subject.
Series: Developments in Heat Transfer, Vol 12
ISBN: 1-85312-858-9
2003 apx 325pp
apx £118.00/US$179.00/€192.05

All prices correct at time of going to press but subject to change.
WIT Press books are available through your bookseller or direct from the publisher.

WIT Press is a major publisher of engineering research. The company prides itself on producing books by leading researchers and scientists at the cutting edge of their specialities, thus enabling readers to remain at the forefront of scientific developments. Our list presently includes monographs, edited volumes, books on disk, and software in areas such as: Acoustics, Advanced Computing, Architecture and Structures, Biomedicine, Boundary Elements, Earthquake Engineering, Environmental Engineering, Fluid Mechanics, Fracture Mechanics, Heat Transfer, Marine and Offshore Engineering and Transport Engineering.

Modelling and Simulation of Turbulent Heat Transfer

Editors: ***B. SUNDÉN****, Lund Institute of Technology, Sweden and* ***M. FAGHRI****, University of Rhode Island, USA*

In this volume the current state-of-the-art in the prediction and control of turbulent heat transfer processes in fundamental and idealized flow circumstances as well as in engineering applications is presented. The contributors cover simple algebraic models, high and low Reynolds number two-equation models based on the eddy viscosity concept, nonlinear eddy viscosity models, EASM (explicit algebraic stress models), RSM (Reynolds stress models) and LES (large eddy simulations) combined with a variety of models for the turbulent heat fluxes.
Topics highlighted include: Generic Transport Processes in Turbulent Convective Flows; Buoyancy Driven Turbulence; Turbulence/Radiation Interactions; Turbulent Reacting Flows; Turbulent Heat Transfer in Gas Turbines; Turbulent Heat Transfer in Industrial Flows; and Turbulent Heat Transfer in Complex Geometries.
Series: Developments in Heat Transfer, Vol 16
ISBN: 1-85312-956-9
2003 apx 300pp
apx £99.00/US$153.00/€161.12

We are now able to supply you with details of new WIT Press titles via E-Mail. To subscribe to this free service, or for information on any of our titles, please contact the Marketing Department, WIT Press, Ashurst Lodge, Ashurst, Southampton, SO40 7AA, UK
Tel: +44 (0) 238 029 3223
Fax: +44 (0) 238 029 2853
E-mail: marketing@witpress.com

Advanced Boundary Elements for Heat Transfer

***M.-T. IBANEZ**, Wessex Institute of Technology, UK and **H. POWER**, University of Nottingham, UK*

"...may be considered as a possible addition to a library that serves the audience interested in advances of numerical methods in engineering."
APPLIED MECHANICS REVIEWS

In this book the authors present an efficient Boundary Element Method scheme for the numerical solution of two-dimensional heat transfer problems. Lacking the major computational difficulties of traditional reinitialization and convolution schemes, this is of the reinitialization type, in which the domain integrals are computed by a recursive relation that depends only on the boundary temperature and flux at the previous time-step.

Contents: Introduction; Integral Representation Formula for Heat Transfer; Non-History- Dependent Convolution Scheme; Recursive Reinitialization Scheme; Numerical Examples for Fixed Boundaries; Thermal Diffusion with Moving Boundary; Numerical Examples with Moving Boundary; Appendices.

Series: Topics in Engineering, Vol 42
ISBN: 1-85312-898-8 2002 144pp
£65.00/US$99.00/€105.79

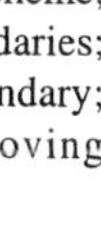

Advanced Computational Methods in Heat Transfer VII

*Editors: **B. SUNDÉN**, Lund Institute of Technology, Sweden and **C. A. BREBBIA**, Wessex Institute of Technology, UK*

Heat Transfer topics are commonly of a very complex nature. Often different phenomena such as heat conduction, convection, thermal radiation and phase change occur simultaneously. Advances in numerical solution methods of nonlinear partial differential equations and access to high-speed, efficient computers have led to dramatic progress in recent years, but there is still a need to develop further innovative approaches for the solution of a variety of problems of current interest.

This book contains edited versions of the papers presented at the Seventh International Conference on Advanced Computational Methods in Heat Transfer. The objective of this conference series is to provide a forum for presentation and discussion of advanced topics, new approaches and application of advanced computational methods to heat transfer problems.

The contributions are divided under the following headings: Diffusion-Convection; Conduction including Nonlinear Problems; Natural and Forced Convection; Phase Change; Metal Casting, Welding, Forging and Other Processes; Heat and Mass Transfer; Advances in Computational Methods; Heat Exchangers; and Modelling and Experiments in Heat Transfer.

Series: Computational Studies, Vol 4
ISBN: 1-85312-906-2 2002 532pp
£184.00/US$285.00/€299.46